World Sustainability Series

Series Editor

Walter Leal Filho, Wissenschaften Hamburg, Hochschule für Angewandte, Hamburg, Hamburg, Germany

Due to its scope and nature, sustainable development is a matter which is very interdisciplinary, and draws from knowledge and inputs from the social sciences and environmental sciences on the one hand, but also from physical sciences and arts on the other. As such, there is a perceived need to foster integrative approaches, whereby the combination of inputs from various fields may contribute to a better understanding of what sustainability is, and means to people. But despite the need for and the relevance of integrative approaches towards sustainable development, there is a paucity of literature which address matters related to sustainability in an integrated way.

Notes on the quality assurance and peer review of this publication

Prior to publication, the works published in this book are initially assessed and reviewed by an in-house editor. If suitable for publication, manuscripts are sent for further review, which includes a combined effort by the editorial board and appointed subject experts, who provide independent peer-review. The feedback obtained in this way was communicated to authors, and with manuscripts checked upon return before finally accepted. The peer-reviewed nature of the books in the "World Sustainability Series" means that contributions to them have, over many years, been officially accepted for tenure and promotion purposes.

Kishore Kumar · Preeti Bhaskar · Seda Yıldırım ·
Monomita Nandy
Editors

Sustainable Value Creation: Corporate Strategies for SDGs Implementation

Sustainable Value Creation and SDGs

Springer

Editors
Kishore Kumar
BU Centre for Sustainable Development
School of Management
Bennett University
Greater Noida, Uttar Pradesh, India

Seda Yıldırım
Tekirdag Namık Kemal University
Tekirdağ, Türkiye

Preeti Bhaskar
College of Banking and Financial Studies
Muscat, Oman

Monomita Nandy
Brunel Business School
Brunel University of London
Uxbridge, UK

ISSN 2199-7373 ISSN 2199-7381 (electronic)
World Sustainability Series
ISBN 978-3-032-24651-6 ISBN 978-3-032-24652-3 (eBook)
https://doi.org/10.1007/978-3-032-24652-3

This Springer imprint is published by the registered company Springer Nature Switzerland AG
The registered company address is: Gewerbestrasse 11, 6330 Cham, Switzerland

Preface

Sustainable development has become a defining priority for governments, institutions, and corporations worldwide. As societies confront climate change, environmental degradation, governance challenges, and increasing demands for corporate accountability, the need for coherent and measurable sustainability frameworks has grown substantially. The Sustainable Development Goals (SDGs) provide a comprehensive roadmap for inclusive and environmentally responsible development; however, translating these commitments into tangible outcomes requires institutional capacity, regulatory coherence, carbon accountability systems, and innovative market mechanisms. This volume brings together interdisciplinary scholarly contributions that examine sustainability from governance, institutional, corporate, environmental, and market perspectives. By integrating policy analysis, carbon accounting frameworks, urban sustainability systems, corporate responsibility structures, and transformative business models, the book offers a multidimensional understanding of sustainability transitions in both emerging and developed contexts.

The book is organized into five thematic parts.

The first part: *SDG Governance, Policy, and Institutional Barriers* lays the foundation for understanding implementation challenges. Chapter "1" identifies structural and systemic barriers that hinder SDG realization. Chapter "2" discusses the localization of global sustainability goals within national and regional policy frameworks. Chapter "3" applies an MCDM-based analytical approach to assess SDG implementation constraints. Chapter "4" examines the role of institutional quality in shaping sustainable development outcomes, emphasizing governance effectiveness and regulatory capacity.

The second part: *Governance, Accountability, and Sustainability Frameworks* focuses on institutional oversight and organizational responsibility. Chapter "5" explores governance mechanisms that guide sustainability agendas through structured accountability systems. Chapter "6" proposes a forward-looking corporate sustainability framework aligned with global standards and long-term value creation. Chapter "7" critically analyzes corporate social responsibility practices in emerging markets, highlighting the influence of governance pressures and stakeholder expectations.

The third part: *Carbon Accounting, Climate Policy, and Market Mechanisms* addresses the technical and regulatory foundations of climate action. Chapter "8" examines carbon accounting as a strategic tool for aligning development strategies with SDG objectives. Chapter "9" discusses the practical challenges of carbon accounting implementation in developing nations. Chapter "10" evaluates carbon markets as mechanisms for promoting sustainable development through market-based incentives. Chapter "11" assesses climate action under SDG 13, particularly in the context of the Paris Agreement and developing economies.

The fourth part: *Urban Sustainability and Development Systems* emphasizes the spatial and systemic dimensions of sustainability. Chapter "12" analyzes the relationship between urbanization and carbon emissions, proposing SDG-aligned strategies for sustainable urban growth. Chapter "13" examines governance innovations in developing economies. Chapter "14" highlights procurement metrics and performance measurement frameworks as tools for optimizing sustainable economic development.

The fifth part: *Sustainable Business Models, Reporting, and Market Transformation* concludes the volume with forward-looking perspectives. Chapter "15" discusses sustainability reporting within the circular economy paradigm. Chapter "16" explores innovative carbon accountability tools that integrate sustainability objectives with inclusive growth. Chapter "17" develops a conceptual framework linking ethical reflexivity with sustainable value creation in business models. Chapter "18" introduces repair-oriented branding as an approach to promoting product longevity and responsible consumption.

Collectively, the chapters provide conceptual clarity, empirical insight, and practical relevance. The book is intended for researchers, policymakers, corporate leaders, sustainability practitioners, and postgraduate students seeking a comprehensive understanding of governance-driven and market-enabled sustainability transitions. This volume reflects the collaborative efforts of scholars from diverse institutional and geographical contexts. We extend our sincere appreciation to all contributors and reviewers for their scholarly commitment, and we acknowledge the publisher's support in advancing interdisciplinary research in sustainability governance and corporate accountability.

As sustainability challenges continue to evolve, meaningful progress will depend on coordinated governance, responsible corporate leadership, effective climate frameworks, and innovation-driven market transformation. It is our hope that this book contributes to informed dialogue and practical action toward building resilient, transparent, and sustainable systems for the future.

Greater Noida, India — Kishore Kumar
Muscat, Oman — Preeti Bhaskar
Tekirdağ, Türkiye — Seda Yıldırım
Uxbridge, UK — Monomita Nandy

Contents

Carbon Accounting, Climate Policy, and Market Mechanisms

Urban Sustainability and Development Systems

Sustainable Business Models, Reporting, and Market Transformation

About the Editors

Dr. Kishore Kumar is Faculty of Sustainability Management and Area Chair (General Management) at the School of Management, Bennett University (Times of India Group). He also serves as Chair of the BU Centre for Sustainable Development (BUCSD) and UN PRME Faculty Lead. He brings strong expertise in sustainability reporting, ESG, and Business Responsibility and Sustainability Reporting (BRSR), along with extensive experience in sustainability training and executive capacity building. His academic and professional work spans corporate sustainability, ESG integration, CSR, and climate change. Dr. Kumar has published over 50 research articles in high-impact journals indexed in Scopus, ABDC, and Web of Science, garnering more than 1750 citations, with an H-index of 18 and an i10-index of 24 (Google Scholar). He is the editor of several Scopus-indexed books on sustainability and climate change and holds multiple editorial roles with reputed international journals. A GRI-certified Sustainability Reporting professional and UGC-NET/JRF qualifier, he is deeply committed to advancing sustainability-driven business education, executive training, and ESG-informed decision-making. e-mail: akishore001@gmail.com

Dr. Preeti Bhaskar is an Assistant Professor at the College of Banking and Financial Studies, Muscat, with over 14 years of academic experience in Human Resource Management. Her research spans Human Resource Management, the adoption of emerging technologies in higher education, and sustainability-oriented practices. She places particular emphasis on digital transformation, responsible innovation, and sustainable institutional development. Her work has been published in reputed Scopus-indexed and ABDC-ranked journals. She has authored books and case studies, and her research excellence has been recognized through Best Research Paper awards. In addition to her research contributions, she actively engages in faculty and student development initiatives at both national and international levels.

Prof. Dr. Seda Yıldırım currently works at Tekirdag Namık Kemal University in Turkey. She has a doctoral degree (Ph.D.) in marketing. She studies on sustainable development, 2030 sustainable development goals, greenwashing, sustainable consumption patterns, consumption patterns, environmental quality and food security. She is married and has two children. She has over 1900 citations by Google scholar. She has many publications in journals indexed by Web of Science (SSCI, SCI-Exp, ESCI, and others) and SCOPUS. She published several book chapters in IGI Global publishing. She won Emerald Literati Award-Outstanding Paper award 2022.

Prof. Monomita Nandy is a globally recognized academic leader in accounting, finance, and sustainability strategy, with over two decades of experience across Europe, Asia, and the US. A dual Ph.D. holder and Fellow of the Academy of Social Sciences, she brings unparalleled depth in corporate governance, financial innovation, and ethical leadership. As a Professor at Brunel University London and former Vice Dean International, she has designed and delivered high-impact executive programs for industry leaders, policymakers, and academic institutions. Her strategic insights have shaped UK Parliament inquiries, EU tax reform, and India's national sustainability reporting framework. She holds advisory roles with EFRAG, IMA USA, and businesses. She has mentored over a dozen doctoral scholars and led practitioner collaborations across MedTech, FinTech, and ESG domains. She is also an award winning academic.

SDG Governance, Policy, and Institutional Barriers

Bridging the Gap: Overcoming Barriers to SDGs Implementation

Durmus Cagrı Yıldırım, Seda Yıldırım, and Rui Alexandre Castanho

Abstract The concept of sustainable development has been presented to world leaders for the continuity of natural resources, energy, food, water and other resources that future generations can access. Although the necessity of sustainable development is accepted, the difficulties encountered in the sustainable and green transition process of both developed and developing countries have led to rethinking sustainability. The 2030 Sustainable Development Goals have guided the world with 17 basic goals and sub-goals for the areas that need to be sustainable soon and that need to change. However, many countries have experienced great difficulties in the progress of the 2030 SDGs adopted in 2015 and some goals are still considered very distant. At this point, the aim of the study is to examine the difficulties and problems encountered in achieving sustainable development goals through investigating individual and organizational challenges.

Keywords Green transition · Sustainable lifestyle · Green economy · Circular economy · Agenda 2030 · SDGs

1 Introduction

The world is a planet that hosts many living things and requires balance for the continuity of the ecosystem (Harvey and Choi 2023). Our planet, which contains many natural resources, fresh water and food sources, has been under the threat of climate change, the effects of which became visible in the 2000s (Chu and Karr 2017; WWF 2016). Although climate change is essentially a natural event, the negative effects on the ecosystem on the planet have caused this natural event to occur faster

D. C. Yıldırım · S. Yıldırım (✉)
Tekirdag Namık Kemal University, Tekirdag, Turkey
e-mail: sedayil1@gmail.com

R. A. Castanho
WSB University and Valoriza-IPP, Portalegre, Portugal
e-mail: acastanho@wsb.edu.pl

K. Kumar et al. (eds.), *Sustainable Value Creation: Corporate Strategies for SDGs Implementation*, World Sustainability Series,
https://doi.org/10.1007/978-3-032-24652-3_1

than expected. Climate change and global warming problems are problems that all world leaders focus on globally. The acceleration of climate change is caused by human-based activities (Hansen et al. 2025; Hausfather 2024) and has turned the eyes to human activities in the solution gap. In 2015, all world leaders were introduced to the 2030 Sustainable Development Goals as seen in Table 1.

Both developed and developing countries started to work around the 17 basic goals that are expected to be implemented globally to solve the sustainability problem. The 2030 SDGs include guiding targets and sub-targets for the slowdown of human activities that accelerate climate change and the green transition. In addition, since the continuity of social welfare and economic development is essential within the scope of a sustainable future, balanced sustainable practices for the protection of economy, social welfare and environmental quality have been stated as targets (The Global Goals n.d).

The emergence of the 2030 SDGs can be considered as a guide showing the right steps for policy makers, organizations and individuals in adapting to the sustainable transition. The fact that the concept of sustainability is still abstract makes the green transition difficult in some areas. The desired words, since every application considered sustainable is new, sometimes the desired goal may not be achieved. The lack of fully clear results for concrete outputs when newly adapting to a sustainable lifestyle can be considered as a difficulty encountered during the journey. Therefore, the adoption of a correct and efficient sustainable practice in the 2030 SDGs targets and green transition is not very fast. Some practices can be misunderstood or deviate from the main purpose over time. For example, greenwashing can be seen because of wrong sustainable policies and practices. If not looked deeply, many wrong practices can emerge in the name of sustainability (Yildirim 2023). However, the way of sustainable future is not so easy to plan in theoretical tasks. In addition to expected challenges, there are unusual crises in the world that prevent ensuring 2030 SDGs efficiently in the short term (United Nations 2023; United Nations 2024; Shulla et al. 2023; Yıldırım 2025; Yıldırım et al. 2025).

Many factors are effective in achieving the green transition and 2030 goals. Some of these factors can be considered as obstacles under normal conditions and some as obstacles under extraordinary conditions. On the other hand, the green transition and 2030 SDGs should be considered both individually and organizationally. Individuals have different roles and responsibilities as citizens and consumers in building a sustainable future. The organizational dimension includes civil society organizations, public institutions and businesses. This study aims to examine the challenges faced at individual and organizational levels in achieving the 2030 SDGs and the green transition. Within the scope of the study, individual and organizational challenge areas will be examined separately.

Table 1 Sustainable development goals

Goals	Main issue	Aim
G1	Poverty	Achieving remedial policies and strategies to reduce poverty
G2	Hunger, food security	Achieving remedial policies and strategies to reduce hunger and adopt sustainable practices towards food security
G3	Health	Adoption of policies and practices towards sustainable public health
G4	Education	Adoption of equal and quality education practices for girls and boys
G5	Gender equity	To prevent gender from being a factor of discrimination
G6	Clean water	Working for access to clean freshwater sources and ensuring that everyone has access to clean drinking water
G7	Energy	Ensuring access to clean and affordable energy and energy security
G8	Employment-economic growth	Ensuring adequate employment of people and sustainability of economic growth
G9	Industry	The industry's transition to green technology, continued investments in production technology and innovation
G10	Inequalities	End and stop all forms of discrimination
G11	Sustainable communities	Creating sustainable communities and transforming cities into sustainable forms
G12	Sustainable consumption-production	Adoption of sustainable consumption, transition to sustainable production
G13	Climate change	Combating climate change
G14	Sea-ocean	Protection and sustainability of sea-ocean ecosystem and biodiversity
G15	Land	Conservation and sustainability of land ecosystem and biodiversity
G16	Peace-justice-strong institutions	Supporting sustainability with peaceful societies and strong institutions
G17	Partnerships	Supporting the green transition through international collaborations, the global fight against climate change

Source Adapted from UN (2015)

1.1 Challenges to Ensuring 2030 SDGs: Individual and Organizational Dimension

Some of the targets in the 2030 SDGS give individual responsibility, while others give organizational responsibility. Therefore, when examining the obstacles to the realization of the 2030 SDGs targets, the obstacles that are reflected on individuals and the obstacles that are reflected on organizations should be considered. (Yildirim 2022). Gaps between environmental attitudes and environmental behaviors have always been investigated by scholars in recent decades. With the visible climate change, sustainability issues have become more prominent. In the early days, gaps between environmental attitudes and environmental behaviors were examined within the framework of consumer behavior. When the literature was examined, it was observed that there was a discrepancy between environmental attitudes and environmental purchasing behavior. The reason for this is that consumers or individuals, when asked, state that they care about the environment, but purchasing environmental or green products or exhibiting sustainable consumption behavior is not seen enough in practice. (Yıldırım and Candan 2015; Malodia and Bhatt 2019; Sundaraja et al. 2021; Geipel et al. 2018). The gap noticed in the organizational green transition can be said to be the difficulties that institutions and brands experience despite wanting to adapt sustainable practices, such as financial-budget, technology, lack of infrastructure, lack of personnel, and lack of information (Pereira et al. 2025; De Haas et al. 2025; Olazabal and Pascual 2015). Individual and organizational difficulty areas will be examined below.

2 Individual Challenges Areas

As viewed from the perspective of a consumer, citizen or individual, it can be said that individual efforts in sustainable development goals are directly related to some goals. Goal-11 as a citizen and Goal-12 as a consumer directly concern individuals.

- Goal 11: Sustainable Communities: Individuals are responsible for the actions they take in the city they live in as citizens. In the creation of sustainable cities, individuals must also adapt to sustainability. Therefore, cooperation between individuals and citizens is a priority in the realization of Goal-11.
- Goal 12: Sustainable Consumption: Adaptation of sustainable consumption behavior is a behavior that directly affects individual consumers. For Goal-12 to be realized, consumers' sustainable transition is a priority.

When literature and current sources are examined, it is seen that some inhibitory factors are effective in consumers' transition to sustainable consumption behavior. For example, individual contribution is also important in the transition to green energy. In their study, Malodia and Bhatt (2019) stated that individuals are skeptical about the energy crisis. In their study in India, they grouped them as "gray, gray-green and green energy consumers". Energy saving was found to be less important for material growth-oriented consumers (grays) (Malodia and Bhatt 2019). Berthiaume (2023) commented on Sphera's study. Sphere conducted a study on inaccessible green products and the purchasing relationship. In the survey, 31% of people stated that the high prices of green products were a major obstacle to sustainable behavior. 30% of people avoid unsustainable brands and 44% of people stated that they could pay more for green products. Although Generation Z is thought to be close to green consumption, very few people between the ages of 18–24 (20%) are seen as close to sustainable consumption. It has been seen that there is a widespread sense of skepticism among consumers regarding sustainability. (Berthiaume 2023). Lipberger (2017) made some comments about the relationships between Generation Y and green brands. In the Ad Age research, more than half of Generation Y did not give a specific brand name for a brand that values the environment. It is thought that there are problems in marketing green products and brands for Generation Y. Generation Y may want to buy green products, but they do not want to put in the intense effort in the process of purchasing these products. On the other hand, the performance of green products, especially cleaning products, has not been found satisfactory for the Millennial generation. In a study conducted by Pew, a very small portion of the US Millennial generation (32%) was determined to be environmentalist. In the study conducted by RetailMeNow, green products were seen as expensive. In studies conducted by Sustainable Brands and Eco Pulse, it was seen that Generation Y does not buy green products very much. (Lıpberger 2017). Vale (2024) identified some factors that prevent sustainable consumption. The uncertainty of these inhibitory factors also creates negativity. Buying green or sustainable shopping is not an accessible consumption behavior for most consumers. Most green products are sold expensively or of low quality. Being expensive discourages most consumers from purchasing. Consumers can buy environmentally friendly products by paying more, but if the usage periods of these products are long, it becomes a biased purchase. Political differences can also keep people away from sustainable consumption (Vale 2024). According to White et al. (2019), there are gaps between wanting to buy green and environmentally friendly products and actual purchasing behavior. The gap between "intention-action" is an obstacle to the spread of sustainable consumption. New policies are still being sought to encourage sustainable consumption. At this point, "social impact" can be an important motivation factor (White et al. 2019). ET Online (2020) commented on the study of the University of Leeds and the University of East Anglia (UEA) and stated that skepticism towards green products in consumers prevents sustainable consumption (ET Online 2020). Viti (2024) commented on the relationship between fashion and sustainability. Citing C. Mair, it was stated that consumers' "Fear of Missing Out-FOMO" and "social impact" factors are effective in purchasing fashion products today. Green fashion products are expensive, but people's psychological influences from a social perspective also affect their purchasing decisions (Viti 2024). Frey et al. (2023) stated that consumers support sustainable consumption but are insufficient in actual purchase. In the McKinsey-2020 survey, more than half of consumers stated that they pay more for green products, and in the NielsenIQ study, more than half of

US consumers considered a sustainable consumption lifestyle important (Frey et al. 2023). Nekmahmud and Fekete-Farkas (2020) examined the sustainable consumption behaviors of Bangladeshi consumers in their study. Consumers stated that green products are beneficial for health and the environment. In this study, inconsistencies were observed between actual actions and intentions in consumers' purchasing behaviors. If the healthy features of green products are emphasized in terms of marketing, sustainable consumption can increase. Tools such as green advertisements, eco-labels, etc. can also encourage sustainable consumption for marketing purposes (Nekmahmud and Fekete-Farkas 2020). Limmer (2023) also compiled the problems related to sustainable consumption. Retailers also do not support the sale of green products that are expensive or of poor quality according to consumers. Some studies have stated that consumers perceive green products as low quality and expensive, creating a negative perception. Creating a sustainable consumption lifestyle that is widespread in society is a difficult task. When a certain part of society buys green products, other parts may oppose it. In some studies, it has been seen that men perceive green products as feminine. This factor is also a factor that prevents sustainable consumption (Limmer 2023). In the compilation of Field (2022), the GreenPrint study interpreted consumer attitudes towards environmentally friendly products. The study stated that most US consumers are aware of climate change and that it affects their consumption. Natural disasters also direct consumers to green consumption. In addition, consumers support transportation with gasoline more. Consumers want to invest in environmentally friendly companies in investments such as stocks. On the other hand, consumers have doubts about the reality of green and environmentally friendly products (Field 2022). Chernev and Blair (2021) focused on how consumers perceive environmentally friendly products. As the sustainable products market grows, determining consumers' perceptions has become more important. Costs are an important factor for companies in the transition to green production. The high perception of product performance will also be important in choosing green products. The Tesla car case indicates the importance of product performance awareness (Chernev and Blair 2021). In the study conducted by Reddy et al. (2023), consumers have concerns about price and accessibility for green products, but they also have positive tendencies towards environmentally friendly products. Traditional media tools such as newspapers are used effectively as a source of information for consumers. In India, sustainability can be achieved if consumers and businesses support environmentally friendly products in harmony during the green transformation period (Reddy et al. 2023). Pankratz et al. (2022) made some comments on sustainable consumption. Environmentally friendly products have a very broad category. Consumers argue that alternatives are not sufficient regarding environmentally friendly products. In addition, high prices are also a source of concern for consumers. High inflation after COVID-19 in many European countries has caused a decrease in sales of environmentally friendly products. While high-income consumers continue to purchase high-priced environmentally friendly products, the decision to purchase high-priced environmentally friendly products has remained distant for low-income consumers. Low-income consumers have to prefer traditional products due to their affordable prices, even if they do not want them very much (Pankratz et al. 2022). Segran

Table 2 Link between sustainable consumption patterns and SDGs

Sustainable consumption patterns	Attitudes and behaviors	SDGs
Zero waste—No waste	Reducing waste and preferring recycling materials to consume	G2, G11
Reduce buying, ending overconsumption	Reduce and end excessive buying attitudes in the short term	G10, G11
Being green	Preferring green and environmental-friendly products	G3, G6, G7, G13, G14, G15
Ethical consumer behavior	Avoid unethical produced products and services Preferring fair traded products	G1, G3, G4, G5, G8, G9, G10, G11
Political consumer behavior	Supporting local producers and farmers etc.	G7, G10, G16, G17

Source Jastrzebska (2017); Yildirim (2022)

(2020) has comments on environmentally friendly products and consumer attitudes. Social media in particular is full of environmentally friendly product promotions. Consumers pay attention to environmentally friendly brands. Prof. Ursey conducted a study and stated that consumers evaluate the same product as having lower performance when the product is environmentally friendly. This finding is negative for sustainable consumption. Consumers not preferring environmentally friendly products will be negative in the long run. Testing the performance of environmentally friendly products with traditional substitute products tests the failure of the product. Environmentally friendly product preferences are also important in categories. While consumers buy brands that they find to have good performance in some categories, they do not prefer environmentally friendly products due to performance in some categories or they approach them with prejudice (Segran 2020).

Table 2 shows the relationship between sustainable consumption and SDGs. The compiled version from the studies of Jastrzebska (2017) and Yildirim (2022) is given in Table 2. It is known that the main areas that sustainable consumption is generally related to are related to other SDGs. Zero waste, being environmentally friendly, reducing purchasing, being an ethical and responsible consumer contribute positively to many SDGs.

3 Organizational Challenges Areas

Organizational green transformation is different from individual green transformation. While the challenges experienced by individuals have a negative impact on the SDGs, the challenges experienced by organizations also have a negative impact on the achievement of the SDGs. While the green transition is happening in many sectors, the challenges faced by some sectors have slowed down the sustainable transition for most businesses. For example, green transformation in energy is one of the most common green transition areas seen globally. As a result of energy security and sustainable energy source research, countries have started to invest in green and renewable energy sources to reduce fossil-energy dependency (Yildirim et al. 2024). However, investment in green and renewable energy sources and green transformation in energy are also struggling with many inhibiting factors. The most important factors inhibiting green energy investments are the high cost of investment and establishment decisions in some renewable energy sources. Since solar and wind energy are the most used renewable energy sources in the literature, there have been lower-cost establishment opportunities today. Using other renewable energy sources, on the other hand, causes higher costs. Factors such as geographical conditions not being suitable for green energy, the country's infrastructure not being suitable, and the lack of green-collar workers are also inhibitors (Kolkowska 2024; Moradiya 2019; Morgan 2025; Qoasja 2024; Pathak 2024).

The green transformation that has come to the fore in the fashion sector has brought sustainable fashion to the agenda. Sustainable fashion emerged with the green transformation in the fashion sector, creating a new sector. The main purpose of sustainable fashion brands is to reduce carbon emissions, ensure efficient use of natural resources and contribute positively to waste management by using recycled materials (Sevik and Yıldırım 2023). Fast fashion brands have also opened separate sub-area for sustainable transformation. For example, H&M is a fast fashion brand, but it markets sustainable fashion under its own brand by adopting green transformation (Yıldırım and Kantarcı 2022). On the other hand, the biggest obstacle to green transformation in the fashion sector is the problem of "Greenwashing". The brands that are most accused of greenwashing are fashion brands. Greenwashing can generally be defined as a deceptive practice that is considered as a false statement and does not have the sustainability feature that brands claim (Yildirim 2023; Tavintharan 2023). Inadequate or overly promotional green transformation efforts by fast fashion brands have been a long-term threat to the SDGs.

The processes experienced by local governments, civil society organizations and businesses within the organizational green transition are different from each other. The COVID-19 pandemic, the Russia-Ukraine conflict and other conflicts and wars have also created more challenging effects on the green transitions of institutions. Sustainable adoption requires institutions to adopt environmental values and make their investments in an environmentally friendly way. However, the health crisis experienced during the COVID-19 pandemic has revealed the vulnerabilities of developed and developing countries. While the European Union has leading countries in the green transition, the food, energy and economic crises during the COVID-19 pandemic have weakened sustainable policies in some areas. It has become clear that the health sector needs to be replanned. The Russia-Ukraine conflict has revealed the food and energy crisis (Yildirim et al. 2024; Leal Filho et al. 2023).

The acquisition of green technology by businesses is an important step in the green transition. However, the acquisition of green technologies involves some challenges. Costs are the primary barrier to institutions adopting green technologies.

High costs prevent many institutions from adopting green technologies. Along with high costs, public skepticism, resistance to change, and infrastructural deficiencies are also barriers. (Lexicon 2024).

Fake labels and fake certificates are also among the factors that negatively hinder the green transition of institutions. Fake labels or wrong labeling practices, which are also included in greenwashing practices, negatively affect most businesses and brands in terms of environmental attitudes (AKEPA 2025; Ecosystems United n.d). In addition, real environmental brands can fall into a suspicious situation due to wrong and fake environmental brands. Many brands under sustainability stress are launching fake or unrealistic environmental products with greenwashing as a quick escape (Fletcher 2022; Fofana 2025; Lyon 2025).

The long-term sustainability of the SDGs will be ensured when all players in an economy (large-scale and SME) experience a green and sustainable transformation. Many factors play a role in green adaptation, such as green technologies, green knowledge, licenses, green-collar workers, and environmental values, and any gaps create a challenge. The European Union attaches great importance to sustainable adaptation and takes the necessary measures for the orderly and sustainable transition of the economy. On the other hand, it transitions to sustainable policies by considering the compelling factors (Eurofi n.d). Ignjatović et al. (2024) studied green transition among the Western Balkans and they stated that institutional capacity and financial difficulties are the main factors hindering the green transition. Green technology and new R&D, which cause difficulties for institutions in green transition (Kekkonen et al. 2023), can also create a kind of dependency. Green technology transfer and costs between countries can make green transition difficult. Different plans may be required for green adaptations of private and public institutions. (Söderholm 2020).

4 Discussion

The difference of this study from other studies in the literature is that it addresses the theme of "Barriers to SDGs" at individual and organizational levels. The emergence of factors that hinder green transition at both individual and organizational levels is a threat to SDGs. This study summarizes both individual and organizational hindering factors as shown in Table 3.

When looking at Table 3, it can be seen that there are individual and organizational challenges for achieving 2030 SDGs. Since the cost of producing green products is reflected in the final price of the product, the cost and price of green products are a cyclical obstacle. As a result of the green product production process that starts with high costs, the price of the final product is also introduced to the market with a high price. This cycle is a factor that prevents the green transition both individually and organizationally due to high price-high cost. High production costs and high sales prices are considered as obstacles that directly affect the realization of Goal-12 in the green product market. Green skepticism and greenwashing are two fundamental

Table 3 Main challenges for green transition and SDGs

Individual challenges	Organizational challenges
Higher prices and lack of affordable	Higher cost of green technology and green investment
Lack of information	Greenwashing trap, fake eco-label, fake green brands
Lack of green product performance	Global Crises; COVID-19 pandemic, wars etc.
Lack of green brands/products availability	Lack of green workers (green skills and info)
Green skepticism	Lack of green corporate values

Source Created by the authors

obstacles that feed each other. Greenwashing practices that businesses and institutions turn to at an organizational level increase green skepticism in consumers. This situation also threatens SDGs practices. When individuals are exposed to too much greenwashing, they may start to be skeptical about real sustainable practices. Businesses may also move away from working for real SDGs due to the greenwashing trap. Although clean and affordable energy is a global dream and an alternative for energy security, the implementation of SDGs is delayed or prevented from reaching the desired level due to reasons such as high energy installation costs, maintenance costs, geographical barriers and lack of green energy workers. Difficulty in green investments and purchasing green licenses and technology are among the factors that slow down the transition to green product production.

5 Conclusion

The achievement of SDGs is essentially possible through the adoption of green transition. Both individuals and organizations contribute to the achievement of SDGs by adopting green transition or sustainable practices. Therefore, the obstacles to the achievement of SDGs can be considered as green transition obstacles.

Individuals do not have the same level of cognitive willingness to be environmentalist. Thinking about buying green products is a question of not actually turning into a purchase. It is important to examine the obstacles to individual green transformation or adoption of sustainable living style. Public awareness, information about green products, making green product prices affordable, and improving green product performance may be useful in accelerating individual contributions to SDGs. Incentives and financial support for green technology and green investments of institutions should be increased and international collaborations should be increased. Solving the high-cost problem can also eliminate other inhibiting factors. It may be possible to end greenwashing by making the sustainability pressure on institutions more positive. However, the greenwashing term maintains its popularity in organizational terms because it is important to appear sustainable.

References

AKEPA (2025) Greenwashing: 18 recent stand-out examples. https://thesustainableagency.com/blog/greenwashing-examples/

Berthiaume D (2023) The No. 1 reason consumers don't buy sustainable products is … https://chainstoreage.com/no-1-reason-consumers-dont-buy-sustainable-products

Chernev A, Blair S (2021) Are customers skeptical of eco-friendly products?. https://insight.kellogg.northwestern.edu/article/are-customers-skeptical-of-eco-friendly-products

Chu EW, Karr JR (2017) Environmental impact: concept, consequences, measurement. Reference module in life sciences. B978-0-12-809633-8.02380-3. https://doi.org/10.1016/B978-0-12-809633-8.02380-3. Epub 2016 Oct 31. PMCID: PMC7157458

Ecosystems United (n.d.) Are you being greenwashed? how to spot fake ecolabels. https://ecosystemsunited.com/are-you-being-greenwashed-how-to-spot-fake-ecolabels/

ET Online (2020) Why don't people buy environment-friendly products like green crackers and cars? Here is a clue. https://economictimes.indiatimes.com/industry/miscellaneous/why-people-dont-buy-environment-friendly-products-like-green-crackers-cars-here-is-a-clue/articleshow/74435995.cms?utm_source=contentofinterest&utm_medium=text&utm_campaign=cppst

Eurofi (n.d) Green transition prospects and challenges. https://www.eurofi.net/current-topics/green-transition-prospects-and-challenges/

Field A (2022) Americans want to buy green, but they don't trust companies. https://www.forbes.com/sites/annefield/2022/06/26/americans-want-to-buy-green-but-they-dont-trust-companies/

Fletcher E (2022) The rise of fake sustainable brands. https://www.sustainabilityforstudents.com/post/the-rise-of-fake-sustainable-brands

Fofana O (2025) EU takes action against fake green marketing. https://www.packaging-gateway.com/news/eu-takes-action-against-fake-green-marketing/?cf-view

Frey S, Am JB, Doshi V, Malik A, Noble S (2023) Consumers care about sustainability—and back it up with their wallets. https://www.mckinsey.com/industries/consumer-packaged-goods/our-insights/consumers-care-about-sustainability-and-back-it-up-with-their-wallets

Geipel J, Hadjichristidis C, Klesse AK (2018) Barriers to sustainable consumption attenuated by foreign language use. Nat Sustain 1:31–33. https://doi.org/10.1038/s41893-017-0005-9

De Haas R, Martin R, Muûls M, Schweiger H (2025) Managerial and financial barriers to the green transition. Manag Sci 71(4):2890–2921

Hansen JE, Kharecha P, Sato M, Tselioudis G, Kelly J, Bauer SE, Ruedy R, Jeong E, Jin Q, Rignot E, Velicogna I, Schoeberl MR, von Schuckmann K, Amponsem J, Cao J, Keskinen A, Li J, Pokela A (2025) Global warming has accelerated: are the United Nations and the public well-informed?. Environ: Sci Policy Sustain Dev 67(1):6–44. https://doi.org/10.1080/00139157.2025.2434494

Harvey A, Choi CQ (2023) Planet earth: everything you need to know. https://www.space.com/54-earth-history-composition-and-atmosphere.html

Hausfather Z (2024) Factcheck: why the recent acceleration in global warming is what scientists expect. https://www.carbonbrief.org/factcheck-why-the-recent-acceleration-in-global-warming-is-what-scientists-expect/

Ignjatović J, Filipović S, Radovanović M (2024) Challenges of the green transition for the recovery of the Western Balkans. Energ Sustain Soc 14:2. https://doi.org/10.1186/s13705-023-00421-4

Jastrzębska E (2017) The responsible consumer as an answer to new sustainable development challenges. Econ Environ 60(1):9. https://ekonomiaisrodowisko.pl/journal/article/view/249

Kekkonen A, Pesor R, Täks M (2023) Stepping towards the green transition: challenges and opportunities of Estonian companies. Sustainability 15(5):4172. https://doi.org/10.3390/su15054172

Kolkowska N (2024) Challenges in renewable energy. https://sustainablereview.com/challenges-in-renewable-energy/

Leal Filho W, Fedoruk M, Paulino, Pires Eustachio JH, Barbir J, Lisovska T, Lingos A, Baars C (2023) How the war in Ukraine affects food security. Foods 12(21):3996. https://doi.org/10.3390/foods12213996. PMID: 37959115; PMCID: PMC10648107

Lexicon (2024) Exploring green technologies: innovations, opportunities, and real-world challenges. https://instituteofsustainabilitystudies.com/insights/lexicon/green-technologies-innovations-opportunities-challenges/#:~:text=Economically%2C%20the%20high%20initial%20investment,and%20implementation%20of%20these%20technologies

Limmer D (2023) Why do consumers demand sustainability while refusing to pay for it?. https://retailwire.com/discussion/why-do-consumers-demand-sustainability-while-refusing-to-pay-for-it/

Lıpberger R (2017) 5 Reasons Why Millennials Don't Buy Green Brands — and a Better Way to Reach Them. https://socapglobal.com/2017/06/5-reasons-millennials-dont-buy-green-brands-better-way-reach/

Lyon J (2025) Misleading labeling and eco-friendly certifications. greenwashing. https://thelyonfirm.com/blog/misleading-labeling-eco-friendly-certifications-greenwashing/

Malodia S, Bhatt AS (2019) Why should I switch off: understanding the barriers to sustainable consumption? Vision 23(2):134–143. https://doi.org/10.1177/0972262919840197. (Original work published 2019)

Moradiya MA (2019) The challenges renewable energy sources face. https://www.azocleantech.com/article.aspx?ArticleID=836

Morgan A (2025) What are the challenges of renewable energy in power markets?. https://montel.energy/resources/blog/what-are-the-challenges-of-renewable-energy-in-power-markets

Nekmahmud M, Fekete-Farkas M (2020) Why not green marketing? Determinates of consumers intention to green purchase decision in a new developing nation. Sustainability 12(19):7880. https://doi.org/10.3390/su12197880

Olazabal M, Pascual U (2015) Urban low-carbon transitions: cognitive barriers and opportunities. J Clean Prod 109(2015):336–346. https://doi.org/10.1016/j.jclepro.2015.08.047

Pankratz D, Novak DR, Rogers S (2022) The cost of buying green. https://www.deloitte.com/us/en/insights/industry/retail-distribution/consumer-behavior-trends-state-of-the-consumer-tracker/sustainable-products-and-practices-for-green-living.html

Pathak S (2024) The challenges and solutions in scaling up renewable energy. https://www.herofutureenergies.com/blog/renewable-energy-challenges-and-solution/

Pereira S, Silva C, Ferreira LMDF (2025) Barriers to energy transition: comparing developing with developed countries. Energy Sustain Dev 87:101753. https://doi.org/10.1016/j.esd.2025.101753

Qosja A (2024) Evaluating barriers to renewable energy adoption. https://www.cyis.org/post/evaluating-barriers-to-renewable-energy-adoption

Reddy KP, Chandu V, Srilakshmi S, Thagaram E, Sahyaja CH, Osei B (2023) Consumers perception on green marketing towards eco-friendly fast moving consumer goods. Int J Eng Bus Manage 2023:15. https://doi.org/10.1177/18479790231170962

Segran E (2020) Green is good, right? Consumers don't always see it that way. https://www.fastcompany.com/90473738/green-is-good-right-consumers-dont-always-see-it-that-way

Sevik N, Yıldırım S (2023) Sustainable enterprises post COVID-19 pandemic: the rise of sustainable fashion enterprises. In: Castanho R (ed) Handbook of research on current advances and challenges of borderlands, migration, and geopolitics. IGI Global Scientific Publishing, pp 343–359. https://doi.org/10.4018/978-1-6684-7020-6.ch017

Shulla K, Filho WL, Policy department for external relations directorate general for external policies of the union (2023) Achieving the UN agenda 2030: overall actions for the successful implementation of the sustainable development goals before and after the 2030 deadline, January 2023–PE 702.576. https://www.europarl.europa.eu/RegData/etudes/IDAN/2022/702576/EXPO_IDA(2022)702576_EN.pdf

Söderholm P (2020) The green economy transition: the challenges of technological change for sustainability. Sustain Earth 3:6. https://doi.org/10.1186/s42055-020-00029-y

Sundaraja CS, Hine DW, Lykins AD (2021) Palm oil: understanding barriers to sustainable consumption. PLoS ONE 16(8):e0254897. https://doi.org/10.1371/journal.pone.0254897

Tavintharan V (2023) Greenwashing: the misleading marketing of sustainable products and initiatives. https://www.sec.org.sg/blog_Greenwashing-The-Misleading-Marketing-of-Sustainable-Products-and-Initiatives.html

The Global Goals (n.d.) The 17 goals. https://globalgoals.org/goals/

UN (United Nations) (2015) Transforming our world: the 2030 agenda for sustainable development, A/RES/70/1. Retrieved from https://sustainabledevelopment.un.org/content/documents/21252030%20Agenda%20for%20Sustainable%20Development%20web.pdf

United Nations (2023) Times of crisis, times of change science for accelerating transformations to sustainable development. https://sdgs.un.org/sites/default/files/2023-09/FINAL%20GSDR%202023-Digital%20-110923_1.pdf

United Nations (2024) The sustainable development goals report 2024. https://unstats.un.org/sdgs/report/2024/The-Sustainable-Development-Goals-Report-2024.pdf

Vale A (2024) Shoppers say they want eco-friendly products, so why aren't they buying them?. https://realstyle.therealreal.com/shoppers-want-eco-friendly-products-but-not-buying-them/

Viti M (2024) Why don't we buy green if we know everything about sustainability?. https://magazine.lineapelle-fair.it/en/green-corner-en/why-dont-we-buy-green-if-we-know-everything-about-sustainability/

White K, Hardisty DJ, Habib R (2019) The elusive green consumer. https://hbr.org/2019/07/the-elusive-green-consumer

WWF (2016) Living planet report 2016. Risk and resilience in a new era. WWF International, Gland, Switzerland. https://awsassets.panda.org/downloads/lpr_living_planet_report_2016.pdf

Yildirim S (2023) Greenwashing: a rapid escape from sustainability or a slow transition? LBS J Manag Res 21(1):53–63. https://doi.org/10.1108/LBSJMR-11-2022-0077

Yildirim S, Candan B (2015) Segmentation of green product buyers based on their personal values and consumption values. Environ Values 24(5):641–661. https://doi.org/10.3197/096327115X14384223590212

Yıldırım S, Kantarcı T (2022) A review on sustainability policies of businesses: recycling and waste reduction. J Recycl Econ Sustain Policy 1(1):1–9. Retrieved from https://respjournal.com/index.php/pub/article/view/1

Yildirim S, Yildirim DC, Erdogan S (2024) Energy security dilemma in OECD countries: the COVID-19 pandemic, Russia–Ukraine War, and energy issues. In: Castanho R (ed) Green economy and renewable energy transitions for sustainable development. IGI Global Scientific Publishing, pp 83–105. https://doi.org/10.4018/979-8-3693-1297-1.ch005

Yıldırım S, Yıldırım DC, Erdoğan S, Aydın K, Ektiren B (2025) The role of digitalization on the way to sustainable future: Turkey case. In: Bridging technology and development for sustainable innovation and geopolitical dynamics. IGI Global Scientific Publishing, pp 363–380

Yildirim S (2022) The consumer role for sustainable development: how consumers contribute sustainable development goals. In: I Management association (ed) Research anthology on measuring and achieving sustainable development goals. IGI Global Scientific Publishing, pp 872–888. https://doi.org/10.4018/978-1-6684-3885-5.ch047

Yıldırım S (2025) Unveiling circularity: challenges and barriers in transitioning to circular practices. In: Singh P, Daga S, Yadav K, Mishra V (eds) Circular economy and environmental resilience. Springer, Cham. https://doi.org/10.1007/978-3-031-93091-1_8

Localizing Global Goals and Policy Implications for Advancement of the SDGs

Obaidur Rahman, Mohd Abdul Moid Siddiqui, and Farah Ahmad

Abstract In 2015, the United Nations introduced Sustainable Development Goals (SDGs), which aims at creating a sustainable world for all by 2030, called as "Agenda 2030". Still, the implementation of SDGs lacks the realistic local consideration. This study focuses on exploring the processes and strategies to translate these global ambitions into place-based realities. On exploration, it is found that the success of the 17 SDGs is intrinsically linked to action at the subnational level. This paper identifies and analyses the primary barriers hindering SDG implementation at local level. These barriers include weak governance and institutional frameworks, significant financial constraints and resource mobilization challenges, a "data paradox" marked by the lack of quality localized data, and critical gaps in local capacity and technical expertise. The study formalizes the key drivers, which drives the transformational change while implementing SDGs at local levels. These drivers include: Comprehensive Sensitization and Engagement of Local Actors, Robust Accountability Mechanisms, Participatory Planning and Service Delivery, Tangible Local Economic Development, and the Formation of Multi-Stakeholder Partnerships. It further identifies the five essential components for successful localization strategies, which include: Establishing Coherent Multilevel Governance, Fostering Inclusive Partnership and Governance Models, Ensuring Flexible and Direct Financing Coupled with Capacity Strengthening, Building Robust Localized Data Ecosystems, and Leveraging Digital Tools for Engagement and Monitoring. The study finally concludes that SDG localization can be empowered by overcoming the identified barriers through the strategic implementation of these drivers and components and by creating an empowerment-focused, co-creative process instead of just top-down implementation. It requires sustained political will, strategic investment, and genuine

O. Rahman (✉)
School of Business, Galgotias University, Greater Noida, India
e-mail: obaidkhan.mba@gmail.com

M. A. M. Siddiqui
School of Management, IILM University, Greater Noida, India
e-mail: mohd.siddiqui@iilm.edu

F. Ahmad
Elizabeth School of London, London, UK

K. Kumar et al. (eds.), *Sustainable Value Creation: Corporate Strategies for SDGs Implementation*, World Sustainability Series,
https://doi.org/10.1007/978-3-032-24652-3_2

collaboration across all sectors of society to unlock sustainable value creation on a global scale.

Keywords Localization · Local · Needs · SDGs · Sustainability · Governance

1 Introduction

In 2015, the United Nations introduced universal Sustainable Development Goals (SDGs), often called Global Goals. According to Bexell and Jönsson (2017), these 17 SDGs together delivers a common development framework, which is designed to remove poverty, protect the planet and ensure peace & prosperity for all by the year 2030. Hariram et al. (2023) highlighted the correlation among these SDGs. It implied that action taken towards one goal, directly or indirectly influences the outcome of others. Therefore, it is important to balance economic, social and environmental pillars. The persistent global challenges increase the urgency of achieving the 2030 Agenda. Global pandemic and disruptions in global trade due to persistent regional conflicts reversed the progress in poverty reduction significantly and now, the world still faces the immense challenge of rising poor population (Buheji 2022; Gayathri et al. 2024). Dutta and Das (2024) mentioned, even in the wealthiest nations, millions of children continue to grow up in poverty. This highlighted the multidimensional and universal nature of this challenge. Barron et al. (2022) underscored the need to invest in improving economic prospects, enhancing education and ensuring social protection to the marginalized, to help in poverty reduction. Moallemi et al. (2020) added that the integrated nature of SDGs represents both a challenge and an opportunity. It makes local-level implementation more crucial, especially in the case of emerging economies where the resources and expertise are traditionally stashed within governmental departments such as health, education, or environment. The holistic approach demanded by the SDGs may be difficult to implement as Institutional silos in emerging economies often hinder integrated SDG implementation (Bringezu et al. 2016; Guha and Chakrabarti 2019; Singh et al. 2024). This study emphasizes the need to overcome traditional departmental boundaries and build effective local strategies, which embrace systems-thinking approach to foster cross-sectoral collaboration and maximize positive ripple effects across the entire SDG framework.

1.1 Defining SDG Localization: Bridging Global Ambitions with Local Realities

Server et al. (2025) explained SDG localization as the process of translating the universal aspirations of the 2030 Agenda into tangible realities within specific cities and regions. Immler and Sakkers (2022) highlighted the need of participatory and community-centred governance for effective localization rather than a purely top-down administrative approach. Server et al. (2025) also identified that this process involves consideration of subnational contexts during goal setting, target determination, selection of operational mechanisms and development of parameters for monitoring progress. Ortiz-Moya and Reggiani (2025) highlighted the fact that two-thirds of the 169 SDG targets necessitate action at the local level. The local government thus, plays a pivotal role in transforming the global vision of Agenda 2030 into ground-level reality as they possess intimate knowledge of individual and collective needs, thus adapting SDGs to local demands.

These studies highlighted the idea of "customization" within the definition of localization. It suggests the requirement of local actors to interpret and adapt global goals while considering their unique circumstances and needs. It helps nations to move beyond just playing a passive role in receiving top-down directives to finally having control over their goals and sustainable growth, which needs a profound understanding of local cultural norms, socio-economic conditions, environmental specificities and existing governance structures (Singh et al. 2025b; Srivastava et al. 2024).

1.2 Relationship Between Localization and Sustainable Value Creation

Lüdeke-Freund et al. (2020) highlighted that sustainable value creation involves integrating economic profit, social and environmental benefits simultaneously. It creates sustainable development and overall well-being for all stakeholders i.e. Sustainability creates shared value. Opoku (2025) underscored integrating the SDGs into organization's culture and strategic operations, may potentially unlock opportunities for innovation, cost savings, risk management and building brand value. Effective localization involves addressing the specific needs of local communities, which can foster local economic development. It creates an enabling environment in which businesses can seamlessly align their operations with sustainable practices. Biloslavo and Edgar (2024) extended this further and stated that local economic development helps the businesses to enter new markets, attract new customers and eventually fuelling growth. Mohieldin et al. (2022) mentioned the relationship between local SDG implementation and corporate value creation as reciprocal. Guarini et al. (2022) analysed the cohesive relationship among local government and the corporate setup. Local government often lack technological innovation, financial resources, expertise; which corporate can help with. This collaborative ecosystem creates shared value proposition, which requires development of clear policy to incentivize corporate action aligned with the SDGs for inclusive sustainable development (Singh et al. 2025c; Tripathi et al. 2022).

This study explores the critical role of localizing the Sustainable Development Goals to cater local sustainability targets. To implement SDGs effectively at subnational levels, it needs to (1) identify and analyse the multifaceted barriers that currently hinder effective SDG implementation at the subnational level, (2) develop strategies and enabling conditions to overcome these obstacles, (3) provide actionable insights for policymakers, local governments, civil society organizations and the private sector, demonstrating how concerted local action is not only indispensable for achieving the broader 2030 Agenda but also crucial for fostering sustainable value creation globally (Kumar et al. 2025a; Kaith and Sachdeva 2024).

2 Literature Review

Yang and Jiang (2017) took the concept of SDG localization from simply implementing SDGs at local level to understanding that "development is place-based". This means, development outcomes depend on local geography & institutions and policies must align with the regional characteristics. Ansell et al. (2022) realised the need of necessary knowledge, skills and resources for policies to effectively fulfil the ambitious 2030 Agenda, within specific local contexts. Leavesley et al. (2022) underlined that SDG localization's success depends on policy coherence and the firm's commitment to strengthen outcomes by exploiting territorial cooperation. Biggeri (2021) stated to achieve sustainable development at the local level, a contextualized approach to SDG localization in every country is demanded. The approach involves carefully considering the specific processes and degrees of decentralization, continuous coordination and negotiation across multiple levels of government-local, regional, national and along with different stakeholders such as states, markets and civil society, within each country (Shivakoti et al. 2021; Srivastava et al. 2018).

SDG localization is the process of designing and implementing local development strategies to achieve the SDGs (Krantz and Gustafsson 2021). Initially, it was perceived only as the application of global goals locally (Bilsky et al. 2021; Srivastava et al. 2021). However, Novovic (2022) helped evolving this concept into a more inclusive process and emphasizes on empowering all local stakeholders. This emphasized that development goals can only be genuinely achieved through participation of local actors at both the execution level and the agenda-setting level (Dulak 2023). This implies that genuine localization is not just about decentralizing administrative tasks but also about decentralizing power and decision-making authority to local communities. Leino and Puumala (2021) underlined that when local actors are actively involved in both agenda-setting and the co-creation of solutions, they develop a stronger sense of ownership which results in better outcomes. Ahmad and Islam (2024) deepened the view of this participatory approach. They mentioned that localization also results to greater accountability and transparency at the local level. These exploration by researchers also helped to find the key challenges that hinders the effective localization of the Sustainable Development Goals (SDGs). First, weak governance framework with insufficient resources, fragmented responsibilities, and misalignment with national policies, dilute accountability and obstruct implementation. The lack of political will or high corruption or both, prevents resources from reaching the intended targets, worsening the existing governance constraints. Second, without adequate funding, local governments are forced to limit work on global goals. A dedicated SDG budget is often absent, which forces local government to rely either on traditional funds or on external aid. The global economic instability can intensify these financial constraints. Third challenge is "data paradox" at local level. It means, when precise, localized data is essential for effective, evidence-based action; the local environments often lack it either due to poor quality, data loss or not disaggregated. This data deficit situation degrades monitoring and evidence-based policymaking. Lastly, a fundamental gap is lack of technical expertise and human capital. Local authorities often lack the skilled personnel to design, execute, and sustain complex SDG initiatives, even when funding is available. It underlines that human capital and technical expertise shortage weakens sustainability.

2.1 Governance and Institutional Frameworks

Lack of robust governance structures at local levels creates a significant challenge in implementing effective SDG localization. Gupta and Sigdel (2024) identified that many local governments often contend with insufficient capacity, limited resources and inadequate political support, necessary to effectively implement SDG initiatives. Guha and Chakrabarti (2019) mentioned that existing institutional frameworks frequently fail to adequately incorporate SDG targets, leading to misalignment between national and local policies. Caiado et al. (2018) stated that fragmentation within governance structures can result in overlapping responsibilities among different entities. It dilutes accountability and undermines the overall effectiveness of efforts to achieve SDG objectives. Arszulowicz (2025) identified another universal challenge i.e. lack of cohesive commitment and trust among different stakeholders. This may result in fragmented efforts towards SDGs, resulting in lack of unified vision. Washaya (2020) underlined that local and regional authorities possess the power and autonomy to make decisions, which when mixed with the political and financial support they receive from national governments, help the local governments to contribute more towards SDG implementation at subnational levels. Rahman (2021) identified the key difficulties that may worsen these institutional weaknesses. It includes the absence of clear allocation of responsibilities, insufficient coordination mechanisms and high turnover. On contrary, in many cases where the financial resources were available but due to unclear responsibility or rampant corruption, these resources may not reach their intended targets or be utilized effectively (Marawu and Utete 2025). These studies together highlight the need for foundational reforms in governance are prerequisites for effective SDG implementation. Without coherent governance system, SDG localization will be significantly hampered, resulting in decelerating the progress towards the 2030 Agenda.

2.2 *Financial Constraints and Resource Mobilization*

Bexell et al. (2025) highlighted that in developing regions, local government face severe budgetary constraints. Thus, financial limitations represent another major impediment to SDG localization. These monetary limits affect the local governments' capability to implement SDG-related programs and initiatives, which could have taken their sustainability goals to next level (Haou et al. 2025). Gustafsson and Krantz (2025) pointed out the misalignment between the command from international agencies and the needs of local communities. This misalignment further intensifies the financial burdens due to inefficient structures and limited local discretion. Clark et al. (2018) highlighted the over-reliance of local levels on external aid and limited dedicated innovative funding mechanisms. Noah and David (2024) demonstrated, using some cases, that SDG budget alignment often relies on reclassifying existing expenditures rather than establishing dedicated funding streams. In short, they lack the explicitly designated budget for the SDGs, reducing the potential outcomes. Choi (2021) used the global economic context to explain that rising inflation and economic downturns further intensifies these financial constraints. This severely impacts the capabilities of local governments to fund ambitious sustainable development project.

Researchers advocated and stressed the need for more flexible and long-term financing mechanisms, especially for local levels to effectively turn SDG localization into a practical reality (Guha and Chakrabarti 2019; Washaya 2020; Choi 2021; Leavesley et al. 2022; Dulak 2023; Gupta and Sigdel 2024; Haou et al. 2025). This also implies the urgent need for innovative financing mechanisms and greater strategic engagement of the private sector to bridge these financial gaps in a sustainable manner.

2.3 *Data Availability, Quality and Monitoring Challenges*

To track the progress of the Sustainable Development Goals (SDGs), effective monitoring and evaluation systems are needed. But data-related constraints like data availability, quality, and granularity, particularly at the local level, hinder these systems substantially. Currently, many developing regions lack socio-economic data, creating constraints for evidence-based decision-making. It also leads to inefficiencies and obstacles in governance and policy formulation.

Anderson et al. (2017) highlighted that developing nations need to build the capacity of their national statistical offices. Enhanced capacity ensures effective monitoring of SDGs implementation by delivering accuracy and timely statistics. Nilashi et al. (2023) explained how inadequate data availability, creates bottlenecks in the monitoring processes. This makes it difficult to assess progress accurately, especially in regions that require it most urgently. Moreover, the variations in methodologies employed in data collection also create discrepancies. Mugellini et al. (2021)

pointed out that varied data collection approaches can lead to challenges in benchmarking i.e. reducing the scope of comparative analyses across different regions and communities. Gennari and Navarro (2019) underscored the absence of standardized methodologies and universally applicable indicators. Such methodologies or indicators are essential to ensure the comprehensive and comparable assessments of SDG implementation across different contexts. Klopp and Petretta (2017); Kharas et al. (2018) mentioned that local administrations often lack the necessary infrastructure and technical expertise to gather accurate data on the diverse range of SDG indicators.

This exhaustive study of literatures, helped us in identifying a notable lack of harmony and standardization in tracking and monitoring systems. Mishra et al. (2024) mentioned that these constraints hinder comparability, especially when attempting to compare progress across different localities. Moreover, the SDG indicators, developed by the UN are designed to serve the global context, highlighting another key gap. Machingura and Lally (2017); Guarini et al. (2022) identified that SDG indicators at local level are not conceptually clear and the necessary data and established methodologies are not be available. Pateman et al. (2021) underlined that this lack of local indicators, forces localities to adapt global indicators, use their proxies, or search for alternative data sources, adding to the complexity for monitoring SDG localization. MacFeely (2019) added that without reliable, granular data, local authorities are unable to accurately assess specific community needs, identify vulnerable populations, tailor interventions effectively, or measure the true impact of their efforts. Fukuda-Parr and McNeill (2019) stated that these constraints lead to misdirected resources, inefficient programs, and a critical lack of accountability.

Various studies identified "Data Revolution" as a key factor to overcome this barrier. Pateman et al. (2021) stressed that local authorities should not only invest in infrastructure and expertise for traditional data collection but also need to actively integrate non-traditional sources, such as citizen-generated data. Grek et al. (2024) added that these integrated non-traditional data sources can provide real-time, context-specific insights and amplify the voices of marginalized populations, transforming SDG monitoring from a bureaucratic exercise into an evidence-based, participatory, and empowering process.

2.4 *Capacity and Technical Expertise Gaps*

Divi et al. (2024) highlighted the limitations faced by local and regional governments, which includes significant limitations in terms of necessary technical capacities and human resources. This scarcity severely hampers their ability to implement SDG initiatives. This problem also leads to a shortage of skilled personnel specifically for data collection and analysis, as well as a general lack of expertise in leveraging new technologies to advance sustainable development goals (Palomares et al. 2021; Rajaonson and Schmitt 2024). Ponce and Escuadra (2024) noticed absence of sufficient training and exchange opportunities within the framework of various regional and global summits. This may lead the local accomplishments to remain unrecognized (Ordonez-Ponce 2023). Masuda et al. (2021) underlined how this lack of resources prevent the dissemination of best practices and inhibits mutual learning among local actors.

Pandey et al. (2022) took the study further and analysed that scarcity of workforce can hinder the execution of SDG goals as by increasing the reliance on importing technologies instead of fostering local innovation. Kotturi et al. (2024) thus, underlined the need for well-planned capacity-building initiatives, knowledge transfer programs and investments in local education to cultivate the required expertise at the local community level. The local governments and stakeholders will otherwise continue to struggle in translating global aspirations into tangible, sustainable outcomes.

3 Research Methodology

This research adopts the systematic literature review (SLR) approach, which is widely recognized as a transparent and rigorous pathway to identify, evaluate and synthesize existing research relevant to the topic. An exhaustive exploration was conducted across key scholarly databases such as Google Scholar, Scopus, Web of Science and specific repositories of organizations like the United Nations, World Bank and leading and leading SDG think tanks; peer reviewed articles, conference proceedings, book chapters and reputable reports from international organizations published primarily in English, focusing on the intersection of localization with at least two of the following domains: the SDGs, Sustainability, and Local governance. To capture the recent advancements, the publications from the last decade (2015–2025) were prioritized alongside some works from earlier periods.

The researcher also used the Compilation and Synthesis method to achieve the objectives of the study. Compilation involves gathering and collecting data from different sources, while synthesis focuses on analysing and integrating this information to identify patterns, relationships, and insights. Collectively, this method is used for combining and interpreting findings from multiple studies or sources to gain a more comprehensive understanding of a research area.

Keywords and phrases used in the search included: *'Sustainable Development Goals', 'SDGs', 'SDG Localization', 'Sustainable Value Creation', 'Systems-Thinking', 'Participatory Approach', 'Local Governments', 'Local Stakeholders', 'Governance', 'Emerging Economies', 'Developing Countries', 'Budgetary Constraints', 'Resource Mobilization', 'Citizen-Generated Data', and combinations thereof. Suitable Boolean operators (AND, OR) were employed for refining search queries* (Alharbi and Stevenson 2020).

Data Collection and Synthesis: Data was derived from relevant research works which included:

(a) The specific SDGs localization initiatives and programs
(b) The SDGs targeted
(c) The geographical contexts, with primary focus on emerging economies

(d) Documented benefits, challenges, and ethical considerations of SDG localization

These integrated research works on SDG Localization helped us to explore patterns, trends, recurring motifs and identify the pertinent research gaps, which sets the stage for further investigation and analysis.

4 Analysis and Synthesis

4.1 Drivers of Transformational Change at the Local Level

This study identified the five key drivers of transformational change, which helps in accelerating SDG implementation at the local level:

Sensitization and Engagement of Local Actors: This study highlights that good governance processes must be inclusive and participatory in nature. The local actors like civil societies, private sector and local councils should be made aware of Agenda 2030 for meaningful outcomes. Lee et al. (2016) stressed the local authorities to understand their roles in SDG implementation. Ningrum et al. (2023) underscored the necessity of making local actors familiarize with the SDGs implications, opportunities and challenges and urging them to fully realize their crucial role.

Accountability Mechanisms: Monaco (2024) clearly state that SDG 16 (Peace, Justice and Strong Institutions), explicitly asks for accountable and participatory governance. It is fundamental to successful implementation of SDGs, which involves transparent and inclusive institutions to reduce all forms of corruption, especially at the local level (Kumar et al. 2025b; Prakash and Tiwari 2025). UNDP has pioneered corruption risk assessments in sectors like education, health and water (Mabry et al. 2024).

Participatory Planning and Service Delivery: Localization is not a novel notion, in fact the UNDP has supported local administrations in providing responsive services for decades (Canton 2021). Taking a cue from the Millennium Development Goals (MDGs), most scholars are of the opinion that SDG progress is heavily dependent on local government actions and active participation of all societal stakeholders (Ansell et al. 2022; Agrawal 2025). This requires the need to ensure that public policies are shared through consultative and participatory mechanisms, involving all relevant stakeholders in decision-making.

Local Economic Development: SDGs will not be realised until communities at the local level receive tangible benefits. UNDP has supported inclusive local economic development efforts that integrate social and environmental standards, reduce territorial disparities, foster social cohesion, generate local business opportunities and jobs and include women and excluded groups (Elayah et al. 2024; Joselin et al. 2025).

Partnerships: Multi-stakeholder partnerships are crucial for transforming global commitments into local action. This involves bringing together national and local governments, the private sector, civil society, foundations, social enterprises, traditional leaders, religious organizations and academia (Bravim et al. 2025). The UN Global Compact, for instance, drives business awareness and action for SDGs by fostering local networks and engagement (Van Tulder et al. 2021).

4.2 Components of Effective Localization

The researchers conducted a thorough review of studies related to role of Localization in attainment of Sustainable Development Goals (SDGs). This section will lay down key components of effective localization, followed by illustration real life examples and discussing the unique considerations for these contexts.

Multilevel Governance: This includes efforts to build national strategies with local priorities. Higher levels of government can create policies and financial incentives for local governments to implement sustainable practices at the grassroot level. Desai (2024); Gupta and Sigdel (2024) integrated local SDG indicators into national dashboards (as implemented in India and Nepal), resulting in better coherence.

Inclusive Partnerships and Governance Models: Lily and Sharma (2025) urged the marginalized sections of the society to participate in policy making for the SDGs to be successful. This involves making digital infrastructure and data generation at lower costs, enabling communities to convey specific needs, thus empowering local sustainability networks.

Direct Financing and Capacity Strengthening: Bisogno et al. (2023) highlighted the need for direct funding and capacity building for local governments and civil society organizations for the realization of SDGs. Almeile et al. (2024) emphasized public–private partnerships, as it may offer additional resources, providing an opportunity to overcome budget constraints.

Localized Robust Data Ecosystems: Bachmann et al. (2022) called for robust data ecosystems, which can be adapted as per the local needs for achieving the SDGs sustainably. This may include disaggregated and community-generated indicators. Oldenburg and Hämäläinen (2025) highlighted that citizen-generated data can amplify local aspirations. As it provides more accurate illustration of marginalized populations, fostering local development.

Digital Tools: ElMassah and Mohieldin (2020) highlighted the importance of employing appropriate digital tools and mechanisms especially at the local level. These tools enable fair participation, facilitate real-time engagement and enable policy feedback. The UNDP's "Localizing the SDGs Toolbox" provides practical tools and strategies for assessing, planning, implementing and monitoring local policies (United Nations Development Programme 2025). These strategies support local ownership and integration into sub-national strategies, promoting SDG progress at both local and global level.

5 Conclusion

This study explores the Sustainable Development Goals' Localization and identifies it not merely as a technical adjustment but as a strategic one. Localization helps in achieving the ambitious "Agenda 2030" and foster sustainable value creation for all. This study highlights that SDGs can be better realised locally, even though the SDGs themselves are universal. Each local level differs from each other in terms of resources, needs etc., there can't be a one size that fit for all. It demands contextualized approaches that empower local authorities and communities. It further emphasizes that empowerment of local bodies and co-creation by public and private sector helps in the evolution of localization from a top-down implementation model to more democratic and effective development pathways. The study proposes five key drivers of this transformational change, which includes- sensitization and engagement, robust accountability mechanisms, participatory planning, local economic development and multi-stakeholder partnerships. The study further explains the critical role of private sector. Private sector is no more seen just as a financier but also as an innovative and strategic partner. It can drive sustainable value creation through CSR, circular economy initiatives and impactful partnerships (Kumar et al. 2025a; Singh et al. 2025a).

Overall, this study analyses the SDG implementation at the local level. It is crucial to accelerate the SDG progress at subnational levels to make communities and processes equitable and sustainable. This will contribute to global well-being and unlock new avenues for sustainable value creation for all. This path forward demands a recognition of the local as the nexus of global ambition, requiring sustained political will, strategic investment and genuine collaboration across all sectors.

References

Agrawal D (2025) A comprehensive review of investing and financing for sustainable tourism projects. Environ, Sustain, Governance Insights 1(1):53–70

Ahmad I, Islam MR (2024) Empowerment and participation: key strategies for inclusive development. In: Building strong communities: ethical approaches to inclusive development. Emerald Publishing Limited, pp 47–68

Alharbi, A, Stevenson M (2020) Refining Boolean queries to identify relevant studies for systematic review updates. J Am Med Inform Assoc 27(11):1658–1666

Almeile AM, Chipulu M, Ojiako U, Vahidi R, Marshall A (2024) Project-focussed literature on public-private partnership (PPP) in developing countries: a critical review. Prod Planning Control 35(7):683–710

Anderson K, Ryan B, Sonntag W, Kavvada A, Friedl L (2017) Earth observation in service of the 2030 agenda for sustainable development. Geo-Spat Inform Sci 20(2):77–96

Ansell C, Sørensen E, Torfing J (2022) The key role of local governance in achieving the SDGs. In: Co-creation for sustainability: the UN SDGs and the power of local partnership. Emerald Publishing Limited, pp 9–22
Arszulowicz M (2025) Sustainable development: challenges and solutions. In: Governance strategies for effective sustainable development. IGI Global, pp 241–260
Bachmann N, Tripathi S, Brunner M, Jodlbauer H (2022) The contribution of data-driven technologies in achieving the sustainable development goals. Sustainability 14(5):2497
Barron GC, Laryea-Adjei G, Vike-Freiberga V, Abubakar I, Dakkak H, Devakumar D, Johnsson A et al (2022) Safeguarding people living in vulnerable conditions in the COVID-19 era through universal health coverage and social protection. Lancet Public Health 7(1):e86–e92
Bexell M, Bornemann B, Moallemi EA, Vien HB (2025) Accelerating the implementation of the sustainable development goals at the local level: unpacking governance processes and tensions. Earth Syst Governance 25:100264
Bexell M, Jönsson K (2017) Responsibility and the United Nations sustainable development goals. Forum Dev Stud 44(1):13–29. Routledge
Biggeri M (2021) A decade for action on SDG localisation. J Hum Dev Capabilities 22(4):706–712
Biloslavo R, Edgar D (2024) Integrating the sustainable development goals in business models. In: The Routledge handbook of accounting for the sustainable development goals. Routledge, pp 152–168
Bilsky E, Moreno AC, Fernández Tortosa A (2021) Local governments and SDG localisation: reshaping multilevel governance from the bottom up. J Hum Dev Capabilities 22(4):713–724
Bisogno M, Cuadrado-Ballesteros B, Rossi FM, Peña-Miguel N (2023) Sustainable development goals in public administrations: enabling conditions in local governments. Int Rev Adm Sci 89(4):1223–1242
Bravim JM, de Carvalho ACV, Segatto AP, Morais-da-Silva RL (2025) Are there key characteristics of social enterprises that foster relationships with government, universities and civil society?. Soc Enterp J
Bringezu S, Potočnik J, Schandl H, Yonglong L, Ramaswami A, Swilling M, Suh S (2016) Multi-scale governance of sustainable natural resource use—challenges and opportunities for monitoring and institutional development at the national and global level. Sustainability 8(8):778
Buheji M (2022) Impact of post-pandemic inflation on global POVERTY: a holistic perspective. Int J Manag (IJM) 13(5):2022
Caiado RGG, Leal Filho W, Quelhas OLG, de Mattos Nascimento DL, Ávila LV (2018) A literature-based review on potentials and constraints in the implementation of the sustainable development goals. J Cleaner Prod 198:1276–1288
Canton H (2021) United nations development programme—UNDP. In: The Europa directory of international organizations 2021. Routledge, pp 179–184
Choi NH (2021) Analyzing local government capacity and performance: Implications for sustainable development. Sustainability 13(7):3862
Clark R, Reed J, Sunderland T (2018) Bridging funding gaps for climate and sustainable development: pitfalls, progress and potential of private finance. Land Use Policy 71:335–346
Desai DJ (2024) Collaborative governance model: an approach to achieve last-mile delivery of SDG 2030 localisation. J Governance Public Policy 14(2):54–73
Divi S, Dash DN, Sahoo MK (2024) Problems and challenges of Indian rural local governments in achieving sustainable development goals: an analysis of the viable perspectives. Int J Sustain Dev Plann 19(7)
Dulak M (2023) Contribution of subnational authorities to multilateralism from the EU perspective—implementation of the SDGs. Global Pol 14:13–21
Dutta S, Das KC (2024) Child poverty and deprivation. In: Mapping sustainable development goals for children in India: progress and present challenges. Springer Nature Singapore, Singapore, pp 13–43
Elayah M, Al-Sameai N, Khodr H, Gamar S (2024) Community-based initiatives and public services delivery in a fragile context: the case of Yemen. Nonprofit Volunt Sect Q 53(1):5–28

ElMassah S, Mohieldin M (2020) Digital transformation and localizing the sustainable development goals (SDGs). Ecol Econ 169:106490

Fukuda-Parr S, McNeill D (2019) Knowledge and politics in setting and measuring the SDG S: introduction to special issue. Global Pol 10:5–15

Gayathri R, Vijayabanu C, Theresa C (2024) Economic disruption and global obscurity–insights and challenges. In: Economic uncertainty in the post-pandemic era. Routledge, pp 1–26

Gennari P, Navarro DK (2019) Validation of methods and data for SDG indicators. Statist J IAOS 35(4):735–741

Grek S, Tichenor M, Bandola-Gill J (2024) Numbers as Utopia: sustainable development goals and the making of quantified futures. Br J Polit Int Relat 26(3):742–758

Guarini E, Mori E, Zuffada E (2022) Localizing the sustainable development goals: a managerial perspective. J Public Budg Account Financ Manag 34(5):583–601

Guha J, Chakrabarti B (2019) Achieving the sustainable development goals (SDGs) through decentralisation and the role of local governments: a systematic review. Commonwealth J Local Governance 22:1–21

Gupta AK, Sigdel TS (2024) Integrating sustainable development goals in local plans: unlocking practices and challenges of local governments in Nepal. Heliyon 10(20)

Gustafsson S, Krantz V (2025) Inside out/outside in—the balance between internal and external localisation of the sustainable development goals in Swedish municipalities. Cities 158:105696

Haou E, Allarané N, Aholou CC, Bondoro O (2025) Integrating sustainable development goals into urban planning to advance sustainability in sub-Saharan Africa: barriers and practical solutions from the case study of Moundou, Chad. Urban Sci 9(2):22

Hariram NP, Mekha KB, Suganthan V, Sudhakar K (2023) Sustainalism: an integrated socio-economic-environmental model to address sustainable development and sustainability. Sustainability 15(13):10682

Immler NL, Sakkers H (2022) The UN-sustainable development goals going local: learning from localising human rights. Int J Hum Rights 26(2):262–284

Joselin D, Singh A, Tiwary KS (2025) The nexus between CSR practices and financial performance: empirical evidence from the manufacturing sector in India. Environ, Sustain, Governance Insights 1(1):1–17

Kaith A, Sachdeva G (2024) Bibliometric insights into green human resource management: a connection with business ethics and values. PURUSHARTHA—J Manage, Ethics Spirituality 17(1):51–63. https://doi.org/10.21844/16202117104

Kharas H, McArthur JW, Rasmussen K (2018) How many people will the world leave behind?. Global economy and development at Brookings, working paper 123

Klopp JM, Petretta DL (2017) The urban sustainable development goal: indicators, complexity and the politics of measuring cities. Cities 63:92–97

Kotturi Y, Hui J, Johnson TJ, Sanifu L, Dillahunt TR (2024) Sustaining community-based research in computing: lessons from two tech capacity building initiatives for local businesses. Proceedings of the ACM on human-computer interaction vol 8, no CSCW1, pp 1–31

Krantz V, Gustafsson S (2021) Localizing the sustainable development goals through an integrated approach in municipalities: early experiences from a Swedish forerunner. J Environ Planning Manage 64(14):2641–2660

Kumar K, Singh R, Choudhary M, Sharma A (2025a) Climate change mitigation strategies in environmentally sensitive industries. In: Climate change and social responsibility, vol 25. Emerald Publishing Limited, pp 185–204

Kumar K, Singh R, Mishra P, Lehri BPS, Kumar A (2025b) What influences companies to go beyond mandatory corporate social responsibility rule? Empirical evidence from India. In: Integrating environmental sustainability and corporate social responsibility: a move beyond profit: corporate social responsibility for environmental sustainability. Springer Nature Switzerland, Cham, pp 45–63

Leavesley A, Trundle A, Oke C (2022) Cities and the SDGs: realities and possibilities of local engagement in global frameworks. Ambio 51(6):1416–1432

Lee BX, Kjaerulf F, Turner S, Cohen L, Donnelly PD, Muggah R, Davis R et al (2016) Transforming our world: implementing the 2030 agenda through sustainable development goal indicators. J Public Health Policy 37(Suppl 1):13–31
Leino H, Puumala E (2021) What can co-creation do for the citizens? Applying co-creation for the promotion of participation in cities. Environ Plann c: Polit Space 39(4):781–799
Lily M, Sharma B (2025) Inclusive society: a pathway to achieving sustainable development goals. Int J Environ Sci 503–510
Lüdeke-Freund F, Rauter R, Pedersen ERG, Nielsen C (2020) Sustainable value creation through business models: the what, the who and the how. J Bus Models 8(3):62–90
Mabry RM, Doctor HV, Khair MN, Abdelgalil M, Rashidian A (2024) Measuring progress on health and well-being in the Eastern Mediterranean region via voluntary national reviews, 2016–2021: what does the data reveal? PLOS Glob Public Health 4(7):e0002838
MacFeely S (2019) The big (data) bang: opportunities and challenges for compiling SDG indicators. Global Pol 10:121–133
Machingura F, Lally S (2017) The sustainable development goals and their trade-offs: case study report. Overseas Development Institute, London
Marawu NM, Utete R (2025) Exploring the lived experiences of citizens on the effects of good governance practices on housing service delivery: a case study of four municipalities in South Africa. SN Soc Sci 5(8):1–26
Masuda H, Okitasari M, Morita K, Katramiz T, Shimizu H, Kawakubo S, Kataoka Y (2021) SDGs mainstreaming at the local level: case studies from Japan. Sustain Sci 16(5):1539–1562
Mishra M, Desul S, Santos CAG, Mishra SK, Kamal AHM, Goswami S, Kalumba AM et al (2024) A bibliometric analysis of sustainable development goals (SDGs): a review of progress, challenges, and opportunities. Environ, Dev Sustain 26(5):11101–11143
Moallemi EA, Malekpour S, Hadjikakou M, Raven R, Szetey K, Ningrum D, Dhiaulhaq A, Bryan BA (2020) Achieving the sustainable development goals requires transdisciplinary innovation at the local scale. One Earth 3(3):300–313
Mohieldin M, Wahba S, Gonzalez-Perez MA, Shehata M (2022) The role of local and regional governments in the SDGs: the localization agenda. In: Business, government and the SDGs: the role of public-private engagement in building a sustainable future. Springer International Publishing, Cham, pp 105–137
Monaco S (2024) SDG 16. Promote peaceful and inclusive societies for sustainable development, provide access to justice for all, and build effective, accountable, and inclusive institutions at all levels. In: Identity, territories, and sustainability: challenges and opportunities for achieving the UN sustainable development goals. Emerald Publishing Limited, pp 157–165
Mugellini G, Villeneuve J-P, Heide M (2021) Monitoring sustainable development goals and the quest for high-quality indicators: learning from a practical evaluation of data on corruption. Sustain Dev 29(6):1257–1275
Nilashi M, Keng Boon O, Tan G, Lin B, Abumalloh R (2023) Critical data challenges in measuring the performance of sustainable development goals: solutions and the role of big-data analytics. Harvard Data Sci Rev 5(3):1–36
Ningrum D, Raven R, Malekpour S, Moallemi EA, Bryan BA (2023) Transformative potential in sustainable development goals engagement: experience from local governance in Australia. Glob Environ Chang 80:102670
Noah AO, David OO (2024) External financing, governance, and infrastructure development: evidence from Sub-Saharan Africa. CSID J Infrastruct Dev 7(3):2
Novovic G (2022) Can agenda 2030 bring about localization? policy limitations of agenda 2030 in the broader global governance system. Dev Policy Rev 40(4):e12587
Oldenburg J-R, Hämäläinen M (2025) Advancing urban sustainability through robust data ecosystem practices. In: Sustainable and smart cities. Routledge, pp 19–36
Opoku A (2025) Synergizing business and sustainability: integrating the SDGs into business strategies in the architecture, engineering and construction sector. Manage Sustain: Arab Rev

Ordonez-Ponce E (2023) The role of local cultural factors in the achievement of the sustainable development goals. Sustain Dev 31(2):1122–1134

Ortiz-Moya F, Reggiani M (2025) Operationalising the follow-up and review of the sustainable development goals at the local level: insights from European cities and their voluntary local review experience. J Urban Ecol 11(1):juaf001

Palomares I, Martínez-Cámara E, Montes R, García-Moral P, Chiachio M, Chiachio J, Alonso S et al (2021) A panoramic view and swot analysis of artificial intelligence for achieving the sustainable development goals by 2030: progress and prospects. Appl Intell 51(9):6497–6527

Pandey N, de Coninck H, Sagar AD (2022) Beyond technology transfer: innovation cooperation to advance sustainable development in developing countries. Wiley Interdisc Rev: Energy Environ 11(2):e422

Pateman R, Tuhkanen H, Cinderby S (2021) Citizen science and the sustainable development goals in low and middle income country cities. Sustainability 13(17):9534

Ponce EJA, Escuadra CJT (2024) Understanding international and regional partnerships for sustainable development goals in higher education: a focus on the Asia Pacific region. Asia-Pacific Soc Sci Rev 24(3):3

Prakash A, Tiwari AK (2025) The role of institutional factors in achieving SDG 16: a thematic review of drivers of corporate anti-corruption action. Environ, Sustain, Gov Insights 1(1):34–52

Rahman MM (2021) Achieving sustainable development goals of agenda 2030 in Bangladesh: the crossroad of the governance and performance. Public Adm Policy 24(2):195–211

Rajaonson J, Schmitt K (2024) AI and data technologies in advancing sustainable development goals. IEEE Technol Soc Mag 43(1):33–38

Sever SD, Tok E, Sellami AL (2025) Sustainable development goals in a transforming world: understanding the dynamics of localization. Sustainability 17(6):2763

Shivakoti R, Howlett M, Fernandez V, Nair S (2021) Governing international regime complexes through multi-level governance mechanisms: lessons from water, forestry and migration policy. Int J Water Resour Dev 37(4):658–675

Singh R, Kumar K, Khan S (2024) A comprehensive view of artificial intelligence (AI)–based technologies for sustainable development goals (SDGs). Artif Intell Enabled Manage: Emerg Econ Perspect 183. https://doi.org/10.1515/9783111172408-012

Singh R, Mushtaq S, Kumar K, Rahman O (2025a) Creating sustainable workplace: enhancing HRM strategies in the digital circular economy. In: Sustainable innovations and digital circular economy. Springer Nature Singapore, Singapore, pp 129–145

Singh R, Sharma A, Kumar K (2025b) Navigating carbon neutrality: strategies for sustainable future and environmental challenges. In: Climate neutrality through smart eco-innovation and environmental sustainability. Springer Nature Switzerland, Cham, pp 69–80

Singh R, Tiwary KS, Kumar K, Kumar A (2025c) Empowering a sustainable future: blockchain's role in economy, organizations, and finance towards smart circular economy. In: Blockchain technologies for smart circular economy and organisational sustainability. Springer Nature Switzerland, Cham, pp 271–283

Srivastava SK, Anshul A, Pathak P, Bansal J (2018) Responsible consumption for curbing food wastage: an exploratory enquiry. PURUSHARTHA—J Manag, Ethics Spirituality 11(2):104–116

Srivastava SK, Singh S, Srivastava M (2021) Food waste at household and social gatherings: drivers and possible remedies. PURUSHARTHA—J Manag, Ethics Spirituality 14(2):117–133

Srivastava SK, Sinha AK, Kumar S (2024) Workplace spirituality empowered by spiritual leadership to promote organizational effectiveness: a bibliometric analysis. PURUSHARTHA—J Manage, Ethics Spirituality 17(1):1–22. https://doi.org/10.21844/16202117101

Tripathi V, Singh K, Parihar S, Srivastava SK (2022) A bibliometric analysis on supply chain collaboration. PURUSHARTHA—J Manage, Ethics Spirituality 15(2):66–103. https://doi.org/10.21844/16202115205

Van Tulder R, Rodrigues SB, Mirza H, Sexsmith K (2021) The UN's sustainable development goals: can multinational enterprises lead the decade of action?. J Int Bus Policy 4(1):1

United Nations Development Programme. Localizing the SDGs toolbox. SDG integration. Accessed 13 Aug 2025. https://sdgintegration.undp.org/localizing-sdgs-toolbox

Washaya W (2020) Contemporary institutional autonomy and SDG implementation: challenges and opportunities. Peace, Justice Strong Institutions 1–10

Yang X, Jiang S (2017) Challenges towards ecological sustainability in China: an introduction. In: Jinan forum on geography and ecological sustainability guangzhou, China. Springer International Publishing, Cham, pp 3–18

Barriers to Sustainable Development Goal (SDG) Implementation: A Structural Analysis Using MCDM Techniques

Remy Jonkam Oben and Shikha Daga

Abstract Environmental challenges and their severe impact have forced economies worldwide to focus on development that is not only focused on growth but is also sustainable. The one goal that binds the world together in this aspect is the attainment of Sustainable Development Goals (SDGs). However, the global commitment towards SDG attainment faces various barriers that delay and/or hinder its adoption. There are various challenges like the cost-related issues, infrastructural hurdles, governance challenges and others. Though literature has provided support for these barriers but still their hierarchy that explains the driving-dependence among the factors is still underexplored. The chapter focuses on identifying the structural relationship across the various barriers that delay the adoption of SDGs. Based on the multi-criteria decision-making technique – Total Interpretive Structural Modelling (TISM) and Cross-Impact Matrix Multiplication Applied to Classification (MICMAC) analysis, the study captures the cause-and-effect dynamics amongst various barriers identified through literature. Subsequently, MICMAC analysis based on the driving and dependence power classifies the barriers into four categories, namely autonomous, dependent, linkage, and driving. This structural insight provides a strategic framework for policymakers, development agencies, and sustainability planners to prioritize efforts and design more effective SDG implementation strategies. The chapter contributes to both theory and practice by offering a systems-based understanding of SDG challenges and demonstrating the utility of interpretive structural modelling in sustainability research.

Keywords Sustainable development goals (SDGs) · Implementation barriers · Total interpretive structural modeling (TISM) · MICMAC analysis · Multi-criteria decision making · Sustainability policy

---

R. J. Oben (✉)
Department of Business Administration, Near East University, Nicosia, Türkiye
e-mail: remyjonkam.oben@neu.edu.tr

S. Daga
PGDAV College, University of Delhi, Delhi, India
e-mail: Shikha.menani@pgdav.du.ac.in

K. Kumar et al. (eds.), *Sustainable Value Creation: Corporate Strategies for SDGs Implementation*, World Sustainability Series,
https://doi.org/10.1007/978-3-032-24652-3_3

1 Introduction

Severe environmental deterioration across economies has led to the initiation of sustainable development beyond economic wealth generation. Change in climate and depletion of natural flora and fauna have led to the adoption of Sustainable Development Goals (SDGs). There has been a recent spike in studies that focus on emission mitigation and the various factors that contribute to it. Whether it be the adoption of clear energy sources, the transition towards green finance, or the adoption of circular principles in production and consumption, all are directed towards the same target of sustainable development. For instance there are studies that have used MCDM techniques to focus on sustainable fashion adoption (Yadav et al., 2025). Promotion of the use of renewable energy backed by institutional support and political will is crucial for the successful implementation of technologies. As highlighted across studies, to promote green growth, advancements in energy storage along with having a more decentralised system can be beneficial for enhancing renewable energy use (Boiko et al. 2025). The recent rise in interest towards mitigating emissions and promotion of sustainable development directly focus the attainment of SDGs. It encompasses 17 interconnected goals and 169 specific targets, thereby addressing a wide range of global issues. The goals though, are important for attaining the sustainable growth are backed by various challenges at the micro as well as the macro level. A study by Matović et al. (2024) highlighted significant challenges in measuring SDG 4 targets in Serbia, including methodological limitations in data collection and others. However, despite wide global acceptance and policy alignment with the SDG framework, progress towards their attainment remains uneven and, in many cases, significantly off-track. A major SDG implementation barrier is (students') limited opportunities to learn about specific targets (Lin 2024). The gap is even more visible across the economies that are struggling with generating and managing decent economic wealth, categorized as low and middle-income countries. Even in relatively advanced economies, the transition toward sustainable practices is hindered by entrenched institutional inertia, short-term political interests, and misaligned economic incentives. Study by Abid et al. (2022) for instance documented how technological innovation is related to the promotion of green growth in Pakistan with other environmental challenges posing a hindrance to the adoption of green growth. There are other studies that highlight how rising geopolitical crises, alarming changes in climatic conditions, and severe economic disruptions the barriers are further exacerbated, making the SDG implementation process increasingly arduous. The results however are found to be heterogeneous depending upon the region, time period and choice of variables. Study by Ahmed et al. (2024) for instance, highlighted how globalization, geopolitical risk, and institutional quality helped in mitigating the environmental challenges while other factors like growth across the economy, trade and prices of the natural resources added to the same.

Literature has discussed multiple factors directly or indirectly that impact the attainment of SDGs at the micro and macro level. There are studies that have discussed the barriers to adoption at the individual SDG level and there are others that have taken the SDGs as a whole. There are certain barriers that are found common to the studies however there is no study that has tried to compile the factors to present an interlinkage between the various barriers. This chapter has done the same by using contemporary techniques of multi criteria decision making that is TISM and MICMAC, which can assist in building a hierarchical structure of the variables and assist in the policy-making.

The chapter's contribution to the SDG is threefold. Firstly it identifies and validates the key barriers to SDG implementation through an extensive literature review and expert validation. Secondly, it makes use of TISM to identify the cause-and-effect hierarchy along with providing its interlinkage. Thirdly, MICMAC classifies the barriers into autonomous, dependent, linkage, and driving barriers based on their driving power and dependence, helping to determine which barriers are foundational and which are outcomes of others within the SDG implementation framework.

The findings of the chapter are relevant, especially for policymakers who can take the clues of the interrelationship amongst the various SDG barriers and formulate relevant policies that can lead to efficient and timely adoption of SDG across the global economy. This research not only enriches academic discourse but also serves as a practical guide for transformative action in the sustainability domain.

2 Literature Review

Attainment of SDGs across all the economies at a consistent pace is pertinent to ensure there is a paradigm shift towards development that is economic and green. There is a vast literature that discusses the necessity to make a shift towards sustainable development and the role that global initiatives like SDGs can play in the same. Even after years of efforts and global support still, we are still lacking in the attainment of the SDGs promptly. The COVID pandemic, geopolitical tussles and economic uncertainty all hinder the progress of sustainability (Singh et al. 2025). Though, literature has documented multiple factors that hinder the achievement of SDGs, a few of them along with its relevance are listed in this section.

2.1 Lack of Awareness (B1)

Attainment of SDGs can be smooth and quick if it is accepted at the micro as well as macro levels. Public awareness hence plays a crucial role in making a smooth transition toward sustainable development. Study by Islam and Jirattikorn (2024) highlight women's under-representation in Bangladesh's STEM education inspite of increased enrolment at the tertiary level. There can be various reasons for the same like the perception in the society, absence of required infrastructure and others. Lack of awareness is also prevalent among other sectors like hospitality and tourism. For instance study by Singh and Dutt (2024) highlighted that though hotels are keen on adopting sustainable initiatives but lack of awareness and understanding hinders the same. The results are supported by Brandão and Verissimo (2024) that provided how lack of awareness in addition to other factors can hinder the adoption of the same in the construction, thereby challenging the SDG 11 progress. It thereby highlights the crucial role of lack of awareness in promoting the SDGs at the micro and macro levels. Community involvement, responsiveness towards development initiatives, and adherence to the regulations all help in promoting the SDGs across economies. Public awareness is in fact one of the significant barriers that has delayed the SDG attainment at the grassroot level. This is because of significant information gaps and asymmetry, especially in the low and middle-income countries. Furthermore, societal apathy and mistrust in institutions often weaken citizen involvement. The significance of the factor in promoting SDG implementation is thus important to be identified along with finding its linkage with other factors.

2.2 Cultural and Social Barriers (B2)

Cultural and social norms have a strong influence on behaviours, attitudes, and development outcomes, and hence play a strong role in SDG implementation. Many economies because of their belief systems and community structures face conflict with the transformative changes required for SDG implementation. Zighan et al. (2026) for instance, highlighted cultural and behavioural factors as foundational constraints that hinder the adoption of circular economy practices in the Gulf Cooperation Council (GCC) region.

For instance, gender biasedness across certain economies may hinder education and/or their participation in the workforce thereby impacting SDG linked to quality education (SDG 4), and gender equality (SDG 5). Similarly, few cultures back big families and are against family planning that hinders progress toward good health and well-being (SDG 3) and responsible consumption and production (SDG 12). Study by Borsatto et al. (2024) for instance studies the role that higher educational institutions can play in SDG attainment through community outreach. This suggest that overcoming such cultural and social barriers is necessary for the attainment of SDGs.

2.3 Data Deficiency (B3)

SDGs focus on goal accomplishment across different sectors within a specific time frame. For this, it is necessary to have data availability to identify the gaps and to take steps to bridge them. However, getting real-time data that is authentic is difficult, which delays the corrective actions, and hinders the SDG goals achievement in a time-bound manner. Study by Fraisl et al. (2025) provided that the lack of data is one of the major challenges in assessing the progress toward the sustainability target attainment, particularly for governance indicator 16.6.2, especially among marginalized populations. Methodological limitations in data collection for instance is found to hamper the accurate assessment of SDG 4 indicators across the Serbian economy (Matović et al. 2024). It highlights how data deficiency is a critical barrier. Many nations struggle with inadequate statistical systems, limited technical capacity, and insufficient funding for data infrastructure. Addressing data deficiency is thus necessary to accomplish the goals of SDG at the local and global levels.

2.4 *Technological Gaps (B4)*

Transition towards improved technology and cleaner sources of energy not only helps in promoting energy efficiency but also helps in promoting the circular economy. Study by Yu et al. (2025) provided that enhancement in the technology, enhanced use of renewable energy along with forest conservation can help in mitigating GHG emissions across the G7 economies. Another study highlighted how digital twin technology though promising, has its barriers that limit their impact on its widespread adoption (Patel et al. 2024). Economies that are lagging in research and development usually face hurdles in meeting the SDG targets. These gaps are more pronounced in least developed and developing countries, where access to the internet, digital literacy, and high-tech manufacturing remains limited. Study by Tahir et al. (2025) for instance, highlighted the role of technological factors in influencing commuters' willingness to use public transit and promoting sustainable transportation. Investment in technological upgradation thus is documented as an important factor that can ensure global sustainable development goals and is also a major barrier in its implementation.

2.5 *Policy Incoherence (B5)*

Policy incoherence is a major hindrance to the attainment of SDGs at the macro as well as the micro level. It refers to the misalignment or contradiction between different policy areas that hampers SDG implementation. Policy incoherence is also documented to erode political trust that makes SDG implementation difficult. A study by Revtiuk and Zelinska (2026) analysed different factors that can impact political trust as it is backed by literature to undermine SDG implementation. Also, economies focusing on enhancing GDP alone may put less focus on environmental sustainability. Such policy contradictions not only create inefficiencies but also promote overall mistrust toward SDGs. For instance, while countries may promote renewable energy to meet SDG 7 (Affordable and Clean Energy), they may simultaneously subsidize fossil fuels, thereby having no net effect on the environment. It is, therefore, necessary to ensure that there is policy integration that ensures efficient SDG implementation. Studies have highlighted how various factors like the quality of the institutions, development in the educational system along with growth in the overall economy amongst others can assist in promoting green energy (Sadiq et al. 2024).

2.6 Capacity Limitations (B6)

In addition to the various barriers listed above, the one factor that integrates all of them together is the capacity constraint. Many economies for instance face issues with experienced staff, technical expertise, and weak administration. These capacity deficits severely and negatively affect all stages of the SDG attainment from planning and coordination to execution, monitoring, and reporting. Strengthening capacities thus requires not just training programs but also long-term investments in knowledge systems, peer learning networks, and resource mobilization. Without overcoming these limitations, even well-funded and well-intentioned SDG strategies may fall short of delivering transformative outcomes. Study by Maity and Sinha (2024) for instance documented how conflict in itself is a major barrier to attainment of any productive goal such as SDG objective of poverty reduction.

2.7 Governance Issues (B7)

In addition to financial constraints, governance issues are also a major hindrance to the attainment of the SDGs. Weak governance and institutional inefficiencies have been documented in the literature to be a major barrier. Study by Bespalyy et al.(2025) highlighted how insufficient institutional support in addition to other barriers, can hinder the integration of SDGs in the educational process. Study by Adam and Fazekas (2021), for instance used a systematic literature review of existing studies to identify how use of information and communication technology (ICT) can support anti-corruption in general by enhancing public scrutiny. The study thus suggests how the same can thus be used for furthering the cause of SDG wherein there is a huge role of transparency and accountability that is needed and is also missing giving rise to exaggerated claims and leading to huge instances of greenwashing (Daga et al. 2025) This is because governance plays a major role in setting policy targets, ensuring accountability, and coordinating stakeholders' targets with the attainment of SDGs. Various economies have adopted stringent regulations for adhering to the goals of sustainability, however many economies especially the ones at the bottom of the development hierarchy still have their issues with the SDGs adherence. These economies are faced with inherent corruption, lack of transparency, and poor rule of law that accompanied by bureaucratic inefficiencies severely hamper the sustainable development efforts. Weak governance also discourages private investment and international aid, compounding challenges related to financing and implementation. It is therefore necessary to ensure a strong governance system integrated with institutional support to sustain progress toward SDGs.

2.8 Financial Constraints (B8)

Sustainable development requires the adoption of cleaner sources of energy along with effective policy implementation, technological innovation, and public awareness to ensure long-term environmental protection, economic resilience, and social equity. Study by Abbass et al (2025) for instance highlighted the important role of technology innovation and green energy sources. It is also indicative of the fact that all these factors have one thing in common and that is availability of requisite finance. However, economies worldwide suffer from a huge gap in funding the SDGs, especially in low and middle-income economies. This is important as LMICs for instance are documented to benefit if they transition towards clean energy as it helps in reducing carbon emissions (Nasrullah et al. 2023). It can be because economies that are in transition phase may focus on promoting economic wealth, limited in-house funding, and excessive reliance on foreign aid often divert funds away from long-term sustainability goals. Another hindrance is the lack of widespread use and availability of green financing options that further aggravate the SDG implementation barriers. Study by Afzal et al. (2022) documented how financial development had a negative impact on the environment and hence suggested for the adoption and promotion of green finance to promote long term sustainability. There are other studies too that suggests that it is therefore essential to identify the financial constraints and their interaction with other factors to ensure there is a quick and smooth transition toward SDG implementation and adoption. Study by Bolatova et al. (2025) identified how infrastructure gaps and financial constraints hindered adoption of clean drinking water and sanitation, thereby highlighting a crucial gap in the adoption of SDGs across the rural areas.

Based on the above-highlighted and literature-identified barriers, this chapter employs multi-criteria decision-making tool—TISM (Total Interpretive Structural Modeling) and MICMAC (Cross-Impact Matrix Multiplication Applied to Classification) analysis. It not only helps in providing interlinkage amongst the various identified factors but also assists in providing a hierarchical structure amongst the variables (Yadav et al. 2025). TISM and MICMAC are powerful tools for identifying complex systems across different variables used in a study. TISM enhances traditional ISM (Interpretive Structural Modeling) by incorporating expert opinions to provide deeper insights into the nature of these relationships, making the model more explanatory and grounded in contextual understanding (Daga et al. 2025). MICMAC goes a step ahead in the MCDM technique as it helps in providing the driving and dependence power of each factor using a cross impact table. It thereafter classifies each of them into categories such as autonomous, dependent, linkage, and driver variables. Such categorization helps in prioritizing strategic actions and in designing effective policy frameworks. Together, TISM and MICMAC provide a robust methodological framework for decision-makers, especially in domains like sustainability, policy planning, and management. A detailed step-by-step analysis of TISM and MICMAC is provided in the next section.

3 TISM and MICMAC Analysis

The first step in TISM and MICMAC analysis is to identify factors of interest either from literature and if there is no base then the Delphi method can be used wherein expert opinions are used to list down the factors required for further analysis. Study by Ametepey et al (2024) used expert opinion to assert how AI can significantly assist in the implementation of SDGs. Here as there is enough literature on barriers to SDG, we have used existing studies to identify the relevant factors. This method also ensures that the model is grounded in existing theory and empirical findings. Based on the literature we have identified 8 factors which are explained in detail in the literature section. The next step in the TISM analysis is to get the factors validated by experts and create a VAXO table. Experts in this chapter consist of seven experts who are familiar with the chapter's domain. A pairwise questionnaire consisting of 56 (8*7) statements is given to each expert who provides their judgment based on the relevance and relation of each factor with the other. For instance, experts are asked whether B1 impacts B2 and so on for all variables based on which a VAXO table is created where V stands for row impacting column, A stands for column influencing row, X stands for both influence each other and O stands for no relationship and influence between the two. Based on the Structural Self-Interaction Matrix (SSIM) (Table 1) all the next steps of TISM are achieved. For instance, the initial reachability matrix which is the binary output of SSIM is created and is presented in Table 2. The IRM captures the direct relationship between the factors and is derived from expert judgments. It consists of only 1 and 0 where 1 stands for the row factor influencing the column factor while 0 stands for no direct influence of the row factor on the column factor. The matrix is thus directional and is derived from expert opinions. All the analysis is done using smart ISM software (Ahmad and Qahmash 2021).

Based on the IRM, transitive links are identified that capture the indirect links, and a new matrix called the final reachability matrix (FRM) is created that captures both direct and indirect links (Table 3). This matrix is used for further classification of the hierarchical matrix.

Table 1 Structural self-interaction matrix

	B1	B2	B3	B4	B5	B6	B7	B8
B1		V	V	V	V	V	V	V
B2			V	V	V	V	V	V
B3				O	A	V	A	O
B4					V	O	V	A
B5						V	V	V
B6							A	A
B7								V
B8								

Aggregated VAXO table from all experts

Table 2 Initial reachability matrix (IRM)

	B1	B2	B3	B4	B5	B6	B7	B8
B1	1	1	1	1	1	1	1	1
B2	0	1	1	1	1	1	1	1
B3	0	0	1	0	0	1	0	0
B4	0	0	0	1	1	0	1	0
B5	0	0	1	0	1	1	1	1
B6	0	0	0	0	0	1	0	0
B7	0	0	1	0	0	1	1	1
B8	0	0	0	1	0	1	0	1

Source Authors' computation using ISM software by Ahmad and Qahmash (2021)

Table 3 Final reachability matrix (FRM)

	B1	B2	B3	B4	B5	B6	B7	B8
B1	1	1	1	1	1	1	1	1
B2	0	1	1	1	1	1	1	1
B3	0	0	1	0	0	1	0	0
B4	0	0	1*	1	1	1*	1	1*
B5	0	0	1	1*	1	1	1	1
B6	0	0	0	0	0	1	0	0
B7	0	0	1	1*	1*	1	1	1
B8	0	0	1*	1	1*	1	1*	1

Source Authors' computation using ISM software by Ahmad and Qahmash (2021)

After the creation of the FRM and identification of diverse relationships among the factors extracted from literature and validated by experts, the next step is to create the interlinkage amongst the variables. This is done using MICMAC wherein the dependency between the variables is organized in a table wherein the factors are classified into four sub-groups namely-dependent, independent, linkage, and autonomous. These classifications are based on two parameters driving and dependence power.

While the dependence provides how much a variable used in the study is influenced by other variables, driving power refers to the opposite that is how much power a variable has in impacting other variables. A variable that is high on dependence is impacted by various other factors and any change in any or all of those factors can significantly impact this factor. A variable that is low on dependence is in fact self-contained as it is not influenced by other variables.

Similarly, a variable that has a greater driving power has an impact on various other variables thereby creating a ripple effect across the system. On the other hand, variables that are low on driving power have limited influence and are passive in the entire model. Based on the combination of dependence and driving powers the factors are then classified into the four subcategories mentioned above: dependent, independent, linkage, and autonomous in the MICMAC table. There are studies that have used TISM and MICMAC methodology across different domains like that in the banking system (Barve and Nayak 2023) and to identify various barriers in the Industry 4.0 (Ben Ruben et al. 2023).

Dependent factors are the ones that are highly influenced by others and have a low driving power. Such variables are thus reactive and depend on other variables. Independent variables are the opposite that have low dependencies and high driving power and are capable of influencing other factors in the system. Linkage factors depict both high dependence and high driving power and thus are the intermediaries in the system. They often play a crucial role as they are the key connectors between different elements. Autonomous are low on both driving and dependence power and are generally isolated from the other factors in the system. They are less influential and less affected by changes in other factors of the system.

After FRM, iterations are repeated to get the levels to which the eight variables belong. It is given in Table 4.

Based on above iterations a digraph showing the levels and its interlinkage is presented below. The diagram is to visually represent the levels and does not discriminate between direct and indirect links here. For the different direct and indirect links, FRM has to be used.

The categorisation of the variables into 5 levels shows the structure and hierarchy of the variables used in this chapter (Fig. 1).

Based on the FRM and the driving and dependence power, the following MICMAC table (Table 5) is drawn, based on which the MICMAC matrix is established (Fig. 2).

Based on the overall driving and dependence power, the eight factors of this chapter are classified into four categories (Table 5). Lack of awareness (B1) and cultural & social barriers (B2) having high driving and low dependence, are classified as independent variables. Data deficiency (B3) and capacity limitations (B6) have the opposite with low driving and high dependence are classified as dependent variables. The other four namely technological gaps (B4), policy incoherence (B5),

Table 4 Iteration table

Factors	Antecedents	Reachability	Intersection	Level
B1	1	1,2,3,4,5,6,7,8	1	Level 5
B2	1,2	2,3,4,5,6,7,8	2	Level 4
B3	1,2,3,4,5,7,8	3,6	3	Level 2
B4	1,2,4,5,7,8	3,4,5,6,7,8	4,5,7,8	Level 3
B5	1,2,4,5,7,8	3,4,5,6,7,8	4,5,7,8	Level 3
B6	1,2,3,4,5,6,7,8	6	6	Level 1
B7	1,2,4,5,7,8	3,4,5,6,7,8	4,5,7,8	Level 3
B8	1,2,4,5,7,8	3,4,5,6,7,8	4,5,7,8	Level 3

Source Authors' computation using ISM software by Ahmad and Qahmash (2021)

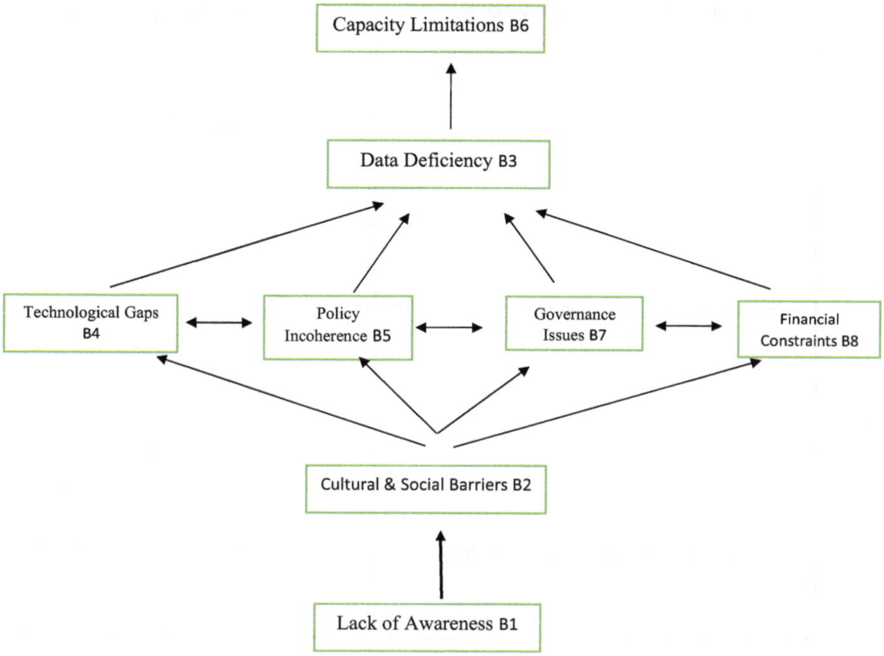

Fig. 1 TISM model

Table 5 MICMAC classification table

Barrier (Code)	Driving power	Dependence power	Classification
Lack of awareness (B1)	8	1	Independent
Cultural and social barriers (B2)	7	2	Independent
Data deficiency (B3)	2	7	Dependent
Technological gaps (B4)	6	6	Linkage
Policy incoherence (B5)	6	6	Linkage
Capacity limitations (B6)	1	8	Dependent
Governance issues (B7)	6	6	Linkage
Financial constraints (B8)	6	6	Linkage

Source Authors' computation using ISM software by Ahmad and Qahmash (2021)

governance issues (B7), and financial constraints (B8) are the linkage variables with high dependence and high driving power. Based on the MICMAC table the variables are classified into different levels and none are in the autonomous category justifying the importance of all the variables in adoption of SDG. TISM is an advancement over the traditional interpretive structural modelling as ISM just helps in identifying and capturing the existence of contextual relationships between the variables, TISM goes

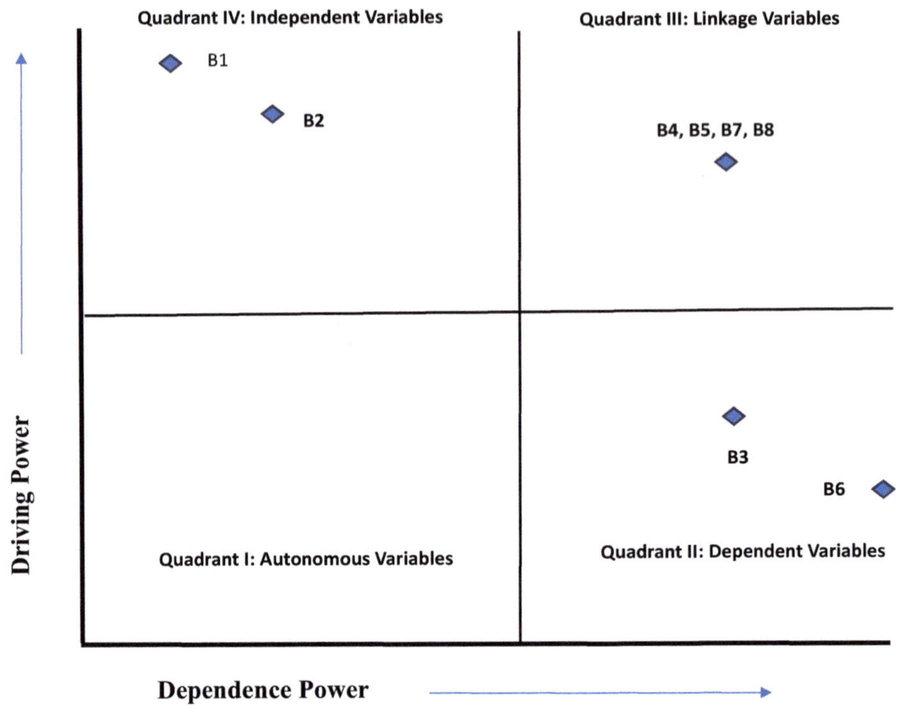

Fig. 2 MICMAC graph. *Source* Authors' Computation using ISM software by Ahmad and Qahmash (2021)

a step further by giving an explanation to each relationship. A detailed TISM with an explanation for each relationship is provided in Table 6.

4 Discussion and Conclusion

Environmental sustainability is the top priority of economies worldwide for which various steps have been undertaken at the micro as well as macro level. Attainment of 17 SDGs in a time-bound manner is one such global effort that can ensure a transition towards growth that is not only economic but also green. However, various barriers make it difficult for economies to make a full-swing shift towards sustainable development. Financial constraints, lack of awareness, infrastructural hurdles, race towards economic wealth generation, and others have a significant negative impact on SDG attainment. Various studies have highlighted various such barriers and their impact on sustainable development, but there is a huge gap in identifying the interlinkage and creation of a hierarchical structure of the multiple barriers. This chapter has tried to bridge this gap by making use of the contemporary multi-criteria

Table 6 Interpretive matrix

	B1	B2	B3	B4	B5	B6	B7	B8
B1		Lack of awareness increases cultural resistance	Awareness can drive data collection initiatives	Awareness helps address tech gaps	Awareness leads to better policy coherence	Awareness can expose capacity issues	Awareness helps address governance concerns	Awareness can highlight financial needs
B2			Cultural attitudes can limit data sharing	Social resistance affects tech adoption	Social norms affect policy priorities	Social norms can hide capacity gaps	Cultural resistance affects governance reforms	Social barriers can restrict financial support
B3						Lack of data hides capacity limitations		
B4			Tech gaps result in poor data systems		Tech issues affect policy implementation	Tech gaps limit operational capacity	Tech infrastructure impacts governance	Tech needs strain financial resources
B5			Policy incoherence undermines data consistency	Policies affect technology deployment		Incoherent policies affect institutional capacity	Policy clarity enables governance	Policy decisions drive financial allocations
B6						–		
B7			Governance failure leads to poor data	Governance affects tech management	Governance affects policy enforcement	Weak governance impacts capacity		Governance determines financial flows
B8			Lack of funds impacts data systems	Funding gaps prevent tech access	Financial issues disrupt policy programs	Funding affects capacity	Financial issues disrupt governance	

Source Authors' compilation from expert opinion

decision-making techniques—TISM and MICMAC. It highlights the complex interrelationship between the various barriers that hinder the effective implementation of sustainable development strategies at the micro as well as macro level. Existing literature has provided eight variables namely: lack of awareness, cultural & social barriers, data deficiency, technological gaps, policy incoherence, capacity limitations, governance issues, and financial constraints. Based on the expert opinion and validation of the factors are further examined that give their level of dominance and interlinkage. The dependence and driving power of the factors classify them into three distinct levels that are dependent, interlinkage, and independent. The hierarchical structure not only reflects the influence each barrier has over the others but also helps determine strategic intervention points for maximum impact.

At the bottom level of the hierarchy, we have two important independent variables namely lack of awareness (B1) and cultural & social barriers (B2). These variables have a high driving power with minimal dependence on other variables and hence can be called the major hindering barriers behind the implementation and attainment of global SDGs. Lack of awareness amongst stakeholders leads to limited efforts toward sustainable practices with poor adoption of clean sources of energy and circular practices thereby putting a significant barrier to SDG adoption. On the same front cultural and social barriers combined with lack of awareness lead to rigidity and resistance to change and adopt new practices. The two being foundational barriers need significant efforts by policymakers to ensure they are directly addressed through targeted campaigns to ensure a ripple-down effect across other variables as well. If these are not handled then efforts at other levels may also produce limited and insignificant results. The intermediate levels also called the interlinkage variables includes four variables namely technological gaps, policy incoherence, governance issues, and financial constraints. These variables have both high driving and dependence power and hence can be classified as the mediators between the other two sets that are dependent and independent. These variables often support the feedback loop in the entire system and hence are crucial to be addressed. Technological gaps (B4) represent barriers in the availability, affordability and adaptability of contemporary technology that promote sustainable development. Economies that are low in economic wealth generation usually struggle with a lack of innovation and poor integration thereby hindering the overall attainment of SDGs. Policy incoherence (B5) refers to policies that are contradictory and not well-aligned to the goal attainment. It often leads to fragmented governance structures and the absence of a common strategic direction leading to barriers in sustainable development. Addressing this requires top-level political will, cross-sectoral collaboration, and alignment of economic, social, and environmental goals. Governance issues (B7) include institutional weaknesses, absence of transparency, and bureaucratic hurdles. It is documented to be a linkage variable, meaning that it is not only a push factor but also has dependence on other factors. Its dual relationship with other factors thus makes it important to ensure there is a systemic transformation. For instance, promoting transparency in governance with enhanced accountability and participation in decision-making can lead to enhanced resilience towards SDG adoption. Another major hurdle in the SDG implementation is the financial constraint (B8), which is also an important linkage variable having an impact on other variables. Handling the financial constraints by promoting green financial instruments, promoting public private partnership along with the integration of sustainability metrics. At the top of the hierarchical structure, there is a barrier related to data deficiency and capacity limitations. The factor at the top level is the dependent variable that is being impacted by all other variables but does not influence others. Data deficiency refers to the absence of real-time data availability that creates obstacles in the policy-making thereby delaying response to SDG implementation and improvement. Capacity limitation (B6) is the constraint at the human and infrastructural levels. It is more prominent in economies with financial constraints. Like data deficiency, capacity limitations are heavily influenced by higher-order variables such as technological access, policy coherence, and governance quality.

Combined together the five-level hierarchy provides a comprehensive understanding of the systemic dynamics that have an impact on the efficient implementation of SDGs across the global economies by mitigating or removing the various hurdles. The use of TISM allowed for not just the hierarchical modelling of barriers but also the integration of interpretive logic, making the analysis more contextually relevant and actionable.

5 Implications

This chapter, with the use of MCDM, has helped in identifying the multiple barriers and has created a structure and interlinkage between the same based on expert opinion. The structure will assist policymakers in identifying the immediate action points along with identifying the cascading effect that they may have on others. Identification of lack of awareness and social and cultural barriers as the important factors underlines how the barriers are context specific and require a macro approach to be adopted.

Awareness campaigns need to be conducted at a large scale to mitigate the issue and to promote sustainable development programs. The multiple linkage factors at the intermediate levels need to be handled with utmost care as they have a ripple effect across the system. For instance, policy coherence needs effective governance structures and its integration across different sectors and units. Similarly, the creation of an innovation ecosystem can help in improving access to cleaner sources of energy that can assist in bridging the technological gap along with mitigating the financial constraints.

The dependent variables at level I also require a supportive system to ensure they can also play their part in promoting sustainability. Encouraging open data initiatives, fostering academic-policy partnerships, and investing in digital infrastructure can significantly enhance data quality and usability.

In conclusion, the chapter provides a structured and interpretive understanding of the barriers to sustainable development through TISM-MICMAC analysis. The classification into independent, linkage, and dependent variables not only reveals the complexity of the issue but also provides a roadmap for targeted and phased interventions. Ultimately, the pursuit of sustainable development requires not only identifying the right barriers but also understanding their interlinkages. This chapter provides a step in that direction, equipping stakeholders with the knowledge and tools to navigate complexity and catalyze transformative change.

References

Abbass K, Amin N, Khan F, Begum H, Song H (2025) Driving sustainability: the nexus of financial development, economic globalization, and renewable energy in fostering a greener future. In: Energy and environment. SAGE Publications Inc. https://doi.org/10.1177/0958305X241305374

Abid N, Ceci F, Ikram M (2022) Green growth and sustainable development: dynamic linkage between technological innovation, ISO 14001, and environmental challenges. Environ Sci Pollut Res 29(17). Springer Science and Business Media Deutschland GmbH. https://doi.org/10.1007/s11356-021-17518-y

Adam I, Fazekas M (2021) Are emerging technologies helping win the fight against corruption? A review of the state of evidence. Inform Econ Policy 57. Elsevier B.V. https://doi.org/10.1016/j.infoecopol.2021.100950

Afzal A, Rasoulinezhad E, Malik Z (2022) Green finance and sustainable development in Europe. In: Econ Res-Ekonomska Istrazivanja 35(1). Taylor and Francis Ltd. https://doi.org/10.1080/1331677X.2021.2024081

Ahmad N, Qahmash A (2021) Smartism: implementation and assessment of interpretive structural modeling. Sustainability 13(16):8801

Ahmed B, Wahab S, Rahim S, Imran M, Khan AA, Ageli MM (2024) Assessing the impact of geopolitical, economic, and institutional factors on China's environmental management in the Russian-Ukraine conflicting era. J Environ Managem 356. Academic Press. https://doi.org/10.1016/j.jenvman.2024.120579

Ametepey SO, Aigbavboa C, Thwala WD, Addy H (2024) The impact of AI in sustainable development goal implementation: a Delphi study. Sustainability (Switzerland) 16(9). Multidisciplinary Digital Publishing Institute (MDPI). https://doi.org/10.3390/su16093858.

Barve P, Nayak S (2023) Strengthening market discipline by depositors : a TISM and MICMAC approach. Indian J Finan 17(10). Associated Management Consultants Pvt. Ltd. https://doi.org/10.17010/ijf/2023/v17i10/170558

Ben Ruben R, Rajendran C, Saravana Ram R, Kouki F, Mesfer Alshahrani H, Assiri M (2023) Analysis of barriers affecting industry 4.0 implementation: an interpretive analysis using total interpretive structural modeling (TISM) and fuzzy MICMAC. Heliyon 9(12). Elsevier Ltd. https://doi.org/10.1016/j.heliyon.2023.e22506

Bespalyy S, Alnazarova G, Petrenko A, Zhanuzakova S (2025) Integration of the SDGs into the educational process of Kazakhstani universities: implementation results and effectiveness assessment. Discover Sustain 6(1). Scopus. https://doi.org/10.1007/s43621-025-02270-x

Boiko O, Murzin I, Akhromkin A, Lukashev S, Kyrylenko S (2025) Impact of renewable energy sources on promoting sustainable development nationwide. J Lifestyle SDG'S Rev 5(1). Editora Alumni In. https://doi.org/10.47172/2965-730X.SDGsReview.v5.n01.pe03389

Bolatova Z, Sharapatova R, Kabiyev Y, Berndtsson R, Tussupova K (2025) Towards sustainable solutions: assessing rural access to safe drinking water and sanitation in Atyrau, Kazakhstan. Water (Switzerland) 17(5). Multidisciplinary Digital Publishing Institute (MDPI). https://doi.org/10.3390/w17050664

Borsatto JMLS, Marcolin CB, Abdalla EC, Amaral FD (2024) Aligning community outreach initiatives with SDGs in a higher education institution with artificial intelligence. In: Cleaner and responsible consumption, vol 12. Elsevier Ltd. https://doi.org/10.1016/j.clrc.2023.100160

Brandão R, Verissimo L (2024) Circular economy adoption in construction: a pathway to sustainable development and UN SDG 11 achievement. In: World sustainable series: Part F3447. Springer Science and Business Media Deutschland GmbH, pp 213–230. Scopus. https://doi.org/10.1007/978-3-031-65909-6_13

Daga S, Yadav K, Singh D, Pamucar D, Simic V (2025) Unveiling greenwashing: analyzing the interaction of factors discouraging ESG greenwashing through TISM and MICMAC. J Environ Managem 380:124850. https://doi.org/10.1016/j.jenvman.2025.124850

Fraisl D, Neves M, Seidu O, Darpoh CK, Basnyat A, Usheva F, Tungbani B, Ankamah D, See L, Gadgil A (2025) Leveraging citizen data to improve public services and measure progress toward sustainable development goal 16. Sustain Developm 33(4):5968–5982. Scopus. https://doi.org/10.1002/sd.3441

Islam N, Jirattikorn A (2024) Breaking gender Barriers in STEM education for achieving the SDG of quality education in Bangladesh. Developm Pract 34(1). Routledge. https://doi.org/10.1080/09614524.2023.2229965

Lin JC (2024) Promoting education about and for sustainable development: assessing the incorporation of SDGs in Chinese higher education. Int J Sustain Higher Educ 25(7):1469–1483

Maity S, Sinha A (2024) Disparities in achieving SDG goal of 'no poverty' for developing countries: technical efficiency estimates across Asian, African and American nations. Int J Sustain Developm World Ecol 31(8). Taylor and Francis Ltd. https://doi.org/10.1080/13504509.2024.2377247

Matović S, Bajrami DD, Gajić T et al (2024) Beyond numbers: challenges in measuring SDG4 targets—Serbia's perspective. In: Sustainability (Switzerland), vol 16(22). Multidisciplinary Digital Publishing Institute (MDPI). https://doi.org/10.3390/su162210006

Nasrullah N, ul Husnain MI, Khan MA (2023) the dynamic impact of renewable energy consumption, trade, and financial development on carbon emissions in low-, middle-, and high-income countries. In: Environmental science and pollution research, vol 30(19). Springer Science and Business Media Deutschland GmbH. https://doi.org/10.1007/s11356-023-26404-8

Patel UR, Ghaffarianhoseini A, Ghaffarianhoseini A, Burgess A (2024) Digital twin technology for sustainable urban development: a review of its potential impact on SDG 11 in New Zealand. In: Cities, vol. 155. Elsevier Ltd. https://doi.org/10.1016/j.cities.2024.105484

Revtiuk Y, Zelinska O (2026) Toward sustainability: examining economic inequality and political trust in EU countries. Sustainability (Switzerland) 18(1). Scopus. https://doi.org/10.3390/su18010210

Sadiq M, Talbi B, Ghosh S, Bashir MF (2024) How does external debt and governance quality impact renewable energy consumption: novel policy insights from BRICS countries. In: Environmental science and pollution research, vol 31(30). Springer. https://doi.org/10.1007/s11356-024-33846-1.

Singh S, Dutt CS (2024) The adoption of the UN sustainable development goals in hotels in Dubai. In: Tourism and hospitality research. vol 24(4). SAGE Publications Inc., pp 515–534. https://doi.org/10.1177/14673584231164941

Singh D, Naeem M, Daga S, Ahmed U (2025) Economic policy uncertainty and the climate challenge: moderating effects on the innovation–emissions nexus in G20 nations. Environ Econ 16(3):14–28. https://doi.org/10.21511/ee.16(3).2025.02

Tahir Q, Riaz MS, Khan MA, Javid MA (2025) Moving towards sustainable urban mobility patterns: addressing barriers and leveraging technology in Islamabad and Rawalpindi, Pakistan. Sustainability 17(21):9776. https://doi.org/10.3390/su17219776

Yadav K, Daga S, Saini A (2025) Structural analysis of sustainable fashion adoption: an integrated TISM and MICMAC approach. J Sustain Market 24:1–24. https://doi.org/10.51300/JSM-2025-149

Yu P, Abusamra OA, Usman M, Makhdum MSA (2025) Leveraging technological innovation, renewable energy, and forest management to overcome barriers and achieve sustainable development goal 13 (Climate Action) in G-7 countries. Sustain Developm 33(6):8821–8839. Scopus. https://doi.org/10.1002/sd.70127

Zighan S, Alkalha Z, Jum'a L (2026) Barriers to circular economy transitions in emerging markets: insights from the GCC and Implications for sustainable development. Sustain Developm 34(S1):210–225. Scopus. https://doi.org/10.1002/sd.70177

Institutional Quality and Its Implications for Corporate Sustainability and Sustainable Development Goals in Selected African Countries

Omobolade Stephen Ogundele and Lethiwe Nzama-Sithole

Abstract Despite the growing global focus on sustainable corporate practices, many African nations continue to grapple with institutional weaknesses that hinder the effective implementation and execution of Sustainable Development Goals (SDGs). Additionally, the pursuit of sustainable development in Africa is closely tied to the quality of institutions overseeing the social, economic and environmental policies. This study explored the longitudinal trends of institutional quality indicators among selected African countries. Four out of six key institutional quality indicators, which comprise of Political Stability and Absence of Violence/Terrorism, Control of Corruption, Government Effectiveness and Rule of Law, were considered. They play a vital role in creating an enabling environment for sustainable corporate practices. Thus, it focuses on their implications for the SDGs, principally SDG 16 (Peace, Justice, and Strong Institutions) and Corporate Sustainability. The study purposely selected two African countries from each region: Southern Africa (South Africa and Botswana) and West Africa (Nigeria and Ghana), allowing for a comparative analysis across different governance contexts. It explores secondary data obtained from the World Bank's World Governance Indicators (WGI) from 2002 to 2023. The base year (2002) was chosen because many international databases, such as the WGI and Transparency International, provide consistent data starting from this year, enabling a robust analysis over two decades. The study employed longitudinal trend analysis to assess changes over time and conduct cross-region comparisons. The study recommends that selected countries should strengthen governance frameworks, which include improving government effectiveness, reducing corruption, and ensuring political stability, which is essential for enhancing corporate sustainability and strengthening SDGs.

O. S. Ogundele (✉)
Department of Commercial Accounting, College of Business and Economics, University of Johannesburg, Johannesburg, South Africa
e-mail: omoboladeo@uj.ac.za

L. Nzama-Sithole
Department of Auditing, College of Accounting Sciences, University of South Africa, Pretoria, South Africa
e-mail: nzamall@unisa.ac.za

K. Kumar et al. (eds.), *Sustainable Value Creation: Corporate Strategies for SDGs Implementation*, World Sustainability Series,
https://doi.org/10.1007/978-3-032-24652-3_4

Keywords Control of corruption · Corporate sustainability · Government effectiveness · Political Stability and Absence of Violence/Terrorism · Rule of law · SDGs

1 Introduction

Recently, conversations around sustainable development have advanced from environmental protection to encompass wider issues, which include governance, institutional quality and corporate sustainability. Corporate sustainability is considered as a company strategy aimed at generating long-term shareholder value by managing risk and embracing opportunities derived from social, environmental and economic development. It is a multidimensional construct that integrates the economic, social and environmental facets into business strategies in ensuring long-term success and contributes to enhancing society and the environment (Millan-Tudela 2023). In the emerging economies, institutional quality plays a very fundamental role in the development and growth of any nation. The role of institutional quality has become crucial to discussions around sustainable economic growth and corporate sustainability, especially across African nations. High institutional quality, featured by sound governance, stable democratic institutions, control of corruption and rule of law is vital for enhancing economic development and growth. It is a many-sided concept that significantly influences economic development, governance and public service. It is affected by several socio-economic factors and requires a balanced approach to improve effectively.

Strong institutions have the potential to accelerate growth and enhance efficiency, specifically when supported by economic freedom. Hunjra et al. (2023) and Uddin et al. (2023) stated that institutional quality is crucial to achieving Sustainable Development Goals (SDGs), specifically SDG 16, which emphasize on providing access to justice, fostering peaceful and inclusive societies and building transparent and accountable institutions. Similarly, good institutions enhance financial inclusion by strengthening access to and use of the formal financial services, which is vital for economic development and growth (Zeqiraj et al. 2022; Karim et al. 2025). The SDGs are a set of 17 global objectives that were instituted by the United Nations in 2015 as components of the 2030 Agenda for Sustainable Development. These objectives aim to tackle a wider array of global issues, including inequality, peace, climate change and justice amongst others (Davim 2025). In emerging economies, institutional quality is an essential determinant of sustainable development. Effective governance, strong institutions and targeted policy interventions are crucial for achieving the SDGs. Effective governance requires clear communication and responsible management of resources, which enhances public engagement and trust (Nzama and Ogundele 2025). Policymakers must give attention to improving and enhancing institutional quality, ensuring transparency and fighting corruption to foster sustainable development (Barbier and Burgess 2021; Sun et al. 2025). Most emerging nations are stuck in low-quality institutional traps, which hinder their economic growth and development and keep them in low-income traps (Kar et al. 2019). These traps make it difficult for these nations to escape poverty and achieve sustainable growth.

Institutional quality indicators are often proxied by exploring several indicators that reflect the political, social and economic environment. Such indicators include the rule of law, control of corruption, government effectiveness, political stability and absence of violence, regulatory quality and voice and accountability (Nzama et al. 2022; Barra and Falcone 2024). Its effect varies across regions. Minimising corruption practices is critical for promoting economic development and strengthening institutional quality (Uddin et al. 2023). Control of corruption has a significant positive influence on economic development and sustainability. Another critical indicator of institutional quality that supports economic development by ensuring justice and minimizing uncertainty in economic transactions is the rule of law. Effective governance is also essential for enhancing institutional quality. It influences many aspects of development, including health outcomes, financial inclusion and sustainable economic growth. Institutional quality significantly influences the long-term financial sustainability of corporations through several mechanisms. Strong institutional quality, featured by several factors such as legal protection, corporate transparency, market capacity, political stability and operational efficiency (Ojeka et al. 2019). This is because improved institutional environments minimize market frictions, such as lower transaction costs, information asymmetry and agency conflicts. Likewise, the combination of strong institutional quality and innovative technology influences financial market development, ensuring long-term financial stability and growth (Khan et al. 2024).

In spite of international support and several efforts to sustain development in African nations. Several nations still face persistent challenges related to institutional inefficiencies (Ebohon et al., 1997). Weak regulatory environments, uneven enforcement of laws, high corruption levels and poor governance have undermined efforts in achieving corporate sustainability, accountability and SDGs. These deficiencies not only threaten the operational effectiveness and efficiency of businesses but also compromise their ability to contribute greatly to the SDGs. Likewise, in Africa, where institutional frameworks are often weak and evolving (Adanu 2017), understanding the trends of institutional quality and their implications on corporate sustainability and the achievement of SDGs is essential. Understanding the relationship is essential for strengthening institutional capacity, enhancing policy reforms and guiding private sector engagement in sustainable development strategies across the African continent.

2 Theoretical Underpinning

Institutional theory is highly relevant as it examines the way in which institutional norms, structures and rules influence organisational outcomes and behaviours. Institutions create a rule of the game that shapes organisational actions. It includes both formal (regulations, laws, policies) and informal (traditions, norms, cultural practices) institutions. The study on the influence of institutional quality as it affects SDGs in African nations highlights the major role of institutions in enhancing sustainable development (Musah, 2024). The proponent of the theory is widely recognized as W. Richard Scott, who made substantial contributions to the understanding and development of the theory in 1995 (Scott 1995). Institutional theory emphasizes the impact of some external social forces, such as movements, social institutions and dominant ideologies on decision making. Institutional theory in the context of institutions, organisations and their environments helps in explaining the varying approaches and effectiveness of sustainability practices. For instance, the ties between top-level managerial institutions and community leaders, as well as regulatory officials, drive corporate sustainability strategies in Sub-Saharan Africa, influencing market performance (Nwoba et al. 2021). The theory is relevant for this study because it explains how formal institutions and informal norms shape strategic choices and organizational behaviour. Additionally, the theory helps to explain cross-country differences by highlighting how organisations react to normative and coercive pressures within specific national governance environments, thereby linking institutional quality to SDG attainment and sustainability outcomes.

3 Methodology

The study purposely selected two African countries from each region: West Africa (Ghana and Nigeria) and Southern Africa (Botswana and South Africa) and allowed for a comparative analysis across different contexts. The countries selected were based on the availability of data, various extents of development and regional representation. The study explored secondary data (such as Control of Corruption, Rule of Law, Political Stability and Absence of Violence, Government effectiveness) sourced from the World Bank's Governance Indicators (WGI) spanning from 2000 to 2023. The year 2002 was chosen as the base because the international database selected for this study provides consistent data from the base year to the terminal year. This has enabled a robust analysis for over two decades. The study employed longitudinal trend analysis in assessing changes over time and conducting cross-country comparisons. It will likewise help in tracking the progression of institutional quality indicators for each country over time. The analysis will assist in identifying stabilities, improvements or declines in the indicators and their various influences on corporate sustainability and SDGs.

4 Results and Discussion of Findings

4.1 Rule of Law

The rule of law is a vital component of institutional quality, influencing various socioeconomic and political outcomes. It encompasses predictable, transparent and enforceable laws applied fairly by the government. It ensures that laws are enforced by an independent judiciary, providing safeguards against deterring corruption and arbitrary decisions (Mehmood 2023). Strengthening the rule of law could lead to more stable, equitable and prosperous societies.

The trend analysis of the rule of law in Fig. 4.1 indicates the extent to which the agents abide by and have confidence in the rules of society, which includes the court, the police, and property rights, among others, in the selected West African countries (Nigeria and Ghana). The dataset showed year-by-year values and percentage changes, allowing us to evaluate the deterioration or improvement of the rule of law as an important factor in Nigeria and Ghana over time. Between 2002 and 2004, both countries (Nigeria and Ghana) witnessed a steady decline (from − 3.36 to 4.21%) in the rule of law. The deterioration in the rule of law is possibly due to poor enforcement mechanisms, weak democratic institutions, or political instability. For Nigeria, this aligns with the post-military transition. A noteworthy improvement was made in 2006 (rising to 22.88%), which was marked by a sharp increase, suggesting political stability and increased international support for institutional development, especially in Ghana and Nigeria's anti-corruption stance under President Olusegun Obasanjo's administration. Inokoba and Ibegu (2011) and Adeyemi (2022) affirmed that the two major anti-corruption bodies in Nigeria were established during his (President Obasanjo) tenure: the Independent Corrupt Practices and Other Related Offences Commission (ICPC) and the Economic and Financial Crimes Commission (EFCC). Though this progress was not sustained, as the following years (2007–2010) witnessed a decline in the rule of law indicators, suggesting that institutional challenges persisted and initial reforms may not be sustained. A notable decrease to − 6.41% in 2016 may be tied to weakening public trust in law enforcement and rising insecurity, particularly in Nigeria. The period from 2018 to 2023 is characterized by minor fluctuations, which indicate that the rule of law improvements remain fragile and inconsistent. These trends highlight the need for better governance enforcement mechanisms and sustaining judicial and legal reforms to promote lasting improvements in the rule of law in the Western region.

In the Southern African countries, the Rule of Law, as displayed in Fig. 4.2, has experienced setbacks and improvements over the past 22 years. Several Southern African nations have undertaken substantial legal reforms that include bills of rights and the adoption of new constitutions that enhance judicial reviews of executive and legislative acts (Isanga 2016). Most especially, the South African nation has witnessed substantial law reforms in an effort to promote a constitutional democracy and dismantle the apartheid structure (Botha and Bekini 2018). The early 2000s

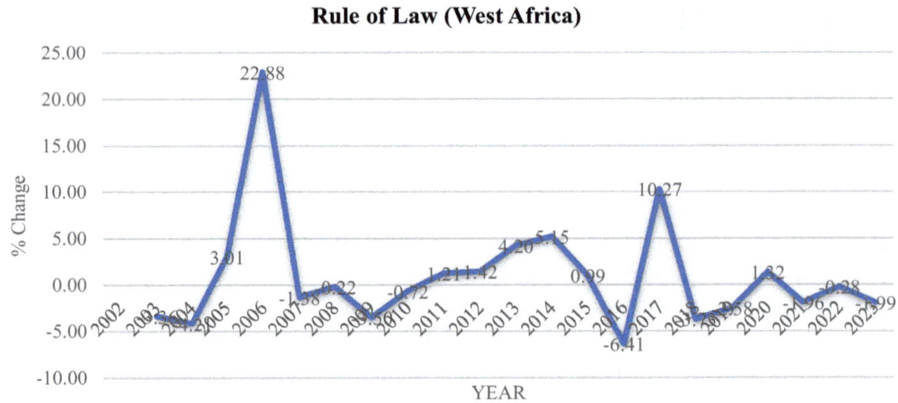

Fig. 4.1 Rule of law (West Africa). *Source* World Bank's WGI

exposed modest fluctuations, with a minor increase in 2003 and 2005 (0.82 and 0.88 percent, respectively).

In 2014, a noteworthy surge occurred when the Rule of Law indicator increased to 9.89 percent, possibly reflecting the influence of governance reforms that were strengthened during that period. Nevertheless, the momentum was short-lived as the subsequent years witnessed consistent declines from 2015 to 2017 (− 6.28 to − 5.60%), driven by factors such as weakened public interest, corruption and political instability. Corruption continues to pose a major challenge, affecting the agencies or institutions responsible for upholding the rule of law. Nzama (2022) affirms that South African municipalities still experience financial mismanagement, fraud and corruption despite reforms. (Cook 2013) also affirmed that corruption, labour unrest, dissatisfaction among youth and political challenges continue to undermine the rule

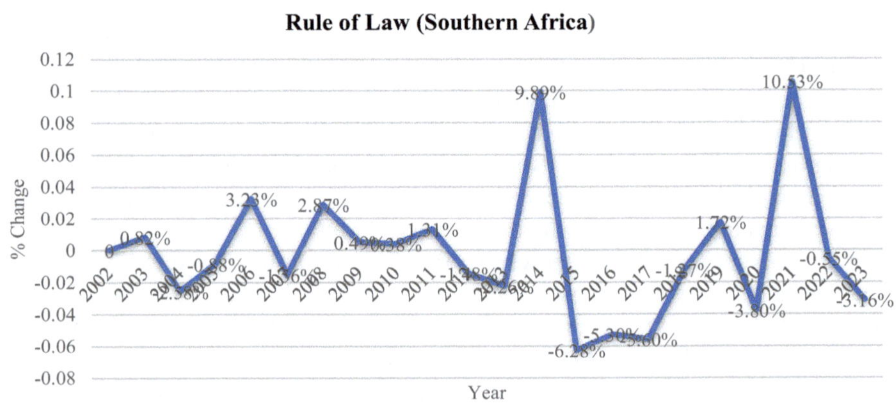

Fig. 4.2 Rule of law (Southern Africa). *Source* World Bank's WGI

of law. Solo (2025) affirmed that there are growing concerns about the use of political tolerance, economic inequality and torture in Botswana.

South Africa and Botswana have made giant strides in improving the rule of law, but continue to face numerous challenges. South Africa's post-apartheid legal reforms and constitutional democracy have brought stronger constitutional protection for human rights and enhanced judicial independence, yet crime and socioeconomic remains a serious concern. Improvements in the legislative reforms and judicial discretion in Botswana are notable, but political intolerance and gender equality still persist (Scribner and Lambert 2010).

4.2 Control of Corruption

The percentage change in the scores for Control of Corruption for the selected West African Countries (Ghana and Nigeria) in Fig. 4.3 spanning from 2002 to 2023 reveals fluctuation in anti-corruption effectiveness, stability in governance and institutional strength. From 2002 to 2006, Ghana and Nigeria experienced steady improvements in their control of corruption, with an annual percentage (from 13.06% to 20.16%) increase until 2006. This upward trend suggests the implementation of anti-corruption reforms. Nigeria's EFCC and ICPC, established in the early 2000s, contribute to the perceived strengthening of institutional quality and integrity.

The improvement actually coincides with the reforms under President John Kufuor's tenure, whose governance emphasized anti-corruption through the Office of Accountability and public sector reforms. From 2007 to 2009, the increase in control of corruption still persisted at a slower pace (from 6.07 to 1.12%). By 2010, the score began to decline, with a notable drop in 2014 (decline from − 2.99% to − 7.43%). Political interference, governance instability and inconsistent policy implementation

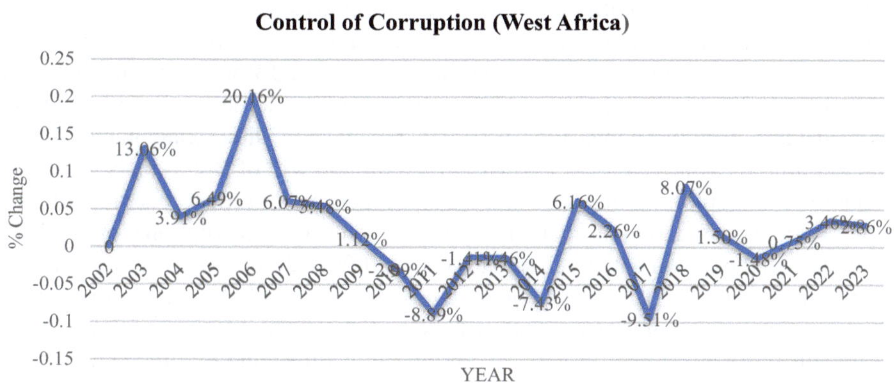

Fig. 4.3 Control of corruption (West Africa). *Source* World Bank's WGI

may have contributed to the downturn. Draman (2019) stated that executive partisanship within the Parliament further hindered effective governance and anti-corruption efforts and a lack of genuine political commitment has rendered their efforts ineffective (Yeboah and Alesu 2016). Political instability and interference in Nigeria have hindered the successful execution of the anti-corruption initiatives (Idemudia et al. 2010; Nwozor and Afolabi 2023).

From 2015 to 2018, the scores reveal a mixed trend. From 2015 to 2016 (6.16% and 2.26% percent for 2015 and 2016, respectively), the result showed that there was a modest rebound, which is likely influenced by renewed political promises to tackle corruption, especially following the changes in Nigerian political leadership. Nevertheless, a sharp decline was witnessed in 2017 (a decline to − 9.51%), which revealed a lack of sustainability in these efforts to control corruption-related issues. Owusu-Mensah and Mathapoly-Codjoe (2023) stated that the emergence of the Office of the Special Prosecutor (OSP) in the year 2017 was seen as a political move rather than a sincere effort to combat corruption, which leads to skepticism about the effectiveness of the control of corruption. The result also showed relative stability from 2019 to 2023 (1.50% to 2.86%), with more fluctuations in the control of corruption scores. The presence of small gains alongside occasional declines points to fragile institutions and the need for more profound structural changes.

In summary, the trend of results displayed shows that while the selected West African countries (Ghana and Nigeria) have made efforts in combating corruption, progress has been uneven. The gains achieved during the specific periods were often reversed, which indicates that anti-corruption reforms and policies have not been fully institutionalized. For lasting impact, the two countries must move beyond politically motivated measures and focus on strengthening the independence of oversight institutions, building a culture of accountability and transparency in the public service.

Figure 4.4 reveals the changes in the Control of Corruption for the Selected South African Countries (Botswana and South Africa). The results show that from 2002 to 2005, the institutional quality (Control of Corruption) showed strong performance and outstanding improvements. This could be as a result of high governance standards and anti-corruption efforts in both South Africa and Botswana. Theobald and Williams (1999) and Gbadamosi (2006) affirm that Botswana has been known for its robust anti-corruption measures, primarily through the Directorate on Corruption and Economic Crime (DCEC). Likewise, the political will of Botswana's leaders in combating corruption, alongside reforms and measures to reduce bureaucratic red tape, has contributed immensely to its success in controlling corruption (Jones 2017).

Although the region witnessed setbacks from 2006 to 2009 which reveals a decline from − 2.24% to − 2.10%. This could be possible for South Africa due to growing concerns around the governance-related issues later in the Mbeki and early Zuma years and some stagnation in anti-corruption innovations and enforcement mechanisms. Nzama et al. (2025) and Charmaine et al. (2024) stated that municipalities are usually challenged due to poor contracts and corrupt practices. South Africa has proven to be vulnerable to corruption, adversely affecting citizens. Mujuzi (2019) and Macdonald (2019) both affirmed that in Botswana, South Africa and other

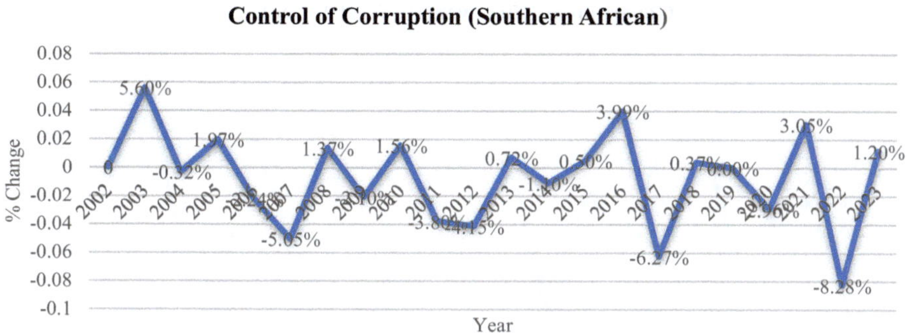

Fig. 4.4 Control of corruption (Southern Africa). *Source* World Bank's WGI

Southern African Development Community (SADC) nations, corruption remains a major concern across various countries in that region. The year 2016 (3.99%) saw a modest improvement, but the rebound was short-lived as 2017 again witnessed a decline. The most dramatic fall occurred in 2022 (fall to − 8.28%), possibly reflecting post-pandemic governance stress or erosion of trust (Nzama et al. 2022). These shifts suggest that the anti-corruption in that region remains fragile and likely prone to institutional instability.

4.3 Government Effectiveness

Figure 4.5 indicates the trend analysis of Government Effectiveness, which is an indicator of institutional quality. Government effectiveness reflects the quality of public services, the competence of the civil service, policy formulation and credibility of government commitments to such policies (Handoyo 2023). It also reflects the efficiency and reliability of public governance.

The data trend spanning from 2002 to 2023 highlights noteworthy differences and fluctuations in government effectiveness for the selected West African Countries (Nigeria and Ghana). The data trend cut across 22 years (2002–2023), highlighting noteworthy differences and fluctuations in government effectiveness for the selected West African Countries (Ghana and Nigeria). Between 2002 and 2004, the results showed a gradual decline (− 2.36 to − 5.73 percent) in government effectiveness, which reflects weak policy implementation frameworks or institutional instability. However, a rebound in 2005 and 2006 (9.39 and 6.57 percent respectively) reveals a renewed effort to improve administrative capacity, probably through improvement in fiscal management or civil service reforms. Nigeria witnessed mixed results in policy performance between 1999 and 2004, with some deteriorations and some improvements. This mixed performance indicates a lack of consistent government effectiveness. Roache affirmed that Ghana also experienced significant aid volatility between

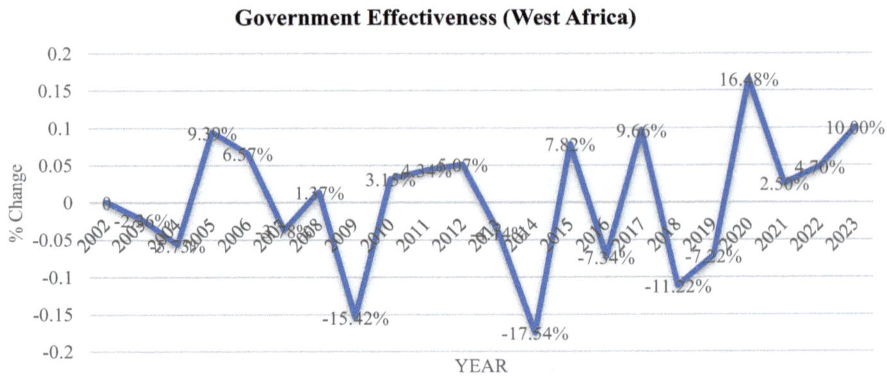

Fig. 4.5 Government effectiveness (West Africa). *Source* World Bank's WGI

2000 and 2003, which led to policy difficulties and hindered governance effectiveness. Results from 2007 to 2009 reveal a renewed volatility. A modest improvement in 2008 was followed by a sharp drop in 2009, which may have reflected domestic policy failure or election-related governance strain (2007 is 3.78%, 2008 is 1.37% while 2009 is 15.42%). The election was held in late 2008 in Ghana, with a presidential runoff concluded in early 2009. Reflecting challenges with the transition of power and new policy directions.

However, 2010 and 2011 witnessed positive growth despite improvements in 2012 (3.15% from 2010 to 5.07% in 2012). The period is marked by severe determination, culminating in a decline in 2014, the worst in the series. This could be linked to bureaucratic inefficiency, increased corruption scandals, or a breakdown in service delivery, which may have also contributed to this erosion of government performance and credibility. Ghana as a country faces severe currency fluctuations and a high inflation rate, contributing to economic instability and public dissatisfaction (Noll and Budniok 2022). Also, in Nigeria, corruption, inefficient resource management and excessive bureaucracy significantly increase the cost of governance, which leads to a budget deficit and heavy debt burden (Titus et al. 2025). The years 2018 witnessed 11.22% and 2019 had 7.22%, also reflect a significant drop in government effectiveness. A notable recovery occurred in 2020, marking the most significant annual improvements in the two-decade span, possibly driven by pandemic-related state interventions, digital governance or donor-supported reforms. Years 2022 and 2023 likewise indicate early signs of improved institutional quality (4.70% in 2022 and 10.00% in 2023).

Figure 4.6 reveals the percentage change result for Government Effectiveness in the selected Southern African region spanning from 2002 to 2023. The result shows insights into government effectiveness over the period. The result showed a consistent decline that culminated in a very sharp decline in the most significant yearly deterioration in the dataset. This could be as a result of weak public services, political instability or economic shock during the period. Despite the end of apartheid, South Africa faced several challenges in addressing social inequality and economic

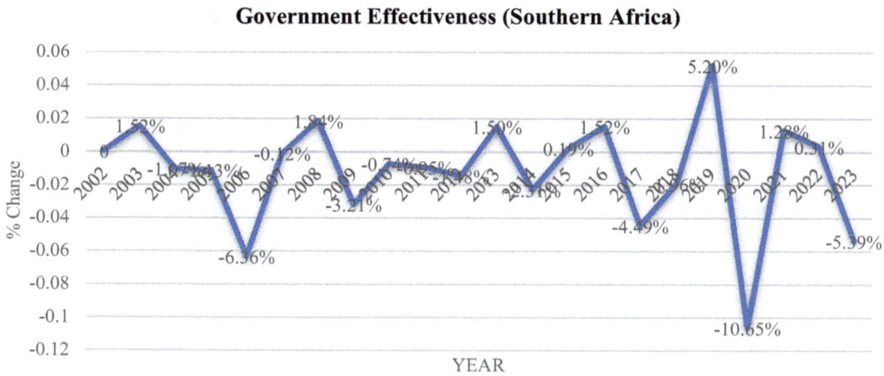

Fig. 4.6 Government effectiveness (Southern Africa). *Source* World Bank's WGI

disparities, which have had a serious effect on governance effectiveness (Nzama et al. 2023). 2007 to 2009 reveals fluctuations in percentage (– 0.12% to – 3.21%), followed by a positive rebound in 2008(1.84%). This period suggests attempts to recover governance effectiveness. 2013 reveals another positive shift after a consistent decline from 2010 to 2012 (from – 0.74% to – 1.48%). The years 2015 and 2016 (0.19% and 1.52% for 2015 and 2016, respectively) witnessed a slight gain; these improvements were sustained, as evidenced by a sharp decline, the second-largest in the time series (Nzama 2022).

From 2016 to 2023, the region witnessed notable gains and declines (1.52 in 2016, while 2023 had – 5.39%). There was a significant improvement in 2016, especially in 2019 (2019 witnessed 5.20%). This suggests periods of governance strengthening or reforms. Although the gains were offset by drastic drops in 2020, likely reflecting the adverse effects of the COVID-19 pandemic on governance and also in 2023. South Africa imposed many lockdowns in 2020 to mitigate the spread of COVID-19. The lockdown exposed poor leadership, the incompetence of state institutions and corruption within the dominant party (Nzama 2022), further weakening government effectiveness (Herbert and Heather 2021). Slight improvements were observed in 2021 and 2022, although they were insufficient to sustain the momentum. Overall, the trend analysis indicates that government effectiveness has been unstable over the past two decades. While there have been seasons of positive changes and improvements, they were often followed by periods of decline, which reveals the underlying vulnerabilities in the governance framework and instability in governance quality. Notwithstanding the progressive corporate governance guidance provided by the King Report on Corporate Governance for South Africa 1994 and the King Report on Corporate Governance for South Africa 2016, which have significantly shaped governance standards and ethical leadership practices in South Africa (Dube 2016).

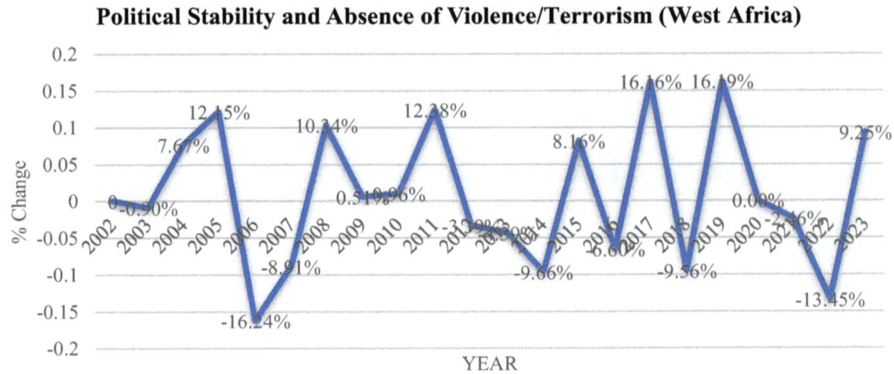

Fig. 4.7 Political stability and absence of violence/terrorism (West Africa). *Source* World Bank's WGI

4.4 *Political Stability and Absence of Violence/Terrorism*

The result displayed in Fig. 4.7 represents the percentage change in Political Stability and Absence of Violence/Terrorism for the selected West African States (Nigeria and Ghana) from 2002 to 2023. This period commenced with slight volatility, which indicates a minor deterioration. However, this was followed by noteworthy improvements in 2004 and 2005, which suggests progress in reducing terrorism and violence or managing political unrest in the region. The democratic system continued to mature in Ghana, marking a peaceful transfer of power and a successful election. The election held in 2004 was part of the process that reinforces political stability (Cook 2010).

2006 and 2007 reveal a sharp decline, especially in 2006 (2006 witnessed 16.24 percent), which indicates a significant increase in political instability, possibly due to rising terrorism threat, conflicts or civil unrest. These years coincide with intensifying conflicts in Nigeria, such as the early signs of Niger Delta militancy and extreme violence in the north. From 2008 and 2011, the results showed gradual improvements in political stability (from − 8.91% to 12.38% in 2011). This period reflects efforts at peaceful transition of power or counter-terrorism. The selected West African countries experienced instability with declines from 2012 to 2014, although a slight improvement in slight but a drop in 2016. Stability stagnated in 2020, probably due to the global effects of the COVID-19 pandemic and also the associated social, political and economic pressures. The modest decline occurred in 2021, which was followed by a sharp decline in 2022, reflecting renewed instability or terrorist resurgence. The year 2023 (9.25%) reveals a modest recovery.

The analysis in Fig. 4.8 reveals the percentage change and volatility pattern in Political Stability and Absence of Violence/Terrorism in South Africa between 2002 and 2023. Over the 21 years, the result alternates between positive and negative changes. Which indicates that while the region suffered significant setbacks in certain years, it has also made progress in others. This inconsistency suggested a fragile institutional environment where episodes of unrest, violence and governance challenges often undermine gains in political stability. The result revealed positive development in quite a few years, especially in 2003, 2005 to 2007, 2010 to 2011 and also from 2018 to 2019. These periods of improvements may be connected with a peaceful transition of power, a decline in violent incidents, a successful democratic process and institutional reforms. Consistent improvements from 2005 (4.51%) to 2007 (6.02%) and from 2010 to 2011 (3.44% in 2010 to 4.43% in 2011) could reflect the efforts to promote conflict resolution and good governance, contributing to a more favourable political climate. South Africa's end of apartheid and subsequent democratic reforms improved political stability. The establishment of democratic institutions and the Truth and Reconciliation Commission helped to address historical injustices and fostered a stable political environment (Clough, 2021). The strong economic development and growth in Botswana have contributed to its stability. The country transitioned from one of the poorest nations at independence to a solid middle-income nation, which helped to mitigate risk with political instability (Mayersen and Stephen 2011).

Conversely, the region experienced significant declines in political stability from 2008 to 2009 (− 4.01% to − 5.05%), 2014, and 2017, and most sharply in 2021 (13.03%). The significant drop in 2021, the largest decline, reflects the widespread political instability and civil unrest in South Africa following President Jacob Zuma's arrest. These events triggered looting and violent protests, impacting the perception of stability and weakening institutional quality indicators in the region. The July riots, also known as the Zuma unrest, were a significant event that highlighted the underlying issues in South Africa (Nzama 2023). These riots were instigated by the arrest and imprisonment of the former South African President (Jacob Zuma) for contempt of court, which incited his supporters to call for his release (Ndlovu et al.

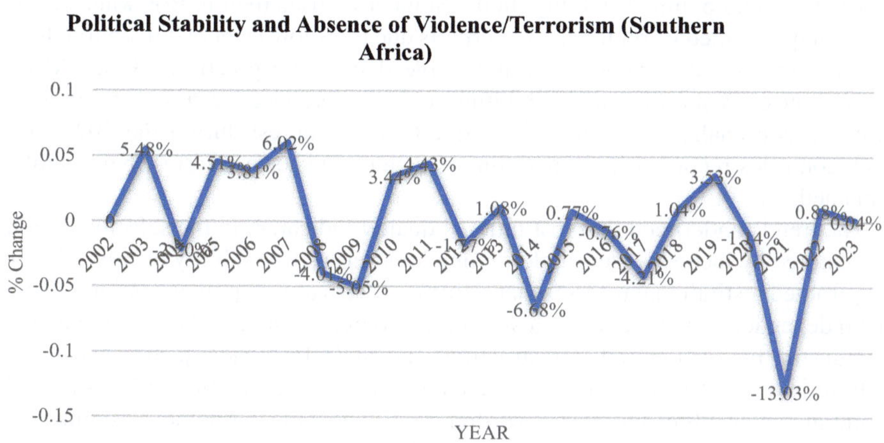

Fig. 4.8 Political stability and absence of violence/terrorism (Southern Africa). *Source* World Bank's WGI

2023). The political instability was also influenced by the government's inability to address the demands of the black majority, which has been a long-standing issue in South Africa (Clough 2021). The political stability of Botswana has been really influenced by developments in South Africa. The unrest and instability in South Africa could have a spillover effect on neighboring nations, including Botswana (Clough 2021). The interconnectedness in the region means that instability in one country could influence the economic and political conditions in other countries. Botswana has also faced several challenges that relate to human rights violations, especially against the San and Bakgalagadi environs (Hitchcock and Babchuk 2010).

In the recent period, modest improvements have emerged, with an increase in 2022 (0.88%) and a marginal gain in 2023 (0.04%). This implies a slow and steady recovery in institutional quality in the selected South African region. Though the minimal nature of these improvements also signals the need for sustainable efforts in political accountability, governance reforms and conflict prevention.

5 Conclusion

Institutional quality is pivotal in promoting corporate sustainability and SGDs by fostering good governance, the rule of law and combating corruption. The trend analysis spanning from 2002 to 2023 across the selected West African countries (Ghana and Nigeria) and Southern African Countries (Botswana and South Africa) reveals fluctuations in key institutional quality indicators explored. Nigeria and Ghana initially experienced some gains in control of corruption, government effectiveness and rule of law following anti- corruption efforts and governance reforms in the early 2000s. However, these gains could not be sustained, with both countries witnessing a decline due to an upsurge in insecurity, especially evident in Nigeria's sharp worsening political stability. In the Southern Africa region, Botswana consistently outperformed its regional counterparts due to its low level of corruption, long-standing institutional frameworks and stable democratic practices. South Africa, while at the early stage initially exhibiting strong governance metrics post-apartheid, faced several challenges in the 2010s due to social unrest during the 2021 riots. Both countries recorded infrequent improvements linked to reforms, but volatility remained.

The trends underlined that institutional quality indicators in the selected regions are featured by inconsistency, undermining long-term sustainability efforts. Without deep-rooted institutional reforms that prioritise transparency, policy continuity, judicial independence and inclusive governance, progress toward SDGs, predominantly 9 (industry, innovation and infrastructure) and SDG 16 (peace, justice and strong institutions) will remain hindered. Enhancing institutional quality and resilience is essential for enabling business environments, enhancing corporate sustainability, inclusive development outcomes and strengthening SDGs.

6 Recommendations

Based on the findings and conclusions, the study suggests the following recommendations:

i. The selected regions and countries should consider and prioritize judicial reforms in enhancing the independence and integrity of their legal systems. Also, collaborations across the regions through ECOWAS and the African Union should support judicial reforms and capacity building.
ii. Selected nations should strengthen governance frameworks, which include improving government effectiveness and ensuring political stability, which is vital for enhancing
iii. The government of the selected countries should invest in public service reforms and address inefficiencies in service delivery by leveraging digital governance and minimizing bureaucratic red tape.
iv. Nations should uphold democratic institutions and encourage political dialogue to enhance long-term stability.
v. Policy must integrate SDGs, especially SDGs 16 and 9, into the budget framework and national development plans.

7 Suggestion for Further Studies

Future studies could consider looking into other regions beyond the selected Western and Southern African countries, which include Central, East and North Africa. Such broader coverage would provide more insights into regional differences and a comprehensive comparative analysis of institutional quality and its implications for SDGs and corporate sustainability. In addition, the study could explore how different sectors (such as manufacturing, financial services, oil and gas, consumer goods and so on) influences of institutional quality on corporate sustainability. One more important avenue for further research is the integration of primary data. Furthermore, the study could also explore how institutional quality influences other SDGs such as SDG 12 (Responsible Consumption and Production) and SDG 13 (Climate Action), which are also critical for long-term sustainability. A multi-SDG approach would provide a better understanding of the impact on sustainability. Utilising qualitative methods such as interviews, focus group discussion and surveys with key shareholders could yield a better understanding of governance dynamics that may not be captured using secondary data.

References

Adanu K (2017) Institutional change and economic development: a conceptual analysis of the African case. Int J Soc Econ 44(4):547–559. https://doi.org/10.1108/IJSE-02-2014-0022

Adeyemi OO (2022) Policy reforms and performance trajectory of anti-corruption agencies in Nigeria. African J Business Econ Res 17:147–165https://hdl.handle.net/10520/ejc-aa_ajber_v2022_nsi1_a7

Barbier EB, Burgess JC (2021) Institutional quality, governance and progress towards the SDGs. Sustainability 13(21):11798.https://doi.org/10.3390/su132111798

Barra C, Falcone PM (2024) Institutional quality and bioeconomy performance in European countries: unveiling the evidence. Ecol Econ 220:108188.https://doi.org/10.1016/j.ecolecon.2024.108188

Botha C, Bekink B (2018). Law reform in South Africa: 21 years since the establishment of a supreme constitutional dispensation. The Theory Pract Legislat 6(2) https://doi.org/10.1080/20508840.2018.1475904

Charmaine MK, Nwosu LI, Enwereji PC, Ogundele OS (2024) Addressing financial mismanagement in state-owned enterprises and municipalities: an integrative review and call for action. Edelweiss Appl Sci Technol 8(6):4735–4748.https://doi.org/10.55214/25768484.v8i6.3025

Clough MW (2021) Southern Africa: descent into Chaos?. In: Global security. Routledge, pp 136–163

Cook N (2010) Ghana, an emergent oil producer: background and US relations. Congressional Research Service, Washington, DC, pp 7–5700

Cook N (2013) In: South Africa: politics, economy, and US relations

Davim JP (2025) Sustainable development goals: a bibliometric analysis. J Sustain Res 7(1):1–4. https://doi.org/10.20900/jsr20250008

Draman R (2019) Weakening parliamentary oversight, increasing corruption: Ghana. Anti corruption evidence: the role of parliaments in curbing corruption. Springer International Publishing, Cham, pp 51–67

Dube ZL (2016) In: The king reports on corporate governance in South Africa: an Ubuntu African philosophy analysis. pp 199–222. https://doi.org/10.1057/978-1-137-56700-0_8

Ebohon OJ, Field BG, Ford R (1997) Institutional deficiencies and capacity building constraints: the dilemma for environmentally sustainable development in Africa. The Int J Sustain Developm World Ecol 4(3):204–213.https://doi.org/10.1080/13504509709469955

Gbadamosi G (2006) Corruption perception and sustainable development: sharing Botswana' santi-graft agency experiences. South African J Econ Managem Sci 9(2):262–276

Handoyo S (2023) Worldwide governance indicators: cross country data set 2012–2022. Data in Brief 51:109814. https://doi.org/10.1016/j.dib.2023.109814

Herbert S, Marquette H (2021) COVID-19, governance, and conflict: emerging impacts and future evidence needs. https://doi.org/10.19088/K4D.2021.029

Hitchcock RK, Babchuk WA (2010) Local justice and legal rights among the San and Bakgalagadi of the Central Kalahari, Botswana. In: Transitional justice: Global mechanisms and local realities after genocide. New Brunswick, NJ, Rutgers, pp 157–178

Hunjra AI, Azam M, Bruna MG, Bouri E (2023) A cross-regional investigation of institutional quality and sustainable development. J Int Finan Market Institut Money 84:101758. https://doi.org/10.1016/j.intfin.2023.101758

Idemudia U, Cragg W, Best B (2010) The challenges and opportunities of implementingtheintegrity pact as a strategy for combating corruption in Nigeria's oil rich Niger Deltaregion. Public Administration and Development 30(4):277–290. https://doi.org/10.1002/pad.576

Inokoba PK, Ibegu WT (2011) Economic and financial crime commission (EFCC) and political corruption: Implication for the consolidation of democracy in Nigeria. The Anthropologist 13(4):283–291

Isanga JM (2016) Rethinking the rule of law as antidote to African development challenges. In *Legitimacy, legal development and change* (pp. 77–102). Routledge.

Jones DS (2017) Combating corruption in Botswana: lessons for policy makers. Asian Educ Developm Stud 6(3):213–226. https://doi.org/10.1108/AEDS-032017-0029

Kar S, Roy A, Sen K (2019) The double trap: institutions and economic development. Econ Modell 76:243–259. https://doi.org/10.1016/j.econmod.2018.08.002

Karim N, Tariq M, Khan MA (2025) Examining the moderated-mediation role of thefinancial sector and institutional quality in the relationship between remittances andeconomic growth in developing countries. J Int Migrat Integrat 26(1):279–316. https://doi.org/10.1007/s12134-024-01184-3

Khan MA, Máté D, Abdulahi ME, Sadaf R, Khan MA, Popp J, Oláh J (2024) Institutional quality, innovation and technologies promote financial market development?. Europ J Int Managem 22(3):484–507. https://doi.org/10.1504/EJIM.2024.136453

Macdonald D (2019) Barriers to reconstruction and development: a brief view of corruption andeconomic crime in Southern Africa. In: Structural adjustment, reconstruction and development in Africa. Routledge, pp 172–178

Mayersen D, McLoughlin S (2011) Risk and resilience to mass atrocities in Africa: a comparison of Rwanda and Botswana. J Genocide Res 13(3):247–269. https://doi.org/10.1080/14623528.2011.606687

Mehmood S (2023) The rule of law approach for more resilient institutions: judicial accountability and independence, and global economic activities (No. 1418). ADBI Working Paper. https://doi.org/10.56506/fghq1674

Millan-Tudela LA, Marco-Lajara B, Martínez-Falcó J, Poveda-Pareja E (2023) Corporate sustainability and long-lived companies: a lesson from the oldest firms in the world. In Corporate sustainability as a tool for improving economic, social, and environmental performance. IGI Global Scientific Publishing, pp 24–39

Mujuzi JD (2019) Investigating, prosecuting and punishing corruption in commonwealth East and Southern African countries. Commonwealth Law Bulletin 45(3):477–525. https://doi.org/10.1080/03050718.2020.1729216

Musah A (2024) The role of institutional efficiency in achieving the SDGs: evidence from Africa. J Business Socio-econ Developm 4(4):359–369. https://doi.org/10.1108/JBSED-02-2023-0008

Ndlovu L, De Kock A, Mkuzangwe NN, Thwala N, Mokoena CJ, Matimatjatji R (2023) Uncovering influential factors of civil unrest in South Africa: a machine learning and OSINT approach. In Machine intelligence and digital interaction conference, December, Springer Nature, pp 180–189. https://doi.org/10.1007/978-3-031-66594-3_19

Noll A, Budniok J (2022) Popular protest, resources, and political opportunities in Ghana: contextualising the case of occupy Ghana. In: Popular protest, political opportunities, and change in Africa. Routledge, pp 39–55

Nwoba AC, Boso N, Robson MJ (2021) Corporate sustainability strategies in institutional adversity: antecedent, outcome, and contingency effects. Business Strategy Environ 30(2):787–807. https://doi.org/10.1002/bse.2654

Nwozor A, Afolabi O (2023) Keeping up appearances? Nigeria's anti-corruption crusade and image dilemma in the global arena. J Finan Crime 30(3):813–827. https://doi.org/10.1108/JFC-02-2022-0039

Nzama L, Sithole T, Kahyaoglu SB (2022) The impact of government effectiveness ontrade and financial openness: the generalized quantile panel regression approach. J Risk Finan Managem 16(1):14. https://doi.org/10.3390/jrfm16010014

Nzama L, Ogundele OS (2025) IPSAS adoption, technological innovations and goodgovernance: catalysts for public sector sustainability. Edelweiss Appl Sci Technol 9(7):2152–2164. https://doi.org/10.55214/2576-8484.v9i7.9130

Nzama L (2022) A governance framework for non-performing municipalities in South Africa (Doctoral dissertation, University of Johannesburg). https://hdl.handle.net/10210/504396

Nzama L (2023) Analysis of local government financial performance: Evidence from South African Metropolitan municipalities. In *ICABR Conference*, June (pp. 775–796). Springer Nature. https://link.springer.com/chapter/https://doi.org/10.1007/978-3-031-461774_42

Nzama-Sithole L, Moloi T, Marx B (2025) A conceptual governance framework for non performing district municipalities in South Africa. Sayıştay Dergisi 36(137):217–253. https://doi.org/10.52836/sayistay.1700555

Ojeka S, Adegboye A, Adegboye K, Umukoro O, Dahunsi O, Ozordi E (2019) Corruption perception, institutional quality and performance of listed companies in Nigeria. Heliyon 5(10):e02569. https://doi.org/10.1016/j.heliyon.2019.e02569

Owusu-Mensah I, Mathapoly-Codjoe E (2023) Judicialization of corruption' in Ghana: an analysis of how Ghanaians view the office of the special prosecutor. Politikon 50(2):119136. https://doi.org/10.1080/02589346.2023.2197295

Scott R (1995) Institutions and organizations, vol 2. Sage, Thousand Oaks, CA

Scribner D, Lambert PA (2010) Constitutionalizing difference: a case study analysis of gender provisions in Botswana and South Africa. Polit Gender 6(1):37–61. https://doi.org/10.1017/S1743923X0999050X

Solo K (2025) Human rights as a moving target in Botswana. In: Legalization of human rights in Africa. Routledge, pp 257–271

Sun X, Meng Z, Zhang X, Wu J (2025) The role of institutional quality in the nexus between green financing and sustainable development. Res Int Business Finan 73:102531. https://doi.org/10.1016/j.ribaf.2024.102531

Theobald R, Williams R (1999) Combating corruption in Botswana: regional role model or deviant case? J Commonwealth Comparat Polit 37(3):117–134. https://doi.org/10.1080/14662049908447799

Titus UM, Idowu OG, Awojobi S, Okon C (2025) High cost of governance and socio economic development in Nigeria. African Ident 1–18. https://doi.org/10.1080/14725843.2025.2462607

Uddin I, Ahmad M, Ismailov D, Balbaa ME, Akhmedov A, Khasanov S, Haq MU (2023) Enhancing institutional quality to boost economic development in developing nations: new insights from CS-ARDL approach. Res Globalizat 7:100137.https://doi.org/10.1016/j.resglo.2023.100137

World Governance Indicator. https://databank.worldbank.org/source/worldwide-governance-indicators

Yeboah-Assiamah E, Alesu-Dordzi S (2016) The calculus of corruption: a paradox of 'strong' corruption amidst 'strong' systems and institutions in developing administrative systems. J Public Affairs 16(2):203–216. https://doi.org/10.1002/pa.1576

Zeqiraj V, Sohag K, Hammoudeh S (2022) Financial inclusion in developing countries: do quality institutions matter?. J Int Finan Markets Institut Money 81:101677. https://doi.org/10.1016/j.intfin.2022.101677

Governance, Accountability, and Sustainability Frameworks

Steering the Sustainability Agenda: Governing, Accounting, and Responsibility

Luzia Arantes and Shivam Bhartiya

Abstract This study undergoes a systematic review of the literature (SLR) with the aim of questioning the sustainability, organisational governance, innovation, and value creation nexus critically. The analysis, based on 49 peer-reviewed articles (presented between 2021 and 2022), outlines three broad thematic clusters, namely, cross-sector partnerships to achieve sustainable value, sustainable capability and strategic innovation, consumer participation in the digital environment, and sustainable governance achieved by innovation and partnership. The findings reveal the growing popularity of ESG practises, the importance of value co-creation, and the challenges associated with the need to uphold legitimacy and stakeholder confidence in collaborative and digital spaces. It gives an integrative conceptual framework that visualises the relationship between the main dimensions and highlights the importance of governance, participation, and innovation as the key agents of sustainable value. The constraints involve the geographic clustering of the empirical studies and the relative lack of research on qualitative or mixed methods. The review provides the in-depth climate of sustainability of organisations, which are congruent with the Sustainable Development Goals (SDG) 9, Industrialisation, Innovation and Infrastructure, and SDG 17, Partnerships for the Goals, which explains their relevance to future research and accountable strategic action.

Keywords Sustainability · Governance · ESG · Value creation · Strategic partnerships · Systematic review of the literature · SDGs

L. Arantes (✉)
Polytechnic University of Cávado and Ave, Barcelos, Portugal
e-mail: lamorim@ua.pt

S. Bhartiya
University of Aveiro, Águeda School of Technology and Management, Águeda, Portugal

Department of Commerce and Management, Centre for Distance and Online Education, Jain University, Bengaluru, India

S. Bhartiya
e-mail: shivam.bhartiya@jainuniversity.ac.in

K. Kumar et al. (eds.), *Sustainable Value Creation: Corporate Strategies for SDGs Implementation*, World Sustainability Series,
https://doi.org/10.1007/978-3-032-24652-3_5

1 Introduction

In recent decades, sustainability has evolved from a more marginal consideration to become one of the key organisational issues, thus affecting the distribution of resources to organisations, strategic precedence, and social accountability (Garcia-Serrano et al. 2025; Štrukelj et al. 2023). Increasing demands toward responsible practises and increased complexity of issues in the world system (climate change and social inequality) have forced companies, governments, and civil society organisations to rethink their value-creating frameworks (Gaba and Madhumathi 2023; Štrukelj et al. 2023). Here, notions like sustainable governance, corporate responsibility, and strategic partnerships have been more popular, as the approach towards more integration and collaboration in the creation of sustainability-oriented values has drifted towards (Carrera 2024).

This tendency is beyond the scholarly literature and, therefore, the thorough study of the interrelationship between sustainability and organisational values has undergone substantial development. Such studies as Ghosh and Singh (2025) and Narula et al. (2025) are coming forward to predict the development of ESG practises as a source of micro-performance, although the outcomes are frequent and not conclusive. Other authors, such as Stadtler et al. (2024), focus on the transformational prospect of cross-sector partnerships, whereas Gomez-Valencia et al. (2021) raise concern over the disjointedness of risks of sustainable development and insist on more integrated structures. Because the authors provide evidence-based arguments on the significance of interdisciplinary and impact-oriented approaches in the scenario of the Sustainable Development Goals (SDGs), Wang et al. (2023) support their arguments.

The current study is in this milieu. Through the systematic literature review (SLR), it is proposed to critically examine the state of the art of the concept of sustainability in the spheres of organisational governance, strategic partnership, and corporate responsibility. The main research question (RQ) is as follows:

RQ1: What has the recent scholarly literature said about the relationship between sustainability, governance, innovation, and value creation?

In line with the nature of the research question posed, the chapter has two main aims: first, to identify the main themes, conceptual axes, and theoretical contributions that have come as a result of the recent scholarship, and second, to develop a model that integrates the perspectives of innovation, responsibility, and sustainable value creation. In practise, the research is useful because it organises an area of knowledge that is fast growing in size, thus providing a more holistic and critical insight into the way organisations are progressing towards sustainability. The results are used in future research agendas by mapping patterns, convergences, and lacunae in the literature to facilitate the design of governance practises that would be more attuned to the modern sustainability problems.

The chapter has the following structure: after the given introduction, the methodology adopted is presented, followed by the results of the descriptive and thematic analysis. Then, the issues of every identified cluster are described with their contributions and limitations. Based on this analysis, an integrative framework is advanced to visualise the connections among the emergent conceptual axes. Lastly, the study implications are considered and the conclusions made.

2 Method

This study follows a rigorous literature review (SLR), the aim of which is to make a critical review of the present level of research on the sphere of sustainability in the context of organisational governance, strategic partnerships, and corporate responsibility. The methodology employed supports the concepts of transparency, replicability, and methodological rigor in the generalization of academic evidence, especially in interdisciplinary settings, including organizational sustainability (De León Pérez et al. 2024). The systemic approach was outlined to exist in four key phases, as shown in Fig. 1.

The bibliographic search was conducted in the first stage at the Web of Science (WoS) database. An elaborate search term was created, which incorporated the terms of sustainability and the idea of governance, responsibility, and value creation. The search was focused on the title, abstract, and keywords, and Boolean operators were used to provide thematic relevance. They used the following expression to do this: sustain AND govern AND (account* OR responsib* OR corporate social responsibility) OR CSR OR stakeholder engagement OR non-financial report* OR sustainability report* or by ESG) AND value creation.

In the second step, the inclusion criteria were used to narrow down the initial findings. Articles that were typed in English and published within the period of two years (20212022) and categorised as scientific, early access, or review papers were considered. Also, only the thematic categories of Management and Business were selected, and the SCI Expanded index and the SSCI indices. The use of these filters cut the sample size of 223 to 54 articles.

The third stage entailed intensive validation of the content. This involved first screening the title and abstract to ensure the relevance of the title to the objective of the review, and a subsequent reading of the articles that were selected to consider the depth and relevance of the articles to the stated objectives. This elaborate sifting resulted in a final list of 49 articles.

An inductive thematic analysis was used in the last phase, where the retained literature was analysed, which helped to determine the conceptual interconnections and the repeated research directions. The result of this analysis is that the studies have been grouped into four thematic clusters, including: (i) cross-sector partnerships as a means of achieving sustainable value; (ii) sustainable capabilities, strategic innovation; (iii) consumer participation, experience in social media; and (iv) sustainable governance and sustainable value creation through innovation and partnerships. Based on this framework, an integrative model was constructed to shed light on the relationships between the four clusters and hence demonstrate conceptual convergence as well as outline areas of future research.

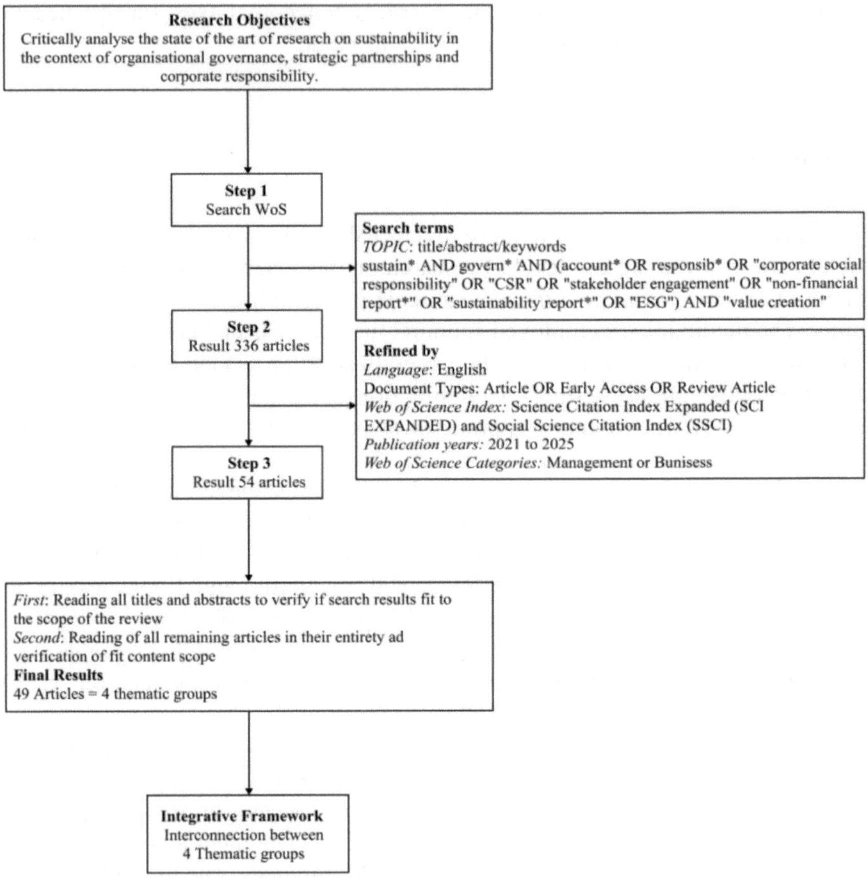

Fig. 1 Stages of the methodological process for the systematic review of literature on sustainability, governance and value creation. *Source* Authors

3 Results

A. Descriptive Results

Figure 2 shows the changes in the annual number of publications included in the sample and the corresponding publication citations in the WoS Core Collection database. This creates an overall upward trend with the number of articles published between 2021 and 2025 (2025 has the highest number of publications, with 15 publications).

Nonetheless, the highest point in citations is 2023; however, even though the number of publications was not the highest that year, the articles have the highest academic impact, with over 400 citations. The difference indicates that the impact of the works can be determined not by the annual production volume, but by the

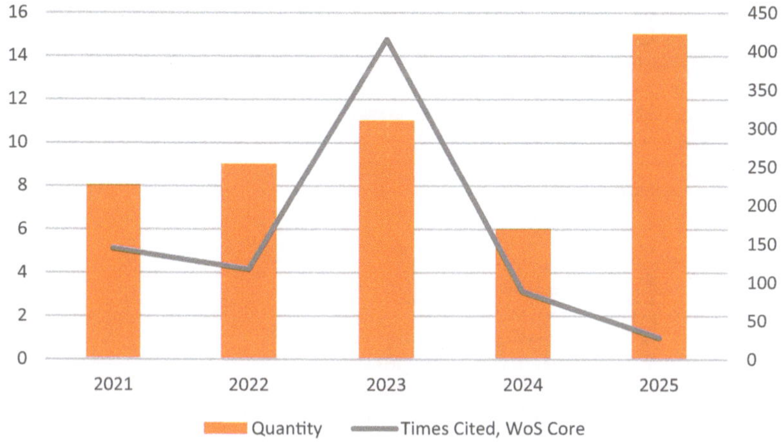

Fig. 2 Temporal distribution of publications and their scientific impact (2021–2025). *Source* Authors

relevance and exposure of the scientific world. This decline in citations in 2024 and 2025 may be attributed to the fact that the articles had shorter circulation times, which explains why the time factor is important in understanding the impact of a scholar.

Figure 3 demonstrates publications distribution of the scientific journals in the sample. A great diversity of journals validates the interdisciplinary character of the issue being studied. However, there are two distinguishing journals, namely Corporate Social Responsibility and Environmental Management, and Sustainability Accounting Management and Policy Journal, which comprise eight and six publications respectively, which highlights the core position of these sources in the sustainability, corporate responsibility, and strategic value database. The other journals, like the Journal of Small Business Management, Strategic Management Journal, and Journal of Business Ethics, are not as frequent but are still relevant to their themes of specialisation. This editorial pluralism indicates that studies undertaken in the field of sustainability and corporate governance are not limited to any field of study but rather are dispersed throughout the fields of business ethics, innovation management, organisation studies, and strategy. All these statistics confirm the thematic arrangement of the results determined in the framework of the qualitative analyses, which promotes the high validity of the review and indicates the further topicality of the subject in the global scientific environment.

B. **Thematic Clusters**

Table 1 highlights the thematic sorting of the systematic literature review (SLR) that identifies the 49 articles chosen according to the systematic literature review in four sorting clusters. With the help of this organisation, it is easier to comprehend the main conceptual areas of focus and discuss convergences in the areas of organisational sustainability, governance, and value creation.

Fig. 3 Distribution of publications by scientific journal. *Source* Authors

Cluster 1 has eight studies that focus on cross-sector relationships to achieve sustainable value. It dwells upon networks among firms, non-governmental organisations, and other social actors as vehicles of co-creation of value. The sample has 20 publications in Cluster 2, which indicates its heart and deals with sustainable capabilities and strategic innovation. The current studies analyse implications of ESG practises to the performance of organisations and implications of governance structures and technological innovation to sustainable value creation. Cluster 3, with five studies, is on the topic of consumer participation/ experience on social media. It puts focus on the role of consumers, NGOs, and digital platforms in the co-construction of value, and advocacy of sustainability, with a focus on justifiable supply chains and relational governance. Lastly, Cluster 4, which consists of 16 articles, is a collection of studies that tie together sustainable governance and value based on innovation and partnerships. This group emphasises the mutual influence of creative organisational activities, the development of new technologies (including artificial intelligence), the legitimacy of the institution, and the normative consistency in developing sustainable value.

Table 1 Thematic grouping of selected studies into clusters

Thematic clusters	Authors
Cluster 1 (n = 8): Cross-sector partnerships for sustainable value	(Al-Tabbaa 2023; Fontana et al. 2021; Joseph and Van Buren 2022; Kotze and Hofmeyr 2022; Lehoux et al. 2021; Muñoz et al. 2022; Robb et al. 2025; Stadtler et al. 2024)
Cluster 2 (n = 20): Sustainable capabilities and strategic innovation	(Abideen 2025; Alkaraan et al. 2023; Atkins et al. 2023; Caratas et al. 2021; Gao et al. 2024; Ghosh and Singh 2025; Gomez-Valencia et al. 2021; Iurkov et al. 2024; Ma 2023; Narula et al. 2025; Sabbaghi 2024; Sandberg et al. 2023; Teti et al. 2022; Tron et al. 2025; Vesal et al. 2025; Wang et al. 2023)
Cluster 3 (n = 5): Consumer participation and experience in social media	(Bennett & Grabs, 2025; Dimitropoulos et al. 2023; Jraisat et al. 2022; Peng et al. 2022; Sardanelli et al. 2022)
Cluster 4 (n = 16): Sustainable governance and value through innovation and partnerships	(Arvidsson, 2022; Barile et al. 2025; Brzustewicz et al. 2021; Cohen & Zhu, 2024; Ecim, 2024; Fahn & Zanarone, 2022; Fosso Wamba et al. 2025; Helfat et al. 2023; Hu et al. 2023; Kavadis et al. 2024; Li et al. 2023; Maroun et al. 2022; McGahan and Pongeluppe, 2023; Rasche et al. 2021; Treepongkaruna et al. 2025; Yin 2021)

Source Authors

Each of the four clusters shall be further discussed in the next segment to enrich their theoretical, methodological, and empirical contributions and fill any gaps found and future research opportunities.

i. *Cross-sector partnerships for sustainable value*

Cluster 1 gathers a line of research that studies the topic of value co-creation and sustainability in cross-sector alliances between non-profit organisations, for-profit companies, and other social entities. At the centre of this argument is the idea of collaborative arrangements providing purposeful strategies according to which organisations are aligned in terms of interests, governance mechanisms, and institutional structures to create mutual value (Al-Tabbaa, 2023). The empirical evidence points to the conditioning aspect of collaborative governance to support such partnerships that strategic alignment and institutional rules define the distribution of benefits among actors (Al-Tabbaa 2023). In line with this view, qualitative studies about the Bangladesh fashion industry have demonstrated the theoretical benefits of intersectoral collaboration in that context; that is, in relation to organisational identity, through empowerment and labour sustainability, especially of the vulnerable groups, when such partnerships are well planned and operated (Fontana et al. 2021). Although these are the contributions that are made, the literature is still theoretically fragmented. Nevertheless, a thorough overview by Stadtler et al. (2024) highlights conceptual overlaps, especially when it comes to the capabilities of impact in intersectoral alliances, but it also suggests challenges related to theorising and measuring impacts over time, thus pointing to relevant avenues that further research should apply.

The framework of Responsible Research and Innovation (RRI) has been applied to investigate the concept of responsible value creation in the sphere of healthcare (Lehoux et al. 2021). Using a qualitative approach that relies on case studies, the authors make the point that co-making of value requires conscious engagement of several stakeholders and special attention to ethical, social, and political consequences (Lehoux et al. 2021).

In a study by Joseph and Van Buren (2022), the dynamics of entrepreneurship in the conflict zones are examined and question the possibility of local entrepreneurs to create value and peace. Based on the group identity theory and the intersection between entrepreneurship and conflict resolution, they argue that whether they can maintain value co-creation under these circumstances rests on the ability of the entrepreneurs to build up trust, create and nurture inclusive social networks, and reduce ethnic and religious tensions (Joseph and Van Buren 2022). Additionally, the process of measurement of impact in new industries is encoded as a negotiation performance following new institutionalist logics (Muñoz et al. 2022).

A resource-based analysis indicates that social impact capabilities have the potential to create a competitive advantage when encompassed in the organisational strategy (Robb et al. 2025). Authentic interaction with local communities is revealed as a prerequisite of creating a mutual value, and purely functional strategies turn out to be counterproductive (Kotze and Hofmeyr 2022).

The cluster methodologically provides a prevalence of qualitative methods and case studies with the inclusion of systematic reviews (Stadtler et al. 2024). These are methodologies that are geared towards a detailed study of the processes and practises, places significantly on contextual richness and interaction with actors. One of the observed trends is the exploration of problematic environments, including conflict zones (Joseph and Van Buren 2022), sectors with a significant amount of job insecurity (Fontana et al. 2021), or revolutionary health systems (Lehoux et al. 2021). The cluster methodological diversity is increased by the introduction of inclusive approaches (Muñoz et al. 2022) and conceptual modelling (Robb et al. 2025).

In principle, it is possible to point out that the literature has narrowed down the valuation of value co-creation as a core construct, although significant differences in operationalisation are observed. Although it is also associated by some researchers with strategic choices and the form of governance (Al-Tabbaa 2023), others investigate its expression in relation to social responsibility, innovation, and institutional change. The most evident theoretical contradiction deals with collaboration and impact relations, especially the issues of measurement and comparison of the long-term effects, which once again appear to be the primary theme in Stadtler et al. (2024).

Overall, cluster one is a developed but developing body of literature on cross-sector or cross-sector collaborations to create sustainable values.

ii. *Sustainable capabilities and strategic innovation*

The key issue of this cluster is the strategic impact of the implementation of environmental, social, and governance (ESG) practices on organisational performance, with the specific focus on innovation mechanisms, the efficiency of organisational operations, and the organisation's structure. Sustainability practises and financial performance are a major group of studies being studied. They show that sustainability enhances the reputation of a company and rates of investor sentiment (Abideen 2025) or examine the disparaging effects of ESG pillars on profitability and liquidity, with positive effects related to the pillars of governance and the environmental pillar and negative effects on the social pillar (Caratas et al. 2021). Similarly, the role of shareholders' organisation and independent board in either an effective sustainable strategy is emphasised (Tron et al. 2025), and the ESG nature's performance in the connection between the investment, its effectiveness, and governance (Gao et al. 2024).

A part of the articles emphasises sustainable efficiency via Data Envelopment Analysis (DEA) and bootstrapping (Lu et al. 2025), the connection between sustainable investments and reducing the risk (Alkaraan et al. 2023), and corporate valuation in multinationals with multidimensional ESG practises (Gomez-Valencia et al. 2021). Some look at the significance of transparency and ESG reporting to investor confidence and risk perception (Atkins et al. 2023; La Torre et al. 2021; Luo et al. 2022; Momtaz and Parra 2024).

Sustainability also seems to be one of the drivers of business innovation, encouraging the new products and processes (Ghosh and Singh 2025). The immediate effect of the ESG performance on financial returns is also examined (Sabbaghi 2024).

The interaction between sustainability and the digitalization approach takes care of the technological aspect (Vesal et al. 2025), whereas ESG practices are operationalized in global chains (Narula et al. 2025).

Institutionally and symbolically, the quality of sustainable practices regarding the reduction of risks is discussed (Wang et al. 2023), the role of authenticity between the ESG discourse and action is emphasized (Ma 2023), and the significance of sustainable alliances in innovation is discussed (Iurkov et al. 2024). Lastly, the internal commitment relevance and national conditions to sustainability impacts on the outcome of ESG ratings are pointed out (Sandberg et al. 2023; Teti et al. 2022).

At the methodological level, there is a preponderance of large-scale quantitative studies based on secondary data, even though there is a variety of statistical methods (e. g., Alkaraan et al. 2023; Iurkov et al. 2024; Narula et al. 2025; Vesal et al. 2025). The qualitative approaches used in a few studies are a gap that can restrict the knowledge of greater processes in an organisation. The high geographical concentration (where multiple studies are dedicated to China) is also one of the weaknesses (Gao et al. 2024; Luo et al. 2022; Ma, 2023), and it casts a concern about generalising the findings. Moreover, financial and operational indicators (Alkaraan et al. 2023; Sandberg et al. 2023; Vesal et al. 2025) are examined in most studies, but there is little analysis regarding the impact on broad social factors and cultural dynamics.

Overall, this cluster displays a coherent idea of the variables that support strategic innovation driven by sustainability, making the internal capabilities, transparency, digital transformation, and an institutional fit the key components of the value creation related to sustainability.

iii. *Consumer participation and experience in social media*

Cluster three is a collection of research papers that explore the co-creation of sustainable value in collaborative settings, which would include non-governmental organisations (NGOs), consumers, and strategic partners.

The former axis dwells upon the significant role of partnerships and NGOs in the development of trust and consideration of sustainability along the lines of supply chains, as it is illustrated by the works of Peng et al. (2022) and Sardanelli et al. (2022). It puts the primary emphasis on relational governance as the cornerstone of launching sustainable practises, which relates well to the literature on sustainable value distribution, which covers supply chains, including fair trade. In such scenarios, Bennett and Grabs (2025) offer alternative processes that will help address asymmetries to foster inclusion.

Simultaneously, academic research has also come up addressing the nature of challenges faced by multinational companies in integrating sustainability into their international supply chain. These difficulties are defined by conflicts between regional demands and international requirements, highlighting the importance of strategic readjustment (Dimitropoulos et al. 2023). The parallel studies on the partnership between for-profit and non-profit organisations reveal institutional tensions that can be resolved with trust, value congruence, and past experiences of cooperation (Jraisat et al. 2022).

In the literature, most qualitative approaches, in particular, case studies, are predominant and fit well with the complexity of the examined contexts (Bennett and Grabs 2025; Jraisat et al. 2022; Peng et al. 2022). However, the lack of mixed-methods and longitudinal research places other restrictions on the generalisability of results. In theory, the recurrent mentioning of co-creation, sustainability, and legitimacy is present in every part of the corpus (Bennett and Grabs 2025; Jraisat et al. 2022; Peng et al. 2022). Nevertheless, the lack of a single conceptual framework indicates a conceptual rupture that should be addressed. There are also some gaps in the literature: minimal representation of consumers as active co-creators, inadequate focus on digital institutions of interaction situations (Bennett and Grabs 2025; Peng et al. 2022; Sardanelli et al. 2022).

Overall, cluster three provides significant insights into the concept of sustainable value within the realm of collaborative networks, but it also indicates a strong necessity for increased theoretical integrity, pluralism of methods, and a better emphasis on the role played by the consumer in the process of co-creation.

iv. *Sustainable governance and value through innovation and partnership*

The cluster unites an array of research works aimed at investigating the complex interconnection between sustainability, value creation, and innovative organisational practises with a specific focus on the assimilation of ESG dynamics in strategic and collaborative terms (Helfat et al. 2023).

A few studies cover sustainability value creation in the multinational companies (Kavadis et al. 2024; Yin 2021), digital platforms (Li et al. 2023), and the joint venture between companies and non-governmental organisations (Brzustewicz et al. 2021). They emphasise non-economic processes like reputation, trust, and value-congruency. The other researchers investigate the applicability of purposeful innovative methods and institutional legitimacy to the situations of big data and sustainability (Arvidsson 2022; Rasche et al. 2021). This variety is also expanded by artificial intelligence and digital innovation integration as the sources of sustainability (Barile et al. 2025; Fosso Wamba et al. 2025).

The cluster can be characterised, methodologically, by its eclectic approach of approaches. There are quantitative fuzzy and set theory-based models (Hu et al. 2023), conceptual and theoretical analyses (Cohen and Zhu 2024; Treepongkaruna et al. 2025), empirical research that is based on secondary research (Li et al. 2023; McGahan and Pongeluppe 2023; Yin 2021), and formal modelling based on relational contracts and incentives (Fahn and Zanarone 2022). In addition, a philosophical or critical view is embraced by some of the investigations, including the exploration of the use of leadership rhetoric in the discourse of sustainability (Arvidsson 2022).

Nevertheless, a general agreement appears despite the theoretical heterogeneity of strategic value of sustainability and alignment of stakeholder interest (McGahan and Pongeluppe 2023) and a necessity to incorporate ethical values and legitimacy standards into governance systems (Ecim 2024; Rasche et al. 2021). The same issue is reported in the proposals of integrated thinking models (Ecim 2024; Maroun et al. 2022) and the transformative role of technology, especially generative artificial intelligence (Fosso Wamba et al. 2025). One of the identified gaps involves the lack of empirical operationalisation of conceptual models (Barile et al. 2025; Treepongkaruna et al. 2025), the immature analysis of data legitimacy related to leadership positions in the integration of sustainable values (Arvidsson 2022), and the descriptive nature of the research on the organisational impacts of artificial intelligence (Fosso Wamba et al. 2025), that requires more solid and interdimensional designs.

Altogether, this cluster is a conglomerate of viewpoints aiming at incorporating the concept of sustainability, innovation, and value creation in a strategic, relational, and technological context.

4 Integrative Framework

Using the results of the observation of the four identified thematic clusters from the systematic literature review, an integrative model was drawn to outline the conceptual interrelationship between the key emerging axes of intersectoral partnership, strategic innovation, consumer participation in digital contexts, and sustainable governance. Figure 4 presents this conceptual framework that encompasses both theoretical and empirical contributions of the discussed studies, hence presenting the relational processes underpinning the formation of sustainable value in the context of a modern, organisational environment.

The framework promotes a relational framework where governance is at the centre of the axis between collaborations that cut across intersectoral and the ability of organisations to focus on sustainable innovations. This association highlights the degree of collaboration interventions amongst various actors to enhance sustainability. On the other hand, innovation acts as the channel that links the strategic capabilities with the creation of the impact-oriented governance models, which underlines the importance of emergent technologies, ESG practises, and digital processes in the transformation of organisations.

This is because of the active engagement of consumers in social media and digital environments, which has been linked to the enhancement of the legitimacy of the sustainability-oriented initiatives, which is one of the key factors of expression between social engagement and institutional actors. The concept of legitimacy, as a dynamic construct, is a result of the consistency between discourse, practise

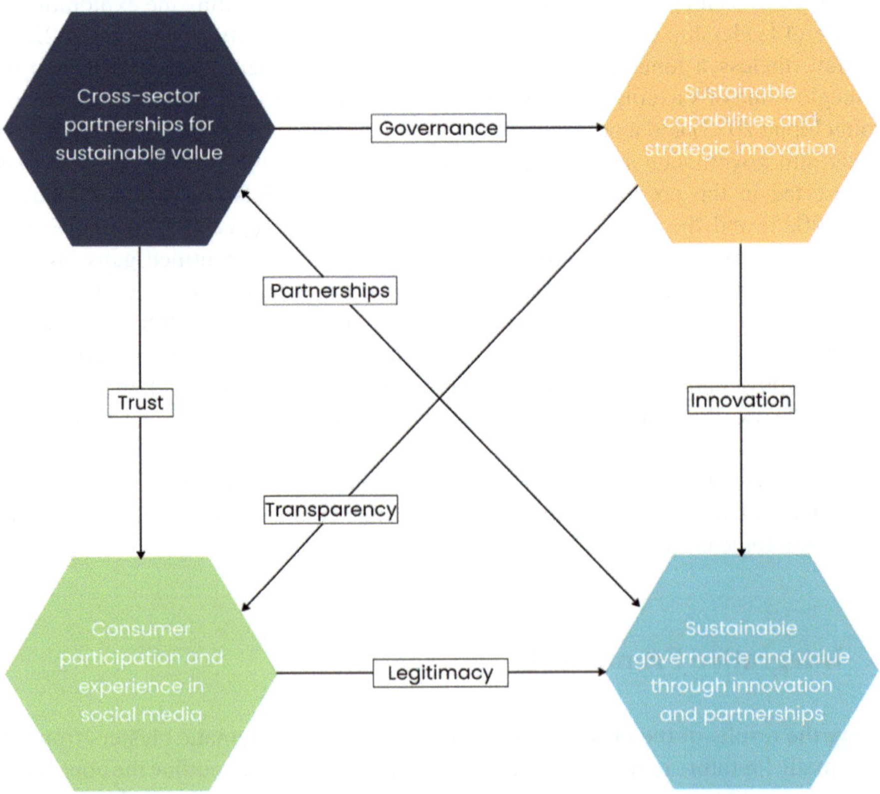

Fig. 4 Integrative model of the dynamics between governance, innovation, participation and sustainable value. *Source* Authors

and values, mediated through the way people perceive the authenticity and moral suitability.

Similarly, trust and transparency are important mediation mechanisms in all the domains of the framework. Relationship Trust forms the relationship basis needed to build long-lasting relationships, and in this situation, it is observed that this is gainfully achieved through cross-sector collaboration or where there is social media activity, and at the same time, consistency is upheld between purposes and outcomes. These factors do not just contribute to the hope of cooperation between actors, but they also confirm the legitimacy of the actions taken.

Finally, the partnership contact between the various blocks of the framework is determining in the mobilisation of the internal abilities within the organisation and movement of the external resources and social legitimacy. The process of partnerships, i.e., among corporations, civil society organisations, consumers, or digital platforms, strengthens the value creation through the encouragement of synergies, collaborative innovation, and a long-term commitment.

As a result, the integrative model presented in this paper is a synthesis of primary convergent ideas found in the recent literature, presenting governance, innovation, participation, and legitimacy as supporting central capstones of value creation that are interdependent on one another. Its usefulness is not only in its organising of extant knowledge but also as the basis of further research to gain a systemic perception of the challenges involved in the sustainability of organisations.

5 Conclusions

Based on an exhaustive systematic review of the literature, the chapter provides a subtle and synthesising understanding of the nexus between sustainability, organisational governance, innovation, and value creation. The study also clarifies how modern organisations are reformatting their strategies and governance processes to address challenges related to sustainability, by highlighting four areas of thematic links, namely the cross-sector partnerships, sustainable capabilities and strategic innovation, engagement with consumers in digital milieus, and sustainable governance. Instead of considering sustainability as a situational managerial issue, the review highlights its fact in the context of relational, institutional, as well as technological processes, which are all involved in formulating sustainable value creation.

In a theoretical approach, the chapter incorporates discontinuous plates of research and drives a more consistent academic framework of sustainability-based governance. Specifically, legitimacy, trust, and involvement of stakeholders are highlighted in the findings as key processes in which sustainability initiatives will be transformed into sustainable value. Combining the findings of the governance theory, studies on innovation and sustainability, the paper supports the necessity to move beyond instrumentalist conceptualisations of the organisational sustainability practises and embrace more relational and systemic approaches towards them.

Besides, the chapter provides relevant practical implications. The findings indicate to managers and policy makers the strategic role of aligning structural governance, innovation processes, and arrangements relating to collaborations with sustainability goals. The cross-sector partnerships are not only a significant rule of operation, but they are a method of governance that allows distributing responsibilities and sharing risks, as well as learning together. Similarly, transparency, moral uniformity, and substantive stakeholder engagement are essential conditions of perpetuating legitimacy and confidence, especially in electronic-mediated space and international value chains.

Despite such contributions, the study has some limitations. The set of reviewed empirical evidence is still geographically centralised, and the prevailing role of quantitative methods restricts the understanding of organisational processes, relations of power, and situational peculiarities. It would be advantageous in future research to have a more methodological pluralism related to qualitative and mixed-method designs, and longitudinal research that will be able to be more focused on the change in sustainability practises and governance arrangements over time.

The review also provides several opportunities for future research. More focus should be given to the stake of the consumers and civil society players as active co-creators of sustainable value, especially in the digital and platform-based environment. In addition, the ever-growing incorporation of artificial intelligence and other sophisticated digital technologies also requires more thorough and empirically based research into what they will mean to sustainable governance, accountability, and ethical decision-making processes. These issues determine that it is necessary to address them to understand how technological innovation may facilitate rather than jeopardise inclusive and responsible sustainability transitions.

Lastly, the chapter has a direct contribution to Sustainable Development Goal 9, Industry, Innovation and Infrastructure, since it prioritises responsible innovation as a source of sustainable value creation, and Sustainable Development Goal 17, Partnerships for the Goals, because it also focuses on cross-sector collaboration as the key element of the organisational changes. These goals make organisations more active in promoting institutional legitimacy and creating more equitable and regenerative ways to achieve sustainable development.

Acknowledgements The authors acknowledge using Grammarly, an AI-assisted tool, exclusively to improve sentence structure and grammar, ensuring greater clarity and accuracy. No content was generated by Artificial Intelligence. The authors assume full responsibility for the content presented.

References

Abideen ZU (2025) The moderating role of investor sentiments between sustainability reporting and firm reputation. evidence from Chinese listed firms. Corporate Soc Respons Environ Managem 32(1):138–159. https://doi.org/10.1002/CSR.2943

Alkaraan F, Elmarzouky M, Hussainey K, Venkatesh VG (2023) Sustainable strategic investment decision-making practices in UK companies: the influence of governance mechanisms on synergy between industry 4.0 and circular economy. Technol Forecast Soc Change 187:122187. https://doi.org/10.1016/J.TECHFORE.2022.122187

Al-Tabbaa O (2023) Nonprofit-business collaboration: an empirical-based framework for value creation. Australian J Managem.https://doi.org/10.1177/03128962231201525/ASSET/052209D5-4067-40D9-9CDA-1537F573721A/ASSETS/IMAGES/LARGE/10.1177_03128962231201525-FIG1.JPG

Arvidsson S (2022) CEO talk of sustainability in CEO letters: towards the inclusion of a sustainability embeddedness and value-creation perspective. Sustain Account Managem Policy J 14(7):26–61. https://doi.org/10.1108/SAMPJ-07-2021-0260/FULL/PDF

Atkins J, Doni F, Gasperini A, Artuso S, La Torre I, Sorrentino L (2023) Exploring the effectiveness of sustainability measurement: which ESG metrics will survive COVID-19? J Bus Ethics 185(3):629–646. https://doi.org/10.1007/S10551-022-05183-1/TABLES/4

Barile D, Secundo G, Del Vecchio P (2025) An artificial intelligence-based innovation ecosystem enabling open innovation and sustainable growth: evidence from a case study. Innovation 1–23.https://doi.org/10.1080/14479338.2025.2514468

Bennett EA, Grabs J (2025) How can sustainable business models distribute value more equitably in global value chains? Introducing "value chain profit sharing" as an emerging alternative to fair trade, direct trade, or solidarity trade. Business Ethics the Environ Respons 34(2):581–601. https://doi.org/10.1111/BEER.12666

Brzustewicz P, Escher I, Hermes J, Ulkuniemi P (2021) Value creation in company–NGO collaboration in corporate volunteering. J Business Indus Market 36(8):1504–1519. https://doi.org/10.1108/JBIM-01-2020-0057/FULL/XML

Caratas MA, Andreea Trandafir R, Iftene C, Cerasela Spatariu E, Gheorghiu G (2021) The impact of sustainability disclosure on companies' performance in healthcare industry. Transform Business Econ 20(2A):593–613. https://openurl.ebsco.com/contentitem/edb:153885466?sid=ebsco:plink:crawler&id=ebsco:edb:153885466&crl=c

Carrera L (2024). Public and Corporate socio-territorial policies. CSR as strategy for a new and enlarged social sustainability. J Emerg Perspect 1:23–32. https://doi.org/10.36253/JEP-16897

Cohen N, Zhu X (2024) Corporate initial responses to COVID-19 and ESG ratings: the role of ESG consistency. Sustain Account Managem Policy J 16(1):44–78. https://doi.org/10.1108/SAMPJ-03-2023-0118/FULL/XML

De León Pérez D, Acosta Vega R, Salazar Galán S, Aranda JÁ, Francés García F (2024) Toward systematic literature reviews in hydrological sciences. Water 16(3):1–25. https://doi.org/10.3390/W16030436

Dimitropoulos P, Koronios K, Sakka G (2023) International business sustainability and global value chains: synthesis, framework and research agenda. J Int Managem 29(5):101054. https://doi.org/10.1016/J.INTMAN.2023.101054

Ecim D (2024) Components of integrated thinking: evidence from South African listed companies. South African J Business Managem 55(1):16.https://doi.org/10.4102/SAJBM.V55I1.4080

Fahn M, Zanarone G (2022) Transparency in relational contracts. Strateg Manag J 43(5):1046–1071. https://doi.org/10.1002/SMJ.3348

Fontana E, Heuer M, Koep L (2021) Cross-sector collaboration and nonprofit boundary work for female workers in developing countries: evidence from Bangladesh. Sustain Account Managem Policy J 12(6):1178–1207. https://doi.org/10.1108/SAMPJ-03-2020-0062/FULL/XML

Fosso Wamba S, Queiroz MM, Randhawa K, Gupta G (2025) Generative artificial intelligence and the challenges to adding value ethically. Technovation 144.https://doi.org/10.1016/J.TECHNOVATION.2025.103235

Gaba N, Madhumathi R (2023). Do pressure-sensitive institutional investors moderate CSR decisions towards value creation of Indian firms? J Finan Report Account ahead-of-print. https://doi.org/10.1108/JFRA-07-2023-0389/FULL/PDF

Gao D, Li S, Zhou Y (2024) Investment efficiency, ESG performance and corporate performance: evidence from Chinese listed enterprises. Chin Manag Stud 19(2):567–599. https://doi.org/10.1108/CMS-06-2022-0210/FULL/XML
Garcia-Serrano I, Central U, Ecuador D, Moran EF, Regina K, Negrão M, Gomes SC, Cesar M, De Carvalho S, Luiz M (2025) Strategic management for sustainability. Revista de Administração Contemporânea 28(6). https://doi.org/10.1590/1982-7849RAC2024240221.EN
Ghosh M, Singh MP (2025) Unveiling sustainable practices in financial institutions: a bibliometric analysis cum meta-analytical review on enhancing financial performance. Corp Soc Responsib Environ Manag 32(2):2295–2316. https://doi.org/10.1002/CSR.3064
Gomez-Valencia M, Gonzalez-Perez MA, Gomez-Trujillo AM (2021) The "Six Ws" of sustainable development risks. Bus Strateg Environ 30(7):3131–3144. https://doi.org/10.1002/BSE.2794
Helfat CE, Kaul A, Ketchen DJ, Barney JB, Chatain O, Singh H (2023) Renewing the resource-based view: new contexts, new concepts, and new methods. Strateg Manag J 44(6):1357–1390. https://doi.org/10.1002/SMJ.3500
Hu X, Zhu W, Zhang C, Zhang T, Zhang C (2023) Research on the rules of ESG performance and value creation based on rough sets. J Bus Econ Manag 24(6):996–1018. https://doi.org/10.3846/JBEM.2023.20631
Iurkov V, Koval M, Misra S, Pedada K, Sinha A (2024) Impact of ESG distinctiveness in alliances on shareholder value. J Business Res 171:114395. https://doi.org/10.1016/J.JBUSRES.2023.114395
Joseph J, Van Buren HJ (2022) Entrepreneurship, conflict, and peace: the role of inclusion and value creation. Bus Soc 61(6):1558–1593. https://doi.org/10.1177/00076503211040238
Jraisat L, Jreissat M, Upadhyay A, Sajjad F, Balodi KC (2022) Paradox of strategic partnerships for sustainable value chains: Perspectives of not-for-profit actors. Bus Strateg Environ 31(7):3491–3508. https://doi.org/10.1002/BSE.3101
Kavadis N, Hermes N, Oehmichen J, Zattoni A, Fainshmidt S (2024) Sustainable value creation in multinational enterprises: the role of corporate governance actors. J World Business 59(1):101503. https://doi.org/10.1016/J.JWB.2023.101503
Kotze R, Hofmeyr K (2022) Effecting successful shared value creation: the role of organisations in fence-line communities. South African J Business Managem 53(1):15. https://doi.org/10.4102/SAJBM.V53I1.2992
La Torre M, Leo S, Panetta IC (2021) Banks and environmental, social and governance drivers: follow the market or the authorities? Corp Soc Responsib Environ Manag 28(6):1620–1634. https://doi.org/10.1002/CSR.2132
Lehoux P, Silva HP, Denis JL, Miller FA, Pozelli Sabio R, Mendell M (2021) Moving toward responsible value creation: business model challenges faced by organizations producing responsible health innovations. J Prod Innov Manag 38(5):548–573. https://doi.org/10.1111/JPIM.12596
Li L, Wang M, Zhou X (2023) Creating value beyond commercial outcomes: the ESG practices of online marketplaces for sustainable development. Electron Mark 33(1):1–19. https://doi.org/10.1007/S12525-023-00682-Z/FIGURES/1
Lu WM, Chou CH, Ting IWK, Liu SM (2025) A new integrated approach for evaluating sustainable development in the electric vehicle sector. Omega 133:103247. https://doi.org/10.1016/J.OMEGA.2024.103247
Luo W, Zhang C, Li M (2022) The influence of corporate social responsibilities on sustainable financial performance: mediating role of shared vision capabilities and moderating role of entrepreneurship. Corp Soc Responsib Environ Manag 29(5):1266–1282. https://doi.org/10.1002/CSR.2268
Ma R (2023) The sustainable development trend in environmental, social, and governance issues and stakeholder engagement: evidence from mergers and acquisitions in China. Corp Soc Responsib Environ Manag 30(6):3159–3173. https://doi.org/10.1002/CSR.2544
Maroun W, Ecim D, Cerbone D (2022) Refining integrated thinking. Sustain Account Managem Policy J 14(7):1–25. https://doi.org/10.1108/SAMPJ-07-2021-0268/FULL/PDF

McGahan AM, Pongeluppe LS (2023) There is no planet b: aligning stakeholder interests to preserve the amazon rainforest. 69(12):7860–7881. https://doi.org/10.1287/MNSC.2023.4884

Momtaz PP, Parra IM (2024) Is sustainable entrepreneurship profitable? ESG disclosure and the financial performance of SMEs. Small Business Econ 64(4):1535–1564. https://doi.org/10.1007/S11187-024-00981-5

Muñoz P, Gamble EN, Beer H (2022) Impact measurement in an emerging social sector: four novel approaches. Acad Managem Discoveries 8(2):298–300. https://doi.org/10.5465/AMD.2020.0044/AMD.2020.0044_WHITEBOARD_VIDEO_AMD_8.2.MP4

Narula R, Rao P, Kumar S, Paltrinieri A (2025) ESG investing and firm performance: retrospections of past and reflections of future. Corp Soc Responsib Environ Manag 32(1):1096–1121. https://doi.org/10.1002/CSR.2982

Peng S, Jia F, Doherty B (2022) The role of NGOs in sustainable supply chain management: a social movement perspective. Supply Chain Manag 27(3):383–408. https://doi.org/10.1108/SCM-05-2020-0191/FULL/XML

Rasche A, Morsing M, Wetter E (2021) Assessing the legitimacy of "Open" and "Closed" data partnerships for sustainable development. Bus Soc 60(3):547–581. https://doi.org/10.1177/0007650319825876

Robb CC, Brownell KM, Brännback M, Rumi S (2025) Applying resource-based theory to social value creation: a conceptual model of contributive advantage. J Small Business Managem.https://doi.org/10.1080/00472778.2024.2309498

Sabbaghi O (2024) The United nations sustainable development goals and global finance: recent evidence. Corp Soc Responsib Environ Manag 31(5):4020–4033. https://doi.org/10.1002/CSR.2785

Sandberg H, Alnoor A, Tiberius V (2023) Environmental, social, and governance ratings and financial performance: evidence from the European food industry. Bus Strateg Environ 32(4):2471–2489. https://doi.org/10.1002/BSE.3259

Sardanelli D, Bittucci L, Mirone F, Marzioni S (2022) An integrative framework for supply chain rating: from financial-based to ESG-based rating models. Total Quality Managem Business Excellence.https://doi.org/10.1080/14783363.2022.2069557

Stadtler L, Seitanidi MM, Knight HH, Leigh J, Clarke A, Le Ber MJ, Bogie J, Brunese P, Hustad O, Krasonikolakis I, Lioliou E, MacDonald A, Pinkse J, Sehgal S (2024) Cross-sector partnerships to address societal grand challenges: systematizing differences in scholarly analysis. J Manage Stud 61(7):3327–3357. https://doi.org/10.1111/JOMS.13053

Štrukelj T, Dankova P, Hrast N (2023) Strategic transition to sustainability: a cybernetic model. Sustainability 15(22):1–32. https://doi.org/10.3390/SU152215948

Teti E, Dell'Acqua A, Bonsi P (2022) Detangling the role of environmental, social, and governance factors on M&A performance. Corp Soc Responsib Environ Manag 29(5):1768–1781. https://doi.org/10.1002/CSR.2325

Treepongkaruna S, Sarajoti P, Korphaibool V (2025) Unlocking synergies: sufficiency economy philosophy for sustainability. Bus Strateg Environ 34(3):3446–3468. https://doi.org/10.1002/BSE.4151

Tron A, Colantoni F, Celentano N, Caputo A (2025) Delving into the influence of sustainability strategy: exploring the influence of sustainability committees on company performance. Bus Strateg Environ 34(2):1551–1573. https://doi.org/10.1002/BSE.4042

Vesal M, Alam MA, Rahmati MH (2025) Blending minds, balancing goals: TMT heterogeneity and sustainable-innovation in family firms. J Small Business Managem.https://doi.org/10.1080/00472778.2025.2495788

Wang N, Pan H, Feng Y, Du S (2023) How do ESG practices create value for businesses? Research review and prospects. Sustain Account Managem Policy J 15(5):1155–1177. https://doi.org/10.1108/SAMPJ-12-2021-0515/FULL/XML

Yin J (2021) How multinational corporations and nonprofits collaborate for sustainability: assessing social partnerships from China. Sustain Account Managem Policy J 12(6):1289–1311. https://doi.org/10.1108/SAMPJ-04-2020-0104/FULL/XML

A 2030 Framework for Corporate Sustainability and Accountability

Neha Malik, Mohit Verma, Aishwarya Vashishtha, and Prashant Ranjan

Abstract Corporate Sustainability 2030 is a model that embraces UN Sustainable Development Goals. Companies have to relate their long-term goals about environmental sustainability and social concerns to the full list of Sustainable Development Goals offered by the UN. The compliance with sustainable development is one of the key business priorities due to the 2030 Agenda by the United Nations and its 17 Sustainable Development Goals (SDGs). The objectives give complex demands to the businesses that require them to consider their environmental and social implications as well as to direct contributions to global sustainability efforts through correct business operations. Company reports are nowadays not only strategic instruments to illustrate how the company operations contribute to sustainable development goals but also a reporting mechanism of environmental and social governance performance measures. Business organizations that aim at aligning these development goals in their reporting systems have made SDGs a priority. Companies should create precise and definite connections between sustainability-related activities and performance indicators, as well as SDG-related outcomes. The matching of sustainability reports with particular SDGs increases their validity and allows the stakeholders to check corporate involvement in the key global challenges, such as climate change, poverty and inequality as well as access to education. The chapter proves that the development of holistic standardized solutions is needed to incorporate SDGs into the corporate

N. Malik
JSBF, O.P. Jindal Global University, Sonipat, Haryana, India

Indian Institute of Technology Delhi, New Delhi, India

M. Verma (✉)
O.P. Jindal Global University, Sonipat, Haryana, India
e-mail: phdmohitv@gmail.com

A. Vashishtha
International Business (Management and Global Business), Rutgers University, Newark, USA
e-mail: aish.vas@rutgers.edu

P. Ranjan
Department of Management Studies, Jamia Millia Islamia, New Delhi, India

K. Kumar et al. (eds.), *Sustainable Value Creation: Corporate Strategies for SDGs Implementation*, World Sustainability Series,
https://doi.org/10.1007/978-3-032-24652-3_6

sustainability reporting practice in an appropriate manner. Regulators and standard-setting bodies, along with policymakers, should collaborate to come up with efficient and similar disclosure of Sustainable Development Goal indicators across various industries.

Keywords SDG · Corporate sustainability · Business · Performance metrics · ESG

1 Introduction

Corporate Sustainability 2030 is developed based on the UN Sustainable Development Goals. Companies must translate their longer-term objectives in the environmental and social spheres into the UN's wide-ranging set of SDGs (Ponce 2023; Sánchez et al. 2020). Business's obligation at the core of the Sustainable Business agenda is to work towards sustainable development compliance based on the United Nations 2030 Agenda and its 17 Sustainable Development Goals (SDGs). The goals create complex demands on businesses, requiring that they address their environmental and social footprints while simultaneously contributing to global development through responsible business practices (Saizarbitoria et al. 2021; Singh et al. 2025a). Currently, company reports are used as tools of demonstration that show how the functioning of a company is in line with sustainable development goals, as well as for reporting environmental and social governance performance indicators (Ponce 2023; Yang et al. 2021; Singh et al. 2024). Business companies have given attention to the SDGs and need to incorporate these development goals into their reporting frameworks (Pizzi et al. 2020, as cited in Stocker et al. 2022). Businesses should also commit to connecting dots between their sustainability efforts and performance metrics, and SDG-related outcomes in clear and tangible ways (Ponce 2023; Singh et al. 2025b).

The alignment of the sustainability reports with SDGs contributes to the credibility of these reports and provides stakeholders with the possibility to check corporations' efforts to interact with the most serious issues of the global community, including climate, inequality and poverty, education, etc. (Sánchez et al. 2020). This study highlights the need for developing full standardized mechanisms through which SDGs can be adequately mainstreamed in corporate sustainability reporting (Živković 2022). Standard setters and regulators, along with policymakers, need to come together to produce meaningful and consistent disclosures of Sustainable Development Goal indicators in disparate sectors (Stocker et al. 2022; Živković 2022).

This article investigates how firms can create legitimate sustainability approaches to address the requirements of Corporate Sustainability 2030. It focuses writing on the need for common reporting, the connection between business and SDG indicators, and the new role for regulators and standard setters in the drive for common standards, increasing comparability, and greater accountability across sectors (Saizarbitoria et al. 2021; Živković 2022). The purpose of this study is to contribute to a deeper understanding of how Corporate Sustainability 2030, based on the UN SDG framework can be practised through corporate strategy and reporting. Specifically, the objectives are:

RO1. To investigate how companies can align their long-term environmental and social goals and objectives with SDG indicators.
RO2. Understand the value of sustainability reporting in demonstrating how a company supports the implementation of the SDGs.
RO3. To understand where and how current ESG and sustainability reporting captures SDG-specific contributions in their existing forms of reporting.
RO4. To consider the arrangements of both institutions and policies that would facilitate effective, common SDG reporting.
RO5. To suggest models for integrating SDG goals across different sectors in comparable and accountable ways.

Questions concerning corporate social responsibility (CSR) have attracted increasing public attention over the past few years, while society has become more aware, and pressures to implement sustainable development models have intensified. Under the UN 2030 Agenda, businesses are under increasing pressure to provide, not only financial value, but also environmental and social value. Firms have been driven by this new-societal value to align their CSR initiatives with the SDGs and thus use the SDGs as a compass to showcase how their activities contribute to issues such as clean energy, gender equality, or sustainable industrialization in quantifiable ways.

Businesses are becoming more concise that CSR is no longer about the act of philanthropy, but a strategic requirement in being competitive and in maintaining the good will of their stakeholders (Srivastava et al. 2018). To attain the SDGs, which is eventually manifested in a system of interdependence where the concept of CSR turns into a collective one of responsibility. Brand image is one of the most significant instruments of such transformations: the company can become competitive by creating a unique, value-driven story, which centers on sustainability, to which consumers could connect. Then CSR becomes central to the marketing communication as well as to the performance measurement, including the emerging rule that makes the implementation of the SDGs indicators of the SDGs be carried by the CSRs.

Corporate Social responsibility (CSR) is a concept that is experiencing rapid development in the modern business environment that its definition goes beyond net revenue. The list of things that fall under this umbrella includes philanthropy and community engagement, ethical work practices to environmental sustainability, and everything that is closely related to the 2030 Agenda of the UN in 17 Sustainable Development Goals (SDGs) (Saizarbitoria et al. 2021; Živković 2022; Tripathi et al. 2022). One should comprehend the impact of CSR activities on consumer perception and brand image because more companies are integrating long-term sustainability policy with the goals of global development (Canobas and Ylagan 2021).

Over the last few years, there has been increased public criticism of corporate responsibility, where consumers want to understand whether their business is doing the job to make a contribution to the overall social and environmental good (Sánchez et al. 2020; Srivastava et al. 2021). By so doing, CSR reporting is being turned into a strategic instrument to not only demonstrate being committed on an ethical basis but also for shedding off responsibilities as envisioned in the SDGs (Ponce 2023). The initial objective of the research is to present the theoretical and practical information to help companies to strive to create or enhance the brand image with the specific references to the corporate social responsibility (CSR) goals, directly associated with the connection between CSR-focused policy of companies and the 2030 agenda and the international regulation. The article The Impact of Corporate Social Responsibility on Brand Image and Consumer Perception is aimed at exploring the dynamic interaction of CSR programs in the context of communicating their strategic alignment with the SDGs. Evaluating the success of CSR in the enhancement of SDG-related brand image and consumer trust. Identify CSR Characteristics, and those are the most influential in causing ESG transparency and social responsibility. Explore Consumer Behaviour after SDG-related CSR with values-based consumption. Research Industry Diversity in the alignment of CSR with the SDGs, particularly in the high-impact industries. Analyse Cultural and Demographic influences on popular awareness on CSR as a sustainability instrument. Future Research Recommendations that can assist firms in making more purposeful links between CSR reporting practices and SDG frameworks. Understand the Significance of the 3ps (People, Planet, Profit) Student SDGs.

By seeing the future of putting CSR into practice with more knowledge of how translational meaning it is as a strategic country competitive advantage in coming up with transparent, responsible and SDGs sensitive business models.

2 Theoretical Framework

Corporate Sustainability and SDGs has been the focus of business strategy and law. Researchers believe the new antitrust law should not merely be efficient from an economic perspective, but also socially beneficial, fair, and responsible in its innovation. In the case of startups, monopolistic ecosystems pose a significant challenge, often driven by digital concentration; therefore, legal frameworks must encourage innovation in line with the SDGs, especially in emerging digital sectors. Sustainability reporting standards that were previously marginal are now at the forefront of measuring corporate legitimacy, particularly SDG 9 (Industry, Innovation and Infrastructure), SDG 10 (Reduced Inequalities), and SDG 12 (Responsible Consumption and Production). Businesses are advised to connect ESG indicators with the SDG impact not as mandatory measures but strategic needs (Goel 2018; Srivastava et al. 2024).

It is increasingly agreed that powerful legal tools and disclosure requirements are needed to help businesses evaluate how their business models contribute or disrupt sustainability (Al-Shaer and Zaman 2018). The SDGs of the UN is a guide on how the corporate behaviour can be brought into the line with the good of the people. The influence of Power over Goal Pursuits The power of Big Tech influences what we do, both directly and indirectly:

SDG 9 (Innovation): The policy encourages open spaces and availability of digital infrastructure.
SDG 10 (Reduced Inequality): Requiring inclusive digital policies and blocking algorithmic discrimination.
SDG 12 (Responsible Consumption): Promoting transparent data practices and minimising the risk of manipulation through algorithmic opacity.

Sustainability approaches must, in any case, be linked to tangible SDG outcomes in terms of business results. This kind of disclosure should be made towards an antitrust ruling, and should constitute a tool to enhance public responsibility. Table 1 presents the sustainability focus areas in the business and the SDGs. Corporate responsibility is not a recent terminology, however, with regard to the 2030 Agenda and the 17 Sustainable Development Goals (SDGs) of the UN, it has become the doctrine of the twenty-first century long-range business planning. These goals have restructured sustainability as a bottom-up, voluntary systemic concept which requires corporate players to declare their environmental and social effects by means of quantifiable measures. Businesses are not merely to be involved in global sustainability programmes, but also value-additions to profitability, as well as, a business plan.

Table 1 Integration of CSR activities with selected UN SDGs

Sustainability focus area	Aligned SDG	Examples of sustainability indicators
Climate action	SDG 13: climate action	Emission reduction targets, renewable energy use, climate risk disclosures
Gender equality	SDG 5: gender equality	Gender pay equity, women in leadership, anti-discrimination policies
Ethical sourcing and production	SDG 12: responsible consumption and production	Sustainable sourcing, circular economy practices, waste management
Inclusive growth and decent work	SDG 8: decent work and economic growth	Fair wages, safe work environments, local employment generation
Innovation for sustainability	SDG 9: industry, innovation and infrastructure	Investment in green tech, digital inclusion, R&D for sustainable products
Equity and access	SDG 10: reduced inequalities	Inclusive policies, access to services for marginalized communities, digital accessibility

Source Author's own creation

With the world economy experiencing a concurrent digital, environmental and social change, businesses need to design comprehensive models that make their business adapt to well-defined SDG objectives. This transformation is embodied in the Corporate Sustainability 2030 model, which is aimed at ensuring compliance between the long-term performance and the overall development outcomes. Here, the role of sustainability reporting is fulfilled twofold: internally to steer the performance and externally to establish trust in investors and credibility among stakeholders (Utami 2024; Mishra and Srivastava 2025).

The overlap between corporate sustainability and SDGs has therefore become an issue of concern in the academic and policy circles. According to research people, sustainability should not be a peripheral objective but a source of corporate legitimacy and value creation in the long run. Nonetheless, even with the growing references to SDGs in corporate reports, alignment is not grounded and competent. Such problems as the symbolic application of SDG icons, unclear targets, and the lack of connection between sustainability and core strategy are common (Piosik and Genge 2019). These considerations provide the basis of our research, and we propose that transparency is a transformative power of companies that are operating in the business-government-society interface (Pálffy and Mihályka 2024; Srivastava et al. 2025). Key themes include:

- Materiality mapping and SDG prioritisation, organisations have to work out which goals are relevant in their industry and stakeholder pool.
- Combining financial, environmental, and social capital as a single business approach.
- Standardization of metrics, ensuring that SDG data is standardized, comparable, and can be verified across different companies and sectors.
- Stakeholder involvement, as being transparent is critical to trust and credibility in sustainability claims.

2.1 Sustainability Reporting and the SDGs

This dualism has made corporate sustainability reporting to become a strategic instrument that informs the internal decision-making process and a social reporting indicator that reflects the willingness of a firm to work towards the society. With firms incorporating SDGs into the sustainability policies, they must outline clear links between the ESG operations and SDG outputs. For example:

SDG 13 (Climate Action) can be integrated with the help of numerated targets on emissions reduction and transition plans.
SDG 5 (Gender Equality) could include a board-level gender diversity measurement, disclosures on equal pay, and anti-discrimination policies.
SDG 12 (Responsible Consumption and Production) to present how they are promoting the circular economy, reducing waste and sourcing ethically.

But one of the greatest obstacles is that there are no common standards around SDG-related disclosures. As most sustainability frameworks are voluntary, companies can typically pick and choose targets or report to various standards, resulting in a lack of comparability and credibility. This study is informed by future dimensions of the interlocked relationship between corporate sustainability and global environmental policy (as articulated by the United Nations Sustainable Development Goals [SDGs]). Theoretical lenses include stakeholder theory, institutional theory and legitimacy theory, to explore how and why firms react towards expectations of sustainability and the framework of the SDGs.

The Stakeholder Theory view to the extent that it argues that corporations operate within the interstices of contracts that reach beyond shareholders to employees, customers, governments, the community, and the environment. This theory does not consider corporate sustainability as a reputational advantage but as a reaction to diverse stakeholders' demands. Connecting corporate strategy with the targets of the SDGs is based on the responsibilities of the companies to serve an extended purpose in societies, to recognise the interdependencies between companies and their societies, and on business of values.

Further, Institutional Theory helps explain how organisations follow rules, norms and culturally congruent practices in their environments. Like this, increasing global attention to sustainable development puts institutional pressure on firms to incorporate sustainability in reporting and long-term planning. Firms may commence their relations with SDGs with different motivations, namely legitimacy seeking, competitive parity or regulation and norm-induced isomorphism (DiMaggio and Powell 1983), which are imposed by global templates, investor community, or peer benchmark mechanisms. This perspective is supported by the Legitimacy Theory, which contends that organisations desire support from their particular societies. The implementation of SDG-related sustainability disclosures reinforces moral legitimacy, especially at a time when corporate influence on social and environmental systems is questioned. Sustainability reporting, therefore, operates as an accountability tool designed to showcase corporate practices that are in line with the values of society enshrined in the SDGs.

All of these theoretical opinions come into play to explain why businesses are getting involved in SDG frameworks and why this involvement is being driven not only by internal strategic choices but also by external accountability forces. This paper aims at using such theories to examine how sustainability reporting can be used as a tool through which the global development agenda can be translated into specific and actual corporate pledges in the example of the Corporate Sustainability 2030 vision. The SDGs-based analysis of CSR is a more systematic approach to determine the effect of this concept on brand value and the trust of consumers. Companies can no longer afford to conduct CSR in the abstract, but directly in connection to the individual SDG goals and indicators. This heightened overlap has in its turn, led to the creation of reporting tools such as the GRI Standards and the SDG Compass, which can be implemented by companies to show actual progress towards the sustainability aims.

Among the theories that have been adopted to explain this change is the Stakeholder Theory (Freeman 1984), which points out the need to consider all stake holders who are impacted by the operations of a company including the communities and the environment. This also quite closely coincides with the emphasis on partnerships and inclusive development of the UN. Mohr et al. (2001) gave Mohr and Webb (2005) the idea that CSR, which appeals to the consumer values, especially those similar to the SDG targets, enhances purchase intention and purchase loyalty to a large extent. It has further been presented by him that the SDG saliency of CSR initiatives vary across sectors, in high ecological footprint sectors. This has been promoted by the sustainability campaign in the food, energy and fast fashion industry in which customer pressure of transparency is the most intense. It has also been presented in literature that the effective implementation of CSR cannot only be superficial campaigns but the integrity of CSR should be enforced on the identified ESG measurements, which can be checked and compared by the interested parties. In this paper, I make use of five major theoretical perspectives that stand at the core of the interconnection between CSR and the SDGs:

Stakeholder Theory (Freeman 1984)—Focus on corporate responsibility to multiple stakeholders in line with Wilke and Vos (2016) SDG values.
Social Identity Theory—Value-congruent CSR and the identification with brands that endorse the sustainable platform.
Attribution Theory—Tests the extent to which perceived genuine intentions of CSR activities (such as SDG reports) affect consumer trust.
Brand Equity Theory—Emphasises the overall value that SDG-aligned CSR contributes to the reputation, perceived quality and emotional brand loyalty of a firm.
People, Planet, Profit (Triple Bottom)—Views CSR through a sustainability lens that aligns with the 17 SDGs, and which allows businesses to evaluate a wider range of impacts than just the financials.

The combination of these theories rationalises the reasoning behind the fact that connecting CSR with SDG indicators does not only make sense ethically, but it is also a strategically good idea. The focus is specially on the concept of 3Ps (People, Planet, Profit); an iconic designation assumed by a significant number of reporting models as a whole that focuses on sustainability as a holistic matter. Move to Impact suggests that this implies that businesses should not continue their close approach to value to the shareholders but they should take into account the environmental and social impact of their actions. Meeting the SDGs with the help of CSR is a business necessity, as it implies ensuring the long-term sustainability, reduces the degree of reputational risk, and creates resilience in the world that is experiencing climate, equity, and governance challenges.

3 Discussion

The CSR has become a major component of corporate sustainability as an ever-growing realization of the interdependence of companies and society. This change shows how businesses are committed but not only to profits, they are committed to responsible actions in both the environmental and the social systems they operate in. More importantly, CSR has become a tool to ensure alignment of the business effects with SDGs, especially in such fields as climate action, decreased inequalities, quality education and responsible consumption.

Companies that take SDG objectives seriously in terms of CSR practice tend to reap the benefits of increased consumer confidence, customer loyalty and brand recognition. The modern customer requires transparency: he or she desires to understand what benefits the brand brings to sustainable development. Therefore, CSR is a kind of required differentiation that is provided by establishing long-term relations with mindful consumers. However, much depends on the reliability and comparison of the performance indicators reported to make CSR successful- and especially the SDGs. The paper aims to add to the critical analysis on the problems of CSR success and failure, based on some case studies illustrating the effect of taking an action or closing the eyes to the global sustainability expectations. As the dynamics of global business continue to evolve, Corporate Social Responsibility (CSR) developed into a mantra of responsible business and there has been a transition beyond the social shareholder value period to an immediate strategic concern. The shift is highly inspired by the growing significance of the 2030 Agenda of Sustainable Development by United Nations that establishes 17 interdependent Sustainable Development Goals (SDGs) to eradicate poverty, safeguard the planet and guarantee prosperity to all. These global goals have reinvented the purpose of business in the society and demand them to be more purposeful to the sustainable development by incorporating social and environmental concerns in their businesses. The current customers, investors and shareholders are demanding more than the good figures of the companies. They require an evident action in the emergent global challenges such as climate change (SDG 13), gender equality (SDG 5), decent work and economic growth (SDG 8), reduced inequalities (SDG 10), and responsible consumption and production (SDG 12). CSR, turned out to be a rather necessary instrument of a company proving its breakdown with the goals and declaring publicly its adherence to ethical, inclusive and sustainable business practices.

Significantly, CSR has also become a brand differentiation instrument, as well as a strategic communication and performance instrument, with the aid of which the companies can show how their operations are contributing to achieving certain SDG indicators. Companies that actively align their CSR efforts with quantifiable SDG indicators, through clean reporting framework and a detailed system of reporting sustainability information are likely to improve their credibility and trust, and may lead to ethical company image during the lifespan of their brand. The introduction of SDGs into the strategies of CSR fosters a better benchmarking process more transparent and comparable both between industries and demonstrates the effect of different enterprises in the marketplace.

Organisations that follow SDG-based CSR initiatives increase in brand image because more and more people prefer companies that really care about global issues. This connection becomes particularly essential in a time of ethical consumerism, whereby purchasing decisions are more determined by the stance of an organisation in regard to environmental sustainability, social justice and corporate ethics. It has been found that the firms embracing have more customer loyalty, improved word-of-mouth advertising, and competitiveness in the market. It is apparent that ironically, the alignment between CSR and SDGs is problematic. Superficial CSR or unrelated CSR activities (CSR activities that do not correspond to the SDG framework) are a manifestation of a negative greenwashing impression of companies (Kumar 2022). To avoid this occurrence, companies should adopt authenticity, transparency and strategy alignment. It involves establishing distinct boundaries between CSR activities and SDGs, convincing arguments to support such relationships and a clear report to the stakeholders about the outcomes.

The theory underpinning this summary is explored on the theoretical basis of theorizing the CSR-brand perception nexus, including Stakeholder Theory, that requires organizations to consider the interest of all parties who might be affected by corporate activity, and the Triple Bottom Line (People, Planet, Profit) and challenges organizations to focus on the profit bottom line, and the planet bottom line and the people bottom line. The findings based on case studies, empirical evidence, and conceptual analysis indicate the significance of the strategically aligned CSR activities with the SDGs in promoting branding, consumer outlook, and resiliency (Kumar et al. 2024).

The paper also identifies the importance of standardized reporting technology towards SDG integration. Similar, at the level disclosure practices, which are supported by policy and regulatory makers to international standard of disclosure setters, are required to allow an opportunity to the Businesses, in all sectors, to publish transparent reports of their activities that are directly contributing to the 2030 Agenda. It not only facilitates the engagement of stakeholders but also seals the gap between corporate aspiration and social influence (Kumar and Prakash 2019).

To conclude, the interplay between CSR and the SDGs creates a robust source through which companies can enhance their legitimacy, be responsible in innovation and take long-term action. When strategically associated with the SDGs, CSR ceases to be reputational management but, on the contrary, an instrument of transformation that precipitates inclusive growth, environmental responsibility as well as sustainable value generation in all markets around the world. To compete in a purpose-driven world, the alignment of the brand identity and operational strategy with the SDGs is not only optional but critical to organisations that wish to compete. Table 2 is a summary of the theories explaining the connection between CSR and SDGs.

Table 2 Theoretical foundations linking CSR to the SDGs

Theory	Key concept	Relevance to CSR and SDG integration
Stakeholder theory	Firms must address the needs of all stakeholders, not just shareholders	Supports inclusive sustainability strategy; justifies SDG alignment as a response to expectations of consumers, communities, employees, and governments
Institutional theory	Firms conform to norms, values, and rules within their environment	Explains why companies adopt SDG-based frameworks to gain legitimacy, meet peer expectations, and align with global governance structures
Legitimacy theory	Organizations seek societal approval through responsible behaviour	SDG-linked disclosures help maintain or restore public legitimacy, especially under scrutiny of environmental and social impacts
Social identity theory	Consumers align with brands that reflect their personal values	CSR initiatives mapped to SDGs enhance brand affinity and consumer loyalty by enabling value-based identification
Attribution theory	Perceived authenticity of CSR affects stakeholder trust	Demonstrating clear SDG impact through measurable reporting increases perceived authenticity of CSR efforts
Brand equity theory	Strategic CSR enhances brand value, trust, and reputation	SDG alignment strengthens brand equity by linking ethical and sustainable values to company identity
Triple bottom line (3Ps)	Sustainability is achieved through balance of People, Planet, and Profit	Reflects SDG structure; reinforces the need for integrated reporting frameworks that show environmental, social, and financial outcomes

Source Author's own creation

4 Gaps and Challenges

A number of structural and institutional difficulties are also barriers to the successful mainstreaming of the SDG agenda in corporate practice. Furthermore, Table 3 presents the impact assessment of different challenges on businesses:

1. ***Diverse Reporting Standards***: Several frameworks (GRI, TCFD, SASB) give disparate guidance, making it challenging to create coherent SDG-aligned reports.
2. ***Shallow Goal Alignment***: Perhaps SDG icons feature in multiple reports but are not underpinned by specific KPIs or targets.
3. ***No Verification and Accountability***: There are few third-party audits of claims made by SDG, making their reliability questionable.

Table 3 Challenges in aligning corporate reporting with SDGs

Challenge	Description	Impact
Fragmented reporting standards	Use of different frameworks (GRI, SASB, TCFD) without harmonization	Lack of comparability across companies
Superficial goal referencing	Use of SDG icons without mapping to KPIs or measurable outcomes	Greenwashing, loss of stakeholder trust
Inadequate verification	Lack of third-party audits for SDG claims	Reduced credibility and transparency
Sectoral misalignment	Applying all SDGs uniformly across industries regardless of relevance	Irrelevant disclosures, missed material issues
Weak regulatory mandates	Few legal requirements to report on SDGs	Low adoption and accountability

Source Author's own creation

4. ***Sector Mismatches***: While not all targets are sector agnostic, not all targets are equally appropriate for all industries; a simple fact that companies often do not recognise or take into account (however, some ratings systems do not account for such).
5. ***Lack of Regulation Frameworks***: In most jurisdictions, there is no legal regulation about reporting on SDGs, which hampers broad acceptance.

5 Policy Recommendation and Concluding Remarks

Their active stance should be utilized by governments and regulatory authorities in order to facilitate the meaningful and massive implementation of the SDG-inspired sustainability practices. The ESGs supports the argument of obligatory SDG reporting of listed companies and larger companies in general and introduces sustainability reporting into practice in priority industries. To be more transparent, it encourages disclosure through incentives, e.g., tax exemptions or ESG-related financial products. Furthermore, the ESGs emphasizes the topicality of the corporate sustainability reporting being more national-focused to facilitate the integration of a coordinated and targeted SD strategy. It is also conducive to encouraging the access of the population to the data, as it demands the provision of sustainability reports in a way that is non-technical and comparative, and at the same time encourages transparency and informed stakeholder involvement.

These measures may make SDG reporting a standard practice in corporate governance, and increase accountability. Twenty-first century corporate sustainability should not be limited to environmental stewardship, but should be responsible and accountable of systematic contributions to social progress. UN SDGs offer a common point of reference and roadmap—but the real challenge would be to move these global goals to the action, disclosure, and impact levels at companies. According to the Corporate Sustainability 2030 framework, business companies are expected to connect their business processes and performance data to some SDG outcomes. It does not just imply the mentioning of the objectives, but incorporating it into the way the companies innovate, govern, the measurement and communication of their value to society. Only credible, comparable and verified SDG reporting can ensure that any company becomes part of the solution, in the creation of a just, inclusive and resilient future in that it allows a real alignment and active contribution to global endeavours.

References

Al-Shaer H, Zaman M (2018) Credibility of sustainability reports: The contribution of audit committees. Bus Strateg Environ 27(7):973–986. https://doi.org/10.1002/bse.2046

Canobas E, Ylagan A (2021) Transformational leadership and work engagement among managers of casual dining restaurants in Batangas province. Int J Res Stud Manag 9(4). https://doi.org/10.5861/ijrsm.2021.m7735

DiMaggio PJ, Powell WW (1983) The iron cage revisited: institutional isomorphism and collective rationality in organizational fields. Am Sociol Rev 48(2):147–160. https://doi.org/10.2307/2095101

Freeman RE (1984) Strategic management: a stakeholder approach. Pitman, Boston

Goel P (2018) Implications of corporate governance on financial performance: an analytical review of governance and social reporting reforms in India. Asian J Sustain and Soc Responsib 3(1). https://doi.org/10.1186/s41180-018-0020-4

Kumar K (2022) Emerging phenomenon of corporate sustainability reporting: evidence from top 100 NSE listed companies in India. J Public Aff 22(1):e2368

Kumar K, Prakash A (2019) Developing a framework for assessing sustainable banking performance of the Indian banking sector. Soc Responsib J 15(5):689–709. https://doi.org/10.1108/SRJ-07-2018-0162

Kumar K, Singh R, Sharma A, Lehri BPS (2024) Climate change adaptation strategies for achieving net-zero economy. In: Transition towards a sustainable future: net zero policies and environmental sustainability. Springer, Singapore, pp 163–183

Mishra A, Srivastava SK (2025) Role of trust and celebrity endorsement in shaping purchase intention towards electric vehicles. In: Implementing ESG frameworks through capacity building and skill development. IGI Global Scientific Publishing, pp 381–400. https://doi.org/10.4018/979-8-3693-6617-2.ch018

Mohr LA, Webb DJ (2005) The effects of corporate social responsibility and price on consumer responses. J Consum Aff 39(1):121–147. https://doi.org/10.1111/j.1745-6606.2005.00006.x

Mohr LA, Webb DJ, Harris KE (2001) Do consumers expect companies to be socially responsible? The impact of corporate social responsibility on buying behaviour. J Consum Aff 35(1):45–72. https://doi.org/10.1111/j.1745-6606.2001.tb00102.x

Pálffy Z, Mihályka L (2024) Cultural influence on sustainability discourse: a comparative analysis of CEO letters in the cosmetics and fashion industry. J Ecohum 3(4):2779–2792. https://doi.org/10.62754/joe.v3i4.3796

Piosik A, Genge E (2019) The influence of a company's ownership structure on upward real earnings management. Sustainability 12(1):152. https://doi.org/10.3390/su12010152

Ponce H (2023) Sustainability as a strategy base in Spanish firms: sustainability reports and performance on the sustainable development goals. Sustain Dev 31(4):3008–3023. https://doi.org/10.1002/sd.2566

Saizarbitoria I, Urbieta L, Boiral O (2021) Organizations' engagement with sustainable development goals: from cherry-picking to SDG-washing? Corp Soc Responsib Environ Manag 29(2):316–328. https://doi.org/10.1002/csr.2202

Sánchez I, Amor-Esteban V, Galindo-Álvarez D (2020) Communication strategies for the 2030 agenda commitments: a multivariate approach. Sustainability 12(24):10554. https://doi.org/10.3390/su122410554

Singh R, Kumar K, Khan S (2024) A comprehensive view of artificial intelligence (AI)-based technologies for sustainable development goals (SDGs). Artificial intelligence enabled management: an emerging economy perspective, p 183. https://doi.org/10.1515/9783111172408-012

Singh R, Sharma A, Kumar K (2025a) Navigating carbon neutrality: strategies for sustainable future and environmental challenges. In: Climate neutrality through smart eco-innovation and environmental sustainability. Springer, Cham, pp 69–80

Singh R, Tiwary KS, Kumar K, Kumar A (2025b) Empowering a sustainable future: blockchain's role in economy, organizations, and finance towards smart circular economy. In: Blockchain technologies for smart circular economy and organisational sustainability. Springer, Cham, pp 271–283

Srivastava SK, Anshul A, Pathak P, Bansal J (2018) Responsible consumption for curbing food wastage: an exploratory enquiry. PURUSHARTHA 11(2):104–116

Srivastava SK, Singh S, Srivastava M (2021) Food Waste at household and social gatherings: drivers and possible remedies. PURUSHARTHA 14(2):117–133

Srivastava SK, Sinha AK, Kumar S (2024) Workplace spirituality empowered by spiritual leadership to promote organizational effectiveness: a bibliometric analysis. PURUSHARTHA 17(1):1–22. https://doi.org/10.21844/16202117101

Srivastava SK, Mishra A, Kumar K, Singh S (2025) Hanuman's Wisdom & Sita's Peace: a journey from divine thoughts to modern learnings through sacrifice in Ramcharitmanas. PURUSHARTHA 18(1). https://doi.org/10.21844/16202118101

Stocker F, Sajjad A, Raziq M, Pacheco L (2022) Editorial: Innovation and business strategy for sustainability. Innov Manag Rev 19(3):174–179. https://doi.org/10.1108/inmr-07-2022-183

Tripathi V, Singh K, Parihar S, Srivastava SK (2022) A bibliometric analysis on supply chain collaboration. PURUSHARTHA 15(2):66–103. https://doi.org/10.21844/16202115205

Utami R (2024) Analysis of green practices for sustainability in the Indonesian banking industry. Proc ISST 3:51–59. https://doi.org/10.33830/isst.v3i1.2328

Wilke R, Vos J (2016) CSR implementation: developing a stakeholder perspective from corporate communications and business ethics. Corp Commun: Int J 21(4):530–544. https://doi.org/10.1108/CCIJ-08-2015-0055

Yang Y, Wen J, Li Y (2021) The impact of environmental information disclosure on the cost of debt: evidence from China. https://doi.org/10.21203/rs.3.rs-572374/v1

Živković S (2022) Sustainability leadership and boards: a conceptual framework. ECMLG 18(1):456–464. https://doi.org/10.34190/ecmlg.18.1.587

Biggies Are Always on the Hit List: Corporate Social Responsibility in an Emerging Market

Barkha Dhingra and Ubais Parayil Iqbal

Abstract The swiftly growing trend of sustainable financing and green consumerism has intensified the stakeholder expectations of a company's performance above the bottom line. As a result, companies have started considering CSR as a strategic tool beyond the outlook of an altruistic act. This argument is supported by stakeholder theory as well as the significant positive findings of CSR and financial performance studies. Further, it has slowly become a social mandate for companies to spend on the social well-being of people and the planet. For example, India has adopted a mandatory CSR policy as per the revised company law passed in 2013. Consequently, the previous relationship patterns between CSR and its determinants have transformed. Hence, the current study examines the impact of firm visibility on CSR spending in a mandatory and developing economy setting. The study also examines non-linearity arguments in the literature in the context of a developing country. For this, a dataset of S&P BSE 200 indexed firms is extracted from the CMIE Prowess database. The study has deployed panel data regression methods for the analysis. The results highlight the positive significant impact of firm visibility on CSR spending. Moreover, the study's findings negate the previous arguments of non-linear relationships in the context of an emerging economy. The study's practical implications and theoretical relevance, along with the major results, have been discussed in detail in the paper.

Keywords Firm visibility · CSR · Non-linear relationship · Panel-data regression · Mandatory CSR

B. Dhingra
Hansraj College, University of Delhi, New Delhi, India
e-mail: drbarkha@hrc.du.ac.in

U. P. Iqbal (✉)
University of Technology and Applied Sciences, Nizwa, Oman
e-mail: ubaisiqbal@gmail.com

K. Kumar et al. (eds.), *Sustainable Value Creation: Corporate Strategies for SDGs Implementation*, World Sustainability Series,
https://doi.org/10.1007/978-3-032-24652-3_7

1 Introduction

Corporate social responsibility (CSR) is gaining momentum as a strategic tool in the era of impact investing and sustainable finance. CSR has been widely depicted as a response to current or expected demands from the stakeholders (Do et al. 2023). Furthermore, it has been defined as a voluntary initiative by the companies beyond the bottleneck in the form of social and environmental interventions (Weber 2008). However, the major chunk of CSR literature has used stakeholder framework to establish the nexus between CSR and firm-specific variables (Chettri and Makan 2025; Iqbal et al. 2025; Oware 2023). In the same line, the study conducted by Yu et al. (2017) has reported that with the increased size and visibility of the business, there will be more stakeholder pressure to provide extra to sustain its legitimacy. According to Sun et al. (2025), companies engage in sustainable practices not just to remain competitive but also to cater to diverse stakeholder expectations and gain stakeholders' loyalty. These arguments are in tandem with the findings of Li et al. (2019), who reported a significant positive impact of corporate visibility/size on their CSR engagements. In their study, Golrida et al. (2019) has observed a non-linear relationship between media visibility and CSR performance. In light of the potential of this topic to offer a diverse spectrum of unique insights into the literature, the current study examines the role of corporate visibility on CSR spending.

In the Indian context, CSR has undergone a profound transformation, especially after the enactment of the mandatory CSR policy for selected listed companies. Section 135 of the Indian corporate legislature requires the qualifying companies to spend at least 2% of their average profit on CSR activities every year. However, there are no penalties for non-compliance. Consequent to this regulatory push, the CSR literature has gained significant traction and resulted in academic enquiries on various aspects of CSR in the Indian context (for example, Ali 2023; Cherian et al. 2019; Iqbal et al. 2025). With the introduction of a mandatory CSR framework, several companies voluntarily spending for CSR have reduced their actual spending as the regulation requires only 2% (Rajgopal and Tantri 2023). Nonetheless, a company might carry forward and modify its expenditures to subsequent years if they exceed the 2% level. Therefore, any excess spending in CSR in a mandatory setting could be intended to satisfy the stakeholder interests, as suggested by stakeholder theory (Parmar et al. 2010). Reinforcing these arguments, it is observed that the stakeholders demand more from the highly visible firms (Yu et al. 2017). This means that when visibility (size) increase, the firm will be more subject to stakeholder scrutiny, resulting in high stakeholders' expectations (Ferri et al. 2022).

Existing literature have studied the role of corporate visibility on various aspects of their CSR performance. However, only limited studies have used the log of total assets as a proxy for measuring the firm visibility. The study conducted by Udayasankar (2008) has reported that the higher the firm's size, the higher the firm visibility. However, he made a distinction between firm size and visibility. Similarly, there are studies which has used media coverage for measuring the media visibility (Golrida et al. 2019; Schreck and Raithel 2018). However, media visibility is short-term in nature and volatile. Conversely, Firm size represented by asset base of the firm act as a more stable, economically meaningful and reliable variable for measuring the firm visibility. Therefore, this study acknowledges that firms with large asset bases are on the radar of mainstream media and channels and are subject to higher scrutiny from governments and other stakeholders. Which means that the size of the firm remains a stronger and tangible variable representing the firm visibility, though largely ignored in the current literature. Hence, the present study has used the log of total assets as a proxy measuring the firm visibility. Moreover, there is a dearth of studies conducted in statutory settings like India, and the major chunk of exiting literature have ignored the actual CSR spending of firms. Hence, the present study is revisiting the impact of corporate visibility on actual CSR spending by extracting data from Prowess IQ. Furthermore, existing studies on CSR have widely used either BSE Sensex or BSE 100 for data construction (Agrawal et al. 2020; Kuzey et al. 2021). On the other hand, the current study has adopted the S&P BSE 200 index to construct the data set. The choice of S&P BSE 200 companies is deliberate as it covers a broader section of the market, which includes small-cap, mid-cap, and blue-chip companies (Iqbal et al. 2025). Moreover, using the log of total assets as a firm visibility proxy is more justified with this set of firms as their asset base reflects their market dominance and stakeholder exposure.

The study enhances the CSR and firm visibility literature in the following ways: Firstly, it explores the impact of firm visibility on actual CSR spending in a mandatory setting, a highly ignored area in the existing literature. Secondly, the study uses a novel and tangible proxy for firm visibility, which makes more sense than the media citations and other scores. Thirdly, the study uses a unique dataset that is representative and has been explored less. Finally, the study results are expected to guide policymakers, industrialists and investors by explaining the interplay between firm visibility and CSR expenditure in a mandatory setting.

2 Literature Review

A firm's CSR involvement is influenced by several firm-level characteristics, for example: Kaymak and Bektas (2017) showed that a company's board structure, particularly board size and independence, significantly impacts its CSR practices. Companies with more external members increase board independence and show higher CSR engagement. Larger boards also enhance social responsibility as they represent diverse stakeholder sets. In their study, Nidheesh (2021) have shown that older and more leveraged firms have superior CSR engagements. Companies seeking to derive strategic advantage from CSR must understand these firm parameters and their influences. The firm's size is one of the primary drivers among the numerous firm-level parameters (Udayasankar 2008; Madden et al. 2006). The bigger the size of a firm, the greater the visibility it will have as a consequence of the stakeholders' attention they will get. Therefore, firm size has been identified as an effective indicator of an important firm-level characteristic, i.e. firm visibility (Brammer and Millington

2006; Zou et al. 2024). The degree of public attention a company receives is referred to as "firm visibility." Highly visible companies are frequently in the news, which makes their actions more transparent and attractive. It is widely believed that highly visible firms must contribute more to CSR activities to satisfy stakeholders, maintain credibility, and effectively manage their social reputation (Kuzey et al. 2025; Udayasankar 2008). Therefore, the visibility of a firm has a positive impact on its CSR engagement (Udayasankar 2008; Campbell and Slack 2006).

2.1 *Theoretical Background*

Several theories approach the firm visibility and CSR nexus, including legitimacy theory, stakeholder theory, and resource-based view (RBV). Stakeholder approach by Edward Freeman (1984) posits that the fundamental indicator of a firm's performance is its ability to create wealth for all the stakeholders. Furthermore, the present study adopts the normative approach of stakeholder theory, which contends that stakeholders have valid claims and rights in a company, and therefore, the company management must meet their needs for intrinsic benefits (Donaldson and Preston 1995). Meeting the expectations of the stakeholders leads to CSR efforts, which in turn provide societal acceptance (Hooghiemstra 2000; Udayasankar 2008). A firm and the society have an implied social agreement (Suchman 1995). As a result, businesses practice social responsibility in an effort to gain or enhance legitimacy (Pareek et al. 2019). Lindblom (1994) defined legitimacy as "A state or position obtained when an entity's value system is congruent with the value system of the larger social system to which the entity is a member". Engaging in CSR activities is a strategy used by businesses to build credibility. This way, legitimacy theory suggests that small firms may also contribute to CSR activities to legitimatize their actions (Udayasankar 2008). In the present study, this theory can help to comprehend how corporate visibility affects CSR spending by acknowledging its influence on perceived legitimacy and societal acceptability. Yu et al. (2017) has reported that with the increased size and visibility of the business, there will be more stakeholder pressure to do extra to sustain its legitimacy. The RBV theory suggests that a company can generate intangible resources by cultivating good connections with its stakeholders, which leads to competitive advantages and higher profitability. In line with the above discussion, the present study adopts stakeholder and legitimacy theory as a theoretical framework.

2.2 *CSR and Firm Visibility*

According to Cowen et al. (1987), big businesses are more likely to engage in socially responsive behavior due to their stronger social impact. Deegan and Rankin (1996) contended that highly visible corporations are more likely to be subject to public social pressure and respond to it with CSR efforts. From the developing country perspective, Pareek et al. (2023) have shown that firms' visibility has a positive impact on the sustainability performance of the organization. Increased stakeholder monitoring of highly visible companies encourages more socially responsive behavior. From a developed country perspective, Brammer and Millington (2006) provided powerful empirical evidence that organizational visibility is essential in firms' philanthropy and social responsiveness behavior. A company is more likely to participate in CSR activities as the public becomes more vested in its operations (Li et al. 2019). On the other hand, Howard and Jaffee (2013) investigated the three major US firms and indicated an inverse relationship between firm size and sustainability commitment. Similarly, Orlitzky (2001) doubted the hypothesis that large firms are more likely to participate in socially responsible activities.

The above-competing results regarding the impact of firm visibility (firm size) on its CSR engagement cast doubt on the direct or linear relationship between the selected variables. As a result, this highlights the possibility of a curve or nonlinear relationship between the variables. The existing studies in the literature also support it. For example, in their study, Udayasankar (2008) suggested a U-shaped relationship between the size of the firm and CSR activities, with small and large firms equally motivated but medium-sized firms being the least motivated. According to Amato and Amato (2007), medium-sized firms are less inclined to engage actively in CSR initiatives than larger and smaller firms. In contrast, Agarwala et al. (2024) and Karyawati et al. (2017) have provided empirical evidence of an inverse U-shape relationship between the concerned variables. This shows that small firms positively impact CSR, while larger firms experience a negative relationship with CSR. The legitimacy theory explains this tendency of smaller firms as they need to build goodwill and reputation in order to expand. Engagement in CSR activities could help growing firms gain a good reputation and credibility. On the other hand, large firms may be less motivated to participate extensively in CSR due to their established reputation. In line with the previous studies, Russo and Tencati (2009) and Wickert et al. (2016) revealed that small businesses are more willing to engage in CSR to strengthen community relations and survive, while large firms are not.

The review of current literature discovers that the selected field of corporate visibility and CSR spending is contentious. Hence, revisiting this area will offer a unique outlook and illuminating insights into the literature.

3 Data and Methodology

3.1 Data

The present study has selected the S&P BSE 200 index for the analysis. The S&P BSE 200 index is selected for the following reasons: first, it covers a wider market segment, including small-cap, mid-cap, and blue-chip companies. Second, it represents a more diverse set of industries. Third, the index provides greater market depth by including stocks that do not rank in the top 30 or 100 by market capitalization. Therefore, the selected index will offer comprehensive insights into a larger spectrum of businesses and their CSR engagements (Iqbal et al. 2025; Jain and Chakrabarty 2020). The present study aims to investigate the impact of corporate visibility on their CSR engagements. For this, a dataset of S&P BSE 200 listed firms is used. However, eight companies have been removed due to non-availability of data. The final dataset covers 192 Indian firms. The balanced panel data for the study includes the yearly time series from 2015–16 to 2022–2023. Data is extracted from the CMIE (Centre for Monitoring Indian Economy Pvt. Ltd) Prowess IQ database.

3.2 Methodology

Model Specification

To investigate the impact of corporate visibility on their CSR spending, the authors have developed the following Model 1:

$$CSR_{it} = \alpha_1 + \beta_1 Firm\_visibility_{it} + \varepsilon_{it} \tag{1}$$

Here, CSR represents a company's Corporate Social Responsibility expenditure.

Further, the following Model 2 augments Model 1 by adding firm-specific control variables. The purpose is to investigate the effect of firm visibility on CSR expenditure in the presence of other variables.

$$CSR_{it} = \alpha_1 + \beta_1 Firm\_visibility_{it} + \beta_2 TobinsQ_{it} + \beta_3 Sales_{it} + \beta_4 Leverage_{it} + \varepsilon_{it} \tag{2}$$

The previous two models assume a linear relationship between firm visibility and their CSR expenditure. Now, Model 3 includes a quadratic term of firm visibility to examine the nonlinear relationship between the selected variables.

$$\begin{aligned} CSR_{it} = {} & \alpha_1 + \beta_1 Firm\_visibility_{it} + \beta_2 (Firm\_visibility)^2{}_{it} + \beta_3 TobinsQ_{it} \\ & + \beta_4 Sales_{it} + \beta_5 Leverage_{it} + \varepsilon_{it} \end{aligned} \tag{3}$$

The variables used in the above models are explained in the following section.

Variables

To investigate the impact of firm visibility on CSR spending, the study has incorporated firm visibility as an independent variable and CSR as the dependent variable.

Independent Variable

Firm Visibility

Firm visibility refers to the extent to which the company is known to its stakeholders. According to Achour and Boukattaya (2021), large firms are subject to greater visibility, which leads to higher stakeholder demands. In other words, the bigger the size, the larger the visibility, and the greater the stakeholder expectations. These propositions are in concert with Hansen and Mitchell (2000), who have used firm size as an indicator-variable for measuring the corporate visibility. Therefore, in line with Kang and Kang (2025), Udayasankar (2008), and Bushee and Miller (2012), the current study has used the firm's asset base as the proxy of firm visibility. The study considers it a reliable and tangible measure of firm visibility against media coverage and news reporting. Hence, the study used the log of total assets to measure the firm visibility.

Dependent Variable

CSR Spending

Section 135 of the Indian Companies Act 2013 has made it mandatory for certain companies to spend on CSR and report their CSR spending in their annual financial statements. The current study is intended to explain the impact of firm visibility on CSR. There are different approaches to measure a company's CSR engagement. However, the monetary measure of CSR is a more appropriate and direct measure of CSR engagement (Baatwah et al. 2022). Hence, the study has used the actual CSR spending reported by companies to measure CSR engagement. This selection of variable measurement is in light of the previous studies (Oware 2023; Bhattacharyya and Rahman 2019; Iqbal et al. 2025; Bhagawan and Mukhopadhyay 2025).

Control Variables

The study has included three control variables, namely Tobin's Q, sales, and leverage, to avoid any possible confounding effect. Tobin's Q is a market value measure of financial performance, which could probably act as a cofounder having a relationship with both firm visibility and the CSR spending (Iqbal et al. 2025; Uyar et al. 2024). Therefore, the current study has deployed Tobin's Q as a control variable. In addition, the study has used the log value of sales as a control variable, as it is related to both firm visibility and CSR. This selection is justified with the support of previous studies (Hillert and Ungeheuer 2025; Tran and Pham 2024). Finally, the study has adopted leverage as a control variable in tandem with the studies by Kuzey et al. (2021) and Iqbal et al. (2025). Financial leverage can be understood as the degree of debt funds used to finance the firm's total assets. Hence, the study uses the debt-equity ratio to measure the financial leverage (Nguyen et al. 2022; Fahad and Rahman 2020).

Estimation Approach

Due to the nature of the data, the panel data techniques were required to examine the relationship between the selected variables. A range of models, including fixed, random effect, and pooled OLS models, are used in panel data approaches based on their assumptions for the intercept, slope, and error components (Baltagi 2008). The first step is to apply a pooled OLS model followed by fixed and random effect models to reduce the endogeneity and unobserved time-invariant effect (Singh et al. 2023). The next step is to apply various tests, including the F test, Lagrange multiplier test (Breusch and Pagan 1980), and Hausman test (Hausman and Taylor 1981) to determine which estimators produced the best regression results.

4 Results and Analysis

4.1 Descriptive Statistics

Table 1 presents the descriptive statistics of the selected variables in the study. The S&P BSE 200 listed companies spent an average of 2.111 on CSR. With a mean value close to the median, most businesses invested a moderate amount of money in CSR activities. It shows that they can still improve their CSR efforts. CSR expenditure has a minimum value of 0 and a maximum of 3.965. In addition, the mean value of firm visibility is 5.312, indicating that the majority of the firms are medium in size. The average performance of businesses (Tobin's Q) is 9.626, a very low value, given that its greatest value is 403.927. This suggests that most sample firms had poor financial performance. With an average leverage of 1.413, the chosen companies had minimal debt financing. The standard deviation of Tobin's Q and leverage is 19.462 and 17,843, respectively, indicating a lot of variability. The sales mean and standard deviation values are 5.007 and 0.647, respectively.

4.2 Correlation Matrix

The correlation matrix and the VIF values are presented in Table 2 to check for the multicollinearity among the selected variables. The correlation matrix shows that a company's CSR expenditure is positively associated with its visibility and level of sales. However, a firm's Tobin's Q and leverage negatively correlate with its CSR expenditure. Furthermore, the correlation coefficient values are below 0.8, suggesting

Table 1 Descriptive statistics

Variable	Obs.	Mean	Std. Dev.	Min	Max
CSR	1344	2.111	0.880	0	3.965
Firm_visibility	1344	5.312	0.765	3.088	7.743
Tobin's Q	1344	9.626	19.462	0.060	403.927
Sales	1344	5.007	0.647	2.249	6.982
Leverage	1344	1.413	17.843	− 28.485	652.047

Source Authors' calculation

Table 2 Correlation matrix

Variable	CSR	Firm_ visibility	Tobin's Q	Sales	Leverage	VIF
CSR	1.000					
Firm_ visibility	0.409	1.000				2.91
Tobin's Q	− 0.289	− 0.466	1.000			2.80
Sales	0.612	0.799	− 0.431	1.000		1.29
Leverage	− 0.004	0.019	− 0.015	0.020	1.000	1.00

Source Authors' calculation

that the data does not exhibit multicollinearity. Chatterjee and Hadi (2015) state that when the Variance Inflation Factor (VIF) value is greater than 10, multicollinearity is apparent. With a mean of 2.00, all of the VIF values in our instance are low, suggesting that multicollinearity is not present.

4.3 Main Results

The results of Models 1 and 2 using the pooled OLS, fixed, and random effects estimators are shown in Tables 3 and 4, respectively. The authors employed the F test, Lagrange multiplier test, and Hausman test to determine which of the pooled OLS, fixed, and random effects estimators produced the best regression results. The significant p-value (significant) of the F test indicates that the fixed effect is appropriate over pooled OLS. The p-value of the Lagrange Multiplier test is less than 5%, indicating that the random effect is relevant. Finally, the Hausman test is used to determine between fixed and random effects. The significant p-value suggests that the fixed-effect estimator is preferable to examine how firm visibility impacts their CSR expenditure.

This research aims to examine how visibility of a company affects its CSR expenditure. Table 3 findings show a significant positive coefficient of corporate visibility in all the three estimators. This indicates that the CSR expenditure increases with

Table 3 Regression results for Model 1

CSR	Pooled OLS		Fixed effect		Random effect	
	Coefficient	Prob.	Coefficient	Prob.	Coefficient	Prob.
C	− 0.389**	0.012	− 1.946***	0.000	− 1.147***	0.000
Firm_visibility	0.471***	0.000	0.764***	0.000	0.613***	0.000
Number of observations	1344		1344		1344	
R-squared	0.1675		0.0957		0.0957	
F-stat/Wald chi-square test	269.96		121.87		154.44	
Prob (F-statistic)	0.000		0.000		0.000	

Source Authors' calculation
Notes *, ** and *** denote significant at 10, 5 and 1%, respectively

Table 4 Regression results for Model 2

CSR	Pooled OLS		Fixed effect		Random effect	
	Coefficient	Prob.	Coefficient	Prob.	Coefficient	Prob.
C	− 1.671***	0.000	− 2.803***	0.000	− 1.907***	0.000
Firm_visibility	− 0.278***	0.000	0.467***	0.000	0.043	0.547
Tobin's Q	− 0.003***	0.006	0.002*	0.074	0.000	0.917
Sales	1.057***	0.000	0.483***	0.000	0.756***	0.000
Leverage	− 0.001	0.457	0.000	0.890	0.000	0.758
Number of observations	1344		1344		1344	
R-squared	0.3957		0.1190		0.1079	
F-stat/Wald chi-square test	219.17		38.76		281.94	
Prob (F-statistic)	0.0000		0.0000		0.0000	

Source Authors' calculation
Notes *, ** and *** denote significant at 10, 5 and 1%, respectively

the visibility of the company as corporates tend to satisfy stakeholders, uphold their credibility and enhance their public image.

Model 2 augments Model 1 by adding firm-specific control variables. The purpose is to observe the impact of the firm visibility on CSR expenditure in the presence of other variables. Based on the fixed effect estimators, firm visibility has a significant positive impact on CSR spending. Moreover, Tobin's Q and sales influence CSR expenditure positively; the former is significant at 10% and the latter at 1%. Additionally, leverage shows an insignificant relationship with CSR spending.

The previous two models assume a linear relationship between firm visibility and their CSR expenditure. Now, Model 3 includes a quadratic term of firm visibility to examine the nonlinear relationship between the selected variables. The coefficient of linear term tells the overall positive or negative trend, while the coefficient of nonlinear term indicates the direction of curvature as concave (inverse U-curve) or convex (U-curve). The findings of Model 3 (Table 5) show a positive coefficient of

Table 5 Regression results for Model 3

CSR	Fixed effect	
	Coefficient	Prob.
C	− 5.210***	0.001
Firm_visibility	1.375**	0.020
Firm_visibility_sq	− 0.087	0.117
Tobin's Q	0.003**	0.018
Sales	0.501***	0.000
Leverage	0.000	0.874
Number of observations	1344	
R-squared	0.1209	
F-stat/Wald chi-square test	31.54	
Prob (F-statistic)	0.0000	

firm visibility but a negative and insignificant coefficient of quadratic term. The analysis depicts an inverse U-shaped relationship between firm visibility and CSR expenditure. However, the inverted U-shaped relationship lacks statistical significance. Overall, the results support a positive impact of firm visibility on CSR expenditure.

5 Discussion and Managerial Implications

Following Bushee and Miller (2012), the present study examines the impact of firm visibility on actual CSR spending by taking asset base as the proxy for firm visibility. The findings of this study reveal that firm visibility is significant in determining the CSR spending of listed corporates in the Indian stock market. The study results are in tandem with the stakeholder theory, which necessitates the firm to respect stakeholder's demands and value their expectations (Freeman 1984). A firm having higher visibility is expected to invest more in its sustainability performance, as a result, firms are spending more on CSR activities (Yu et al. 2017). Moreover, this study supports the argument that the mandatory CSR dampens the discretionary behavior of companies by discouraging their intrinsic motivation to spend more than what is required (Rajgopal and Tantri 2023). Therefore, companies are forced to spend more on CSR due to increased stakeholder expectations fueled by increased firm visibility. However, these arguments challenge the findings of Bhagawan and Mukhopadhyay (2025), who reports CSR spending as a discretionary decision by companies in India in light of the "comply or explain" flexibility option. Despite this, the current study supports the stakeholder theory and propounds that any increase in the stakeholder expectation due to increased firm visibility will impact the managerial decision on CSR spending. In simple words, in light of the findings of this study, it can be interpreted that CSR spending is a strategic response to stakeholder pressure rather than a voluntary virtuous choice (Carroll and Shabana 2010). The study results corroborate the findings of Gamerschlag et al. (2011), who reported firm visibility as a determinant of CSR disclosure.

Further, Model 3 has been developed to investigate the possibility of a curve or nonlinear relationship between the selected variables. The outcome of Model 3 rejects the assumption of a non-linear relationship between firm visibility and CSR spending. This result contradicts the previous results on the CSR topic, which reported a non-linear interaction. For example, Udayasankar (2008) has reported a non-linear interaction between firm visibility and sustainability engagement. Another study by Schreck and Raithel (2018) has reported that the marginal cost and benefits of sustainability reporting vary with its visibility. This means that in an altruistic CSR setting, smaller firms spend higher amounts on CSR to gain a competitive edge, while larger firms do so as a result of stakeholder pressures. This leaves a U-shaped CSR spending against the firm visibility, indicating that small and large firms spend more while medium-sized firms lack the motivation to spend extra. On the contrary, the current study reveals that in a mandatory setting, smaller companies lack the motivation to spend extra on CSR, though their firm characteristics remain a significant predictor of CSR. This could be due to the lack of resources by smaller companies to outperform beyond the statutory requirements similar to large corporates. Therefore, it can be understood that mandatory CSR setting could discourage smaller firms from engaging in higher CSR performance, whereas larger firms with higher financial resources can spend more on CSR. Adding on, the study result also indicates that mandatory CSR nullifies the existence of any competitive or strategic determinants of CSR in the case of smaller firms and at the same time, CSR gets amplified by firm characteristics such as firm visibility (Li et al. 2019). Conversely, the study conducted by Agarwala et al. (2024) reported an inverted U-shaped relationship, where medium firms are spending more on CSR.

Interestingly, the study raises questions in light of its findings on the CSR performances of smaller firms due to the absence of strategic intent and stakeholder pressures. Therefore, the current study suggests implementing monitoring mechanisms to ensure CSR compliance by the less visible firms. Moreover, companies must focus on integrating CSR into their strategic planning rather than keeping it as a regulatory requirement. In essence, the study findings confirm that the CSR regulation is effective as it ensures larger firms spend proportionately more than smaller firms. In conclusion, the present study acknowledges the stakeholder theory by reporting a significant and positive role of firm visibility and CSR spending. The linear relationship between firm visibility and CSR spending negates the assumption that small firms spend more on CSR to get a competitive edge. The absence of U-shaped results indicates that the CSR regulation is effective, where large firms are spending proportionately more on CSR.

6 Limitations and Direction for Future Studies

Despite the above observations, the current study has several important limitations as well. Firstly, the study has used only the log of total assets as the proxy for firm visibility. However, the log of total assets alone may not fully represent the visibility characteristics of a firm. Therefore, future studies may focus on measuring firm visibility by incorporating multi-dimensional constructs. Furthermore, a comparison of developing or developed economy settings and mandatory vs voluntary CSR settings would yield a better outlook on the topic. Hence, future research may pursue a cross-country comparison between diverse economic units. Adding on, future research could use more meaningful control variables and econometric techniques to avoid any possible influence of unobserved firm-specific factors. Finally, future research could deploy a multi-stage study using a mixed-method approach. Qualitative enquiries will be useful in digging out the reasons for the linearity or non-linearity, if any, observed in the data. Moreover, it helps explain how the interaction between firm visibility and CSR is happening.

References

Achour Z, Boukattaya S (2021) The moderating effect of firm visibility on the corporate social responsibility-firm financial performance relationship: evidence from France. In: Corporate social responsibility. IntechOpen

Agarwala N, Pareek R, Sahu TN (2024) Do firm attributes impact CSR participation? Evidence from a developing economy. Int J Emerg Mark 19(12):4526–4542

Agrawal OP, Bansal P, Kathpal S (2020) Effect of financial performance on corporate social responsibility and stock price: a study of BSE listed companies. Int J Emerg Technol 11(1):286–291

Ali J (2023) Does mandatory corporate social responsibility expenditure affect the financial performance of food and agribusiness firms? Evidence from India. Eur Bus Rev 35(4):520–533

Amato LH, Amato CH (2007) The effects of firm size and industry on corporate giving. J Bus Ethics 72(3):229–241

Baatwah SR, Al-Qadasi AA, Al-Shehri AM, Derouiche I (2022) Corporate social responsibility budgeting and spending during COVID–19 in Oman: a humanitarian response to the pandemic. Financ Res Lett 47:102686

Baltagi BH (2008) Forecasting with panel data. J Forecast 27(2):153–173

Bhagawan P, Mukhopadhyay JP (2025) Does mandatory expenditure on CSR affect firm value? Empirical evidence from Indian firms. J Account Lit 47(3):525–546

Bhattacharyya A, Rahman ML (2019) Mandatory CSR expenditure and firm performance. J Contemp Account Econ 15(3):100163

Brammer S, Millington A (2006) Firm size, organizational visibility and corporate philanthropy: an empirical analysis. Bus Ethics: Eur Rev 15(1):6–18

Breusch TS, Pagan AR (1980) The Lagrange multiplier test and its applications to model specification in econometrics. Rev Econ Stud 47(1):239–253

Bushee BJ, Miller GS (2012) Investor relations, firm visibility, and investor following. Account Rev 87(3):867–897

Campbell D, Slack R (2006) Public visibility as a determinant of the rate of corporate charitable donations. Bus Ethics: Eur Rev 15(1):19–28

Carroll AB, Shabana KM (2010) The business case for corporate social responsibility: a review of concepts, research and practice. Int J Manag Rev 12(1):85–105
Chatterjee S, Hadi AS (2015) Regression analysis by example. Wiley
Cherian J, Umar M, Thu PA, Nguyen-Trang T, Sial MS, Khuong NV (2019) Does corporate social responsibility affect the financial performance of the manufacturing sector? Evidence from an emerging economy. Sustainability 11(4):1182
Chettri N, Makan LT (2025) The impact of excess CSR expenditure on firm value and dividend payout in India: an analysis using firm age and size dynamics. Int J Soc Econ 52(6):931–946
Cowen SS, Ferreri LB, Parker LD (1987) The impact of corporate characteristics on social responsibility disclosure: a typology and frequency-based analysis. Acc Organ Soc 12(2):111–122
Deegan C, Rankin M (1996) Do Australian companies report environmental news objectively? An analysis of environmental disclosures by firms prosecuted successfully by the Environmental Protection Authority. Account Audit Account J 9(2):50–67
Do TK, Huang HH, Shan L, Tsang A, Yu L (2023) Stakeholder demands and corporate social responsibility: evidence from the Olympic Games. China J Account Res 16(3):100307
Donaldson T, Preston LE (1995) The stakeholder theory of the corporation: concepts, evidence, and implications. Acad Manag Rev 20(1):65–91
Fahad P, Rahman PM (2020) Impact of corporate governance on CSR disclosure. Int J Discl Gov 17(2):155–167
Ferri LM, Pedrini M, Minciullo M (2022) Corporate social responsibility and stakeholder dialogue under institutional voids: decoupling the role of corporate motives, ethics, and resources. J Manag Governance 26(1):159–188
Freeman RE (1984) Strategic management: a stakeholder approach. Cambridge University Press, Cambridge
Gamerschlag R, Möller K, Verbeeten F (2011) Determinants of voluntary CSR disclosure: empirical evidence from Germany. RMS 5(2):233–262
Golrida KP, Muliani M, Joshi PL (2019) Re-examining firm size and corporate social responsibility: the visibility approach. Emerg Markets J 9(1):1
Hansen WL, Mitchell NJ (2000) Disaggregating and explaining corporate political activity: domestic and foreign corporations in national politics. Am Polit Sci Rev 94(4):891–903
Hausman JA, Taylor WE (1981) Panel data and unobservable individual effects. Econometrica: J Econom Soc 1377–1398
Hillert A, Ungeheuer M (2025) The value of visibility. Available at SSRN 2689652
Hooghiemstra R (2000) Corporate communication and impression management—new perspectives why companies engage in corporate social reporting. J Bus Ethics 27(1):55–68
Howard PH, Jaffee D (2013) Tensions between firm size and sustainability goals: fair trade coffee in the United States. Sustainability 5(1):72–89
Iqbal UP, Mathew Jose S, Daniel TM (2025) Levels of mandatory CSR compliance and the payoff: evidence from India. Soc Responsib J 21(2):378–390
Jain S, Chakrabarty SP (2020) Does marginal VaR lead to improved performance of managed portfolios: a study of S&P BSE 100 and S&P BSE 200. Asia-Pacific Finan Markets 27(2):291–323
Kang J, Kang Y (2025) Capital market scrutiny and strategic distinctiveness of the firm: evidence from a natural experiment. Br J Manag 36(1):461–476
Karyawati PG, Angesty K, Joshi PL (2017) A study of inverted U-shape relationship between firm size and corporate social responsibility—the case of Indonesia. Int J Account Audit Perform Eval 13(3):308–323
Kaymak T, Bektas E (2017) Corporate social responsibility and governance: information disclosure in multinational corporations. Corp Soc Responsib Environ Manag 24(6):555–569
Kuzey C, Uyar A, Nizaeva M, Karaman AS (2021) CSR performance and firm performance in the tourism, healthcare, and financial sectors: do metrics and CSR committees matter? J Clean Prod 319:128802

Kuzey C, Uyar A, Al-Shaer H, Karaman AS (2025) Do firms with more powerful CEOs and greater visibility have better social reputation? Int J Account Inf Manag

Li F, Morris T, Young B (2019) The effect of corporate visibility on corporate social responsibility. Sustainability 11(13):3698

Lindblom CK (1994) The implications of organizational legitimacy for corporate social performance and disclosure. In: Critical perspectives on accounting conference, New York

Madden K, Scaife W, Crissman K (2006) How and why small to medium size enterprises (SMEs) engage with their communities: an Australian study. Int J Nonprofit Volunt Sector Mark 11(1):49–60

Nguyen CT, Nguyen LT, Nguyen NQ (2022) Corporate social responsibility and financial performance: the case in Vietnam. Cogent Econ Finance 10(1):2075600

Nidheesh KB (2021) Determinants of CSR disclosure: an evidence from India. J Indian Bus Res 13(1):110–133

Orlitzky M (2001) Does firm size confound the relationship between corporate social performance and firm financial performance? J Bus Ethics 33(2):167–180

Oware KM (2023) Mandatory CSR reporting and disability employment. Evidence from India. Cogent Bus Manag 10(2):2204601

Pareek R, Pandey KD, Sahu TN (2019) Corporate governance, firms' characteristics and environmental performance disclosure practices of Indian companies. Indian J Corp Governance 12(2):142–155

Pareek R, Sahu TN, Gupta A (2023) Gender diversity and corporate sustainability performance: empirical evidence from India. Vilakshan-XIMB J Manag 20(1):140–153

Parmar BL, Freeman RE, Harrison JS, Wicks AC, Purnell L, De Colle S (2010) Stakeholder theory: the state of the art. Acad Manag Ann 4(1):403–445

Rajgopal S, Tantri P (2023) Does a government mandate crowd out voluntary corporate social responsibility? Evidence from India. J Account Res 61(1):415–447

Russo A, Tencati A (2009) Formal vs. informal CSR strategies: evidence from Italian micro, small, medium-sized, and large firms. J Bus Ethics 85(Suppl 2):339–353

Schreck P, Raithel S (2018) Corporate social performance, firm size, and organizational visibility: distinct and joint effects on voluntary sustainability reporting. Bus Soc 57(4):742–778

Singh J, Singhania S, Aggrawal D (2023) Does board gender diversity impact financial performance? Evidence from the Indian IT sector. Soc Bus Rev 18(1):51–70

Suchman MC (1995) Managing legitimacy: strategic and institutional approaches. Acad Manag Rev 20(3):571–610

Sun G, Sulemana I, Agyemang AO (2025) Examining the impact of stakeholders' pressures on sustainability practices. Manag Decis

Tran HT, Pham HST (2024) How environmental, social, and governance disclosure promotes sales? Empirical evidence from global firms. J Strateg Mark 32(1):1–19

Udayasankar K (2008) Corporate social responsibility and firm size. J Bus Ethics 83(2):167–175

Uyar A, Lakhal F, Kuzey C, Karaman AS (2024) Do stockholders appreciate CSR? The role of firm visibility, financial slack, and monitoring. Manag Int 28(special):92–111

Weber M (2008) The business case for corporate social responsibility: a company-level measurement approach for CSR. Eur Manag J 26(4):247–261

Wickert C, Scherer AG, Spence LJ (2016) Walking and talking corporate social responsibility: implications of firm size and organizational cost. J Manag Stud 53(7):1169–1196

Yu J, Lo CWH, Li PHY (2017) Organizational visibility, stakeholder environmental pressure and corporate environmental responsiveness in China. Bus Strateg Environ 26(3):371–384

Zou H, Yang X, Wang R (2024) Keeping pace with competitors in CSR: from a perspective of competitive dynamics. Chinese Manag Stud

Carbon Accounting, Climate Policy, and Market Mechanisms

Leveraging Sustainable Development Goals Through Carbon Accounting

Ajai Prakash and Nyaribo Misuko

Abstract There are various ways and means adopted for leveraging the SDGs to make this world a better place to connect and live in. Carbon accounting is one such creation. It is a means where greenhouse emissions are measured, which helps in bringing out a systemic relationship with different activities carried out, with the intent to develop. The alignment with the seventeen overarching goals of United Nations SDGs have setup the agenda for the 2030. The framework of SDGs specially the SDGs 17 (Partnerships for the Goals), 15 (Life on Land), 13 (Climate Action), 12 (Responsible Consumption and Production), and 9 (Industry, Innovation and Infrastructure) connect with how carbon emission can be spotted on the SDGs canvas. The Greenhouse Gas Protocol, the Carbon Disclosure Project, creates a setup to work together, to enhance accountability and transparency with respect to carbon emission. This study examines how carbon accounting serves as a critical marker for implementing and monitoring the Sustainable Development Goals (SDGs), enabling more effective measurement and tracking of contributions from industrial sectors and other stakeholders. The following work is an attempt to connect how carbon accounting can be a marker in implanting connecting the study highlights the supporting role of carbon accounting for a positive climate action for a sustainably developed world.

Keywords Carbon accounting · SDGs · Transparency · Emission · Sustainable development

A. Prakash (✉)
Department of Management, University of Lucknow, Lucknow, India
e-mail: ajaiprakash1@gmail.com

N. Misuko
School of Business, KCA University, Nairobi, Kenya
e-mail: drnyaribo@kcau.ac.ke

K. Kumar et al. (eds.), *Sustainable Value Creation: Corporate Strategies for SDGs Implementation*, World Sustainability Series,
https://doi.org/10.1007/978-3-032-24652-3_8

1 Introduction

People often crave progress, yet saying exactly "how much" feels beyond reach. So far, attempts to improve life come from new ideas, yet killing inventions when they are unwanted happens rarely. With every advance, more carbon adds to the growing stack. Without balance, life cannot settle on the planet. Over decades, carbon output has grown fast—tied closely to factory work rising globally. People once hoped change could come through policy. Now different standards and steps exist, shaped by effort, aiming toward cleaner air and land. Carbon accounting is one such creation. Many of the objectives of the SDGs are curated along the pillars of the triple bottom line approach, but it has become difficult to meet the expectations of the same (Nica and Georgescu 2025).

In fact, it is challenging to retard reckless production and reduce, the unprecedented increasing consumption behaviors in every walk of life, which has an increased impact on the environment. The rise of capitalism and better technology has resulted in rapid population growth, unchecked industrialization, and sprawling urbanization. This damages the efforts towards sustainable development (Nica and Georgescu 2025). Among the SDGs, SDG 13 (Climate Action) stands out as crucial response to the growing environmental crisis. Various indicators, such as extreme weather, water shortages, air pollution, biodiversity loss, societal health crises, economic upheavals, and the predicted ocean acidification, support forecasts on environmental degradation (Kumar et al. 2024). According to Roy et al. (2024), these challenges can only be improved by making people's lives more sustainable through improved work habits. Behavioral shift is strongly required in the educational progression of producers and consumers.

The United Nations released the 17 interconnected Sustainable Development Goals (SDGs) in 2015 in response to this global need, helping to create a more sustainable and just future, for the next generations (Tosun and Leininger 2017). However, the transition to sustainable lifestyles is slow, despite people's growing awareness. Societal behavior drives people towards common actions. The same is true in case of sustainable approaches for a better common future. Higher the number of individuals attempting to adopt sustainable lifestyles the more influences are gathered at individual behaviors working towards adoption of sustainable actions. The lack of infrastructure and people's inability to utilize the current infrastructure causes problems, and widespread adoption becomes gradual and low (Butar et al. 2025).

The following section is an attempt to understand how different efforts align with creating a sustainable future. It has been observed that electronic media serves as a catalyst in the pursuit of sustainable development. This is demonstrated by actions such as substituting e-books for printed books, virtual or online meetings for in-person meetings, digital banking, digital medical reports, online reservations, restaurant menu readers, digital photos substituted by physical photos, and numerous other actions. However, the dark side of this shift is seen in greenhouse gas (GHG) emissions emanating out of these activities. GHGs include multiple gases such as carbon dioxide, methane, nitrous oxide, fluorinated gases like hydro fluorocarbons, per fluorocarbons, and sulphur hexafluoride. GHGs must be balancers for environment and cannot be released in quantities that adversely leverage the atmosphere. There are multiple ways in which GHGs are emitted such as energy production, transportation, industry and manufacturing, infrastructure development, burning waste, dump yards, data centers and cloud computing, streaming services etc. All these activities add to the climatic conditions in a negative way. When emissions exceed, they harm natural ecosystem including forest, marine life and human life.

People connect with many more actions to create an even workplace to live in, but we fall short of indicators that can identify the extent of each activity. Carbon accounting is one such mechanism, that has been created to look at the efforts pursued by the various measuring systems, which finally aggregates into a common accepted system.

1.1 Carbon Accounting Meaning

Figuring out where greenhouse gases emanate whether from individual actions or from within the organizations, it is imminent to account for the same. Carbon accounting helps in tackling climate shifts as accounting helps in a better management of such emissions. Keeping track of carbon levels helps shape clearer ways to cut emissions say researchers like Nica and Georgescu in 2025. Since carbon makes up most % of gases emitted, about 79% (according to EPA data from 2023), calling it "carbon accounting" makes sense. Although, what counts as carbon accounting is not always clear i.e. how people see the shifts based on where they stand (Saraswati 2020; Wright et al. 2011). Still, the version offered by Stechemesser and Guenther in 2012 shows up more often than most would expect. According to them, "Carbon accounting comprises the recognition, the non-monetary and monetary evaluation and the monitoring of greenhouse gas emissions on all levels of the value chain and the recognition, evaluation and monitoring of the effects of these emissions on the carbon cycle of ecosystems". The definition provides a broader perspective of carbon accounting as it includes how evaluating and monitoring carbon emissions impact business functions. Most greenhouse gas measurements report carbon dioxide equivalents in tons. The related numbers indicate how much warming each gas can cause, using CO_2 as a base reference. It helps compare gases like methane or fluorinated gases. Scientists often write this as CO_2e, simply shortening the full term. Peter Drucker's observation "what counts shows up in decisions. Whatever an executive does, it is a decision. If it is not a decision, it is not an executive action." (Drucker 1967). Climate shift leans heavily on tracking carbon because without numbers, control fades. Measuring emissions gives companies a clearer view, shaping choices through reporting and follow-up. Goals take shape only when real data point the way. Progress moves into view once metrics track each step forward. Adjustments happen more easily when impacts come into focus. Global climate aims, gain strength when individual actions align with observable results.

Acts and actions committed by organizations can now be associated with harm to nature, firms face stronger demands to share how they affect ecosystems. Earning profits is no more an indicator of development as they do not capture the need to act ethically, protect nature, while matching world goals like the UN's Sustainable Development Goals today (Kumar et al. 2021). Open reporting, shapes frameworks on how to track emissions, as seen when methods like carbon banking reveal real outcomes. Numerous sustainability reporting mechanisms have been designed at the global and national levels like The Global Reporting Initiative (GRI) (Stefanescu 2022), Sustainability Accounting Standards Board (SASB) (Parfitt 2024), Carbon Disclosure Project (CDP) (Siew 2015), ISO 26000 (Castka and Balzarova 2008). These are some of globally utilized frameworks. In the Indian context the National Voluntary Guidelines (NVGs), Business Responsibility and Sustainability Report (BRSR), Companies Act 2013 (Section 135) are sustainability guidelines that have been created in the same intent as above (Kumar and Prakash 2019a, b).

The EU Taxonomy and the Corporate Sustainability Reporting Directive (CSRD) are two European frameworks that work in the same way (Hummel and Bauernhofer 2024). Other frameworks that are worth considering are the Task Force on "Climate Related Financial Disclosure of Japan" (Board 2017), Guidelines presented by "China Securities Regulatory Commission" (CSRC) of China (Dong et al. 2021) or King II Report (2002) and King IV Report (2016) Report of South Africa, which are prominent. These are specific and focused on certain demands of a specific countries. These reporting rules aim to push companies toward handling greenhouse gases more carefully. Though the reporting rules are linked together, how each one applies shifts depending on the respective context. The central goal is to link firms to meet their duty while tracking emissions. With such a setting and rules in place, clarity grows which helps to evaluate what is expected in terms of sustainable development and how far a specific operation could achieve from what is expected. Trust builds slowly through such consistency, shaping how others view the effort. This also imbibes a sense of fairness which gives credence to a wider approval among the recipients of development i.e. all the stakeholders. This simultaneously also raises the confidence among those who invest, especially when the obvious patterns from actions become decipherable.

Firms, while acting with integrity, track carbon emissions. They consider themselves as forward thinking, caring, and ready to face climate hurdles. Moving forward often means involving all the stakeholders—investors, rule makers, shoppers, civilians—to shape what comes next. Being held responsible for nature's toll on business has become more pressing (Kumar and Prakash 2019a, b). In the current world, a company is evaluated on the basis of their financial and non-financial reports (sustainability report). A significant part of sustainability reporting is the data on carbon emissions. With this, carbon disclosure has ceased to be a voluntary measure instead it is becoming an obligatory part of corporate governance. By delivering clear, user-friendly, and reliable carbon reporting to stakeholders, organizations demonstrate their social and legal preparedness (Abeysekera 2022). International collaborations are made possible by carbon reporting, which draws organizations from all over the world to express interest in the policy of collective climate action.

Organizations address sustainability challenges, and the need for dynamic sustainability approaches by arranging sustainability action and implementation through training courses, sponsoring carbon accounts research and even collaborating for the professional development of employees on climate related issues (Lovell and MacKenzie 2011). Such initiatives show the adoption of a higher climate-sensitive corporate culture. In spite of the higher awareness, the domain of carbon accounting is not at realms of industrial practice yet (Bui and Fowler 2019; Hettler and Graf-Vlachy 2024). Barriers to adoption include unskilled trainers, regulatory decisions, unpredictable reporting systems, and economic resource limitations, particularly within small and medium enterprises (Sharma and Kumar 2021).

2 Contemporary Frameworks for Carbon Accounting

A widely known framework The Greenhouse Gas Protocol (GHG Protocol)- is being utilized to overcome these obstacles. This tool is prevalent and has been developed by the World Resources Institute (WRI) along with The World Business Council for Sustainable Development (WBCSD). WBCSD categorizes carbon emissions into three areas: Scope 1 (direct emissions), Scope 2 (indirect emissions purchased energy), and Scope 3 (other indirect emissions value). By offering a variety of reporting options and information on the so-called emission hotspots, each of these organizations enables businesses to develop targeted plans (Hettler and Graf-Vlachy 2024).

Researchers have tried to measure GHG emission activities through metrics. These metrics are characterized and point towards the utilization of carbon accounting systems by companies. These metrics are considered significant indicators of company's utilization of carbon accounting system such as reporting Scope 1, 2 and 3 emissions, tracking carbon intensity (e.g. CO_2 per unit of product or service) and setting emission reduction targets. Age and ownership of the organization related standards matter a lot, when it comes to effective carbon accounting (Solikhah et al. 2021; Singh et al. 2025a). Different issues (domestic) can change how businesses respond and how they measure carbon emissions. Carbon reporting is also important. It helps companies understand their negative or positive contribution to the environment. Carbon reporting helps companies look at and evaluate how climate challenges are addressed by them and what actions globally are being taken by them Rankin et al. 2011). Tracking carbon emission helps firms stay on course, nudging them toward greater duty and answerability during output decisions. Step by step, these entities shift not only responsibility solely to shareholders but look at broader social roles tied to varied interests. With the growing awareness their actions spread not just for profits, but they look at the effects on communities, ecosystems, and individuals.

3 Carbon Accounting and Its Extent

Nowhere was the shift clearer than in how businesses track emissions, once tucked into academic corners by a few specialists. Today, conversations about climate show up in issues that nobody expected. Researchers link finance records to environmental shifts (Stechemesser and Guenther 2012). As awareness grows about lasting resource challenges, companies find themselves turning toward methods like carbon tracking not just as theory, but as daily practice for steering toward more resilient futures. Now the scrutiny is stronger on sustainability, and researchers and companies alike turn towards it as a measure for existence than ever before (Schaltegger and Csutora 2012). Still, there is a bridge required in a shared meaning for carbon accounting. Various tools and measures track emissions, yet diversity leads us to understand it in separate ways (Chen et al. 2019; Ascui and Lovell 2011).

In some cases, emissions data is a "carbon commodity" which is a measurable and a tradable commodity with financial value (Ascui and Lovell 2012). Carbon Accounting is especially significant because of its potential to effectively support practical solutions regarding climate change. Carbon Accounting can successfully evaluate effective climate mitigation initiatives and provide guidance for policy decisions when implemented properly (Ascui and Lovell 2011). Carbon accounting as a recent trend has appeared to be more of window dressing in many cases. Many a times there are overstatements, knowingly or unknowingly with respect to carbon accounting. According to Turnhout et al. (2015), inaccuracies or overstatements in reporting are followed, as such expression leverage the obtaining of loans or other contemporary benefits for businesses. Carbon tracking guides company choices even as efforts for zero emissions meet resistance in different ecosystems. Plenty of companies struggle to rely on carbon data effectively. Gains show up in different spheres but the habit of adopting such practices hasn't taken root.

What happens often comes from multiple forces too. Many frameworks have been developed for carbon accounting but among these many frameworks are context and country specific. This leads to poor reporting and lack of consistency in global reporting. Potential investments in businesses are discouraged by discrepancy between the company's carbon emissions and offsets due to the fragmented reporting system (Ascui and Lovell 2011). Companies assist the SDGs in a number of ways, including through carbon accounting. Despite the potential of carbon accounting to positively connect with SDGs, it has been given the least importance till now. Many approaches are being widely adopted, including as ESG frameworks, sustainability reporting, circular economy models, and employee awareness campaigns to enhance the benefits from Carbon accounting. This chapter focuses on investigating theoretically and exploiting the use of carbon accounting as a tool to advance SDGs and how it corresponds with several SDGs. This object is closely tied to Sustainable Development Goals (SDGs) 9, 12, 13, and 15. Literature has been gathered through a focused method, aiming at the objectives of various SDG and connecting them to the issue. Using key terms, databases like Scopus, Web of Science, and Google Scholar were explored. The emphasis was placed on newer peer-reviewed articles discussing shifts, obstacles, and usage in carbon tracking. Work centered on how different industries put these methods into practice this revealed key details about the subject while supporting long-term environmental efforts.

4 SDGs and Climate Synergies

UN set the SDGs with targets meant to wrap up by the close of 2030 and each shaping progress down the path of growth. Among seventeen aims, one stands out clearly on climate action as a core priority under SDG 13. Just creating an ecosystem around SDG 13 cannot slow global warming. Other targets of SDGs 12, 9, 15, and 17 are also linked in ways that stretch beyond single issues and give a comprehensive framework. These aims spotlight distinct challenges shaped partly by climate shifts. Together, they show how joint actions can start weaving together lasting and resilient environments. The chapter further discusses each of the SDGs with respect to web created towards understanding carbon accounting and environment.

4.1 SDG 9

This focuses on "building resilient infrastructure, promote inclusive and sustainable industrialization and foster innovation" (United Nations General Assembly 2015) with 8 targets. 3 of the targets are implementation (like expanding access to technology and finance for business and infrastructure) and 5 are outcome targets (building high-quality infrastructure, encouraging inclusive industrialization, raising the industry's GDP share, and fostering technological advancement), Out of these targets, target 9.4 emphasizes improvement in infrastructure and update industries to be more sustainable by using resources efficiently and adopting clean, eco-friendly technologies. This is based on the extent to which each country can manage the preceding issues in falling in line with carbon accounting, by assisting firms in gauging its carbon data. The carbon accounting reports facilitate the firms in taking informed and responsible investment decisions.

In response to climate change, many firms are transitioning toward sustainable technologies such as solar panels, energy-efficient machinery, electric vehicles, and carbon mineralization systems. Although adoption of related technologies is not enough as many times they are just in the namesake and do not provide enough effectiveness. This is where Carbon Accounting plays the pivotal role. It is a means to assess the actual environmental alignment of technologies by expressing the effectiveness of given technology towards specific carbon accounting system in play.

4.2 SDG 12

Responsible consumption sits at the center of SDG 12, which spans eleven targets laid out through three action-oriented and eight result-driven goals. This focus pushes people toward choices that balance needs with impact. Progress marks a shift toward mindful use of resources across daily habits. Reducing excess food loss becomes one path forward alongside managing chemical emissions from enterprise activities. The key target, 12.6, highlights why companies need to include eco-conscious methods and transparent updates in their reports. Tracking carbon helps organizations stay aware about environmental effects from operations. Frequently, analysts turn to Life Cycle Assessment which is just one part of carbon tracking. It has been witnessed that operations like transport to even wrapping goods add to a lot of environmental damage and add to pollution. As people look at these flaws, choices of consumers might shift towards picking items with greener materials or avoiding travel with heavy carbon footprints. Support for planet-minded businesses could follow too. Over time, habits shape up and tend to stick around as thoughtful actions in everyday life (Srivastava et al. 2018; Tripathi et al. 2022).

4.3 SDG 13 (Climate Action)

Global efforts to slow heating up are the focus of SDG 13. Three goals tie directly to putting things into practice, while other pair helps to make that happen—all part of broader work laid out by Zhu and team in 2024. Resilience against climate shocks becomes one of the aims. Tracking carbon ties closely to meeting goal 13.2, which asks for clear strategies to be developed around the known climate realities (Singh et al. 2025b). Tracking carbon involves collecting and storing emissions data through clear methods that shape how climate rules are created. By doing so, organizations along with public agencies gain insights into their progress in reducing gases and creating targets rooted in facts. Firms often secure carbon credits alongside purchases from outside sources to prove output levels at or under zero. Usually, businesses do this to boost how much people trust them and their actions. This helps win deals with public agencies and global organizations. By tracking carbon data, anyone can easily check details tied to buying or selling carbon credits (Srivastava et al. 2021; Kaith and Sachdeva 2024).

4.4 SDG 15 (Life on Land)

SDG 15—Life on Land- places a lot of emphasis on protecting ecosystems, boosting biodiversity, and using them sustainably. There are twelve important targets for fixing damaged land. They focus on stopping deforestation and protecting local wildlife. Out of the different goals of SDG 15 the two goals 15.1 and 15.5 look at how to protect the biodiversity on earth and use our land resources wisely. Carbon accounting provides a good platform while considering the greenhouse gas emission and measuring the value of carbon stored in soil and plants. This includes all forms of ecosystems. The data on emission emanating from deforestation is of utmost importance. The environmental impact assessment, which is a tool for understanding the principle of protection used in understanding sustainable development, comes into consideration. Carbon accounting is a sophisticated tool to measure the same. It can account for the carbon offset process, aid in ecosystem restoration as well as achieving a balance between lowering carbon emissions and preserving biodiversity. Carbon accounting integrates land related emissions data into carbon inventories (Srivastava et al. 2024; Mishra and Srivastava 2025).

4.5 SDG17 (Partnerships)

The SDGs in essence are a 16-pillar structure with the SDG 17 as a networking system between many variables. It connects with networking among the countries, systems that emanate within these countries, similar systems that exist in different countries and the like. In fact it is a set of 19 networking systems, i.e. the targets of SDG 17. Target 17 keeps track of carbon emissions and directs to use accurate data to check how well the policies and development plans align between different countries and regions. The different reports on carbon emission help countries work together and in an integrated manner. The sustainability standards used by these reports are the same hence they align in the constructs and measures. Hence carbon accounting can also be considered as a universal expression working towards a common measure. This builds trust and gives a lot of room for sharing information, especially in the multinational arena. The targets of SDG 17 also direct at teaching people; issues related to sustainability and sharing information. In addition, one of the targets is keeping track of emissions. It directs at using accurate data to check how integrated do our polices and development plans work. This SDG is directed at facilitating countries to work in tandem with others and achieve the SDGs. SDG17 creates a ground for partnerships and carbon accounting creates a unison to get all on the same page this builds trust and encourages sharing of information. This action with similar standards helps to achieve the goals for Earth's ecosystems.

The direction that projects should follow is to create financial plans to address issues of climate change. It helps in making the society, government and businesses work together. Fresh ideas often emanate and succeed where cleaner technologies meet tougher climate goals. Easier funding paths open when past results become witness to the tools that tackle emissions best. Across regions, new methods take shape by adopting better ways to track carbon regularly. A worldwide team called the Partnership for Carbon Accounting Financials runs from the Netherlands, bringing together over 400 banks across countries. Every continent plays a role here. Information flows easily within such groups as they share data. This helps build clarity and the joint efforts run smoother because of centrality in the idea and unison in action by the members. Similar programs already exist, as shown by recent data from PCAF.

Funding sits at the center, tied to lending and investing alongside it. Tracking emissions becomes clearer when banks share data through a linked system. Gains appear without exaggeration as lower emissions from operations and such reporting show real impact over time. Goals tied to long-term sustainability gain momentum and move forward due to the shift. Clearer methods build confidence among users who see outcomes more openly. More people tend to join spaces where honesty leads to stronger results. Pushing eco-friendly spending choices grows easier when systems align behind shared aims. Collaboration across sectors gains strength through shared effort across borders.

5 Challenges and Limitations of Carbon Accounting

Carbon accounting is important for managing sustainable development. But there are challenges that companies and governments face. Universal usage of carbon accounting appears because of the following issues. The absence of a standard framework is the primary problem firms face. There is currently no standard method for accounting for carbon, which makes reports difficult to compare and influences green investment decisions. Even though there are numerous national and international laws for ecosystem protection and that businesses use various frameworks as part of their voluntary disclosure (Saraswati 2020; de Bortoli et al. 2023; Yan et al. 2023; Mengis et al. 2023) there is lesser cohesion among all. Moreover, there is lack of clear policies for carbon data sharing, therefore many firms which are practicing carbon accounting do not share their data. This limits transparency and affects collaboration from different parties like other businesses, MNCs and government (Yan et al. 2023). Another major issue is the underdeveloped measurement tools. As Bowen and Wittneben (2011) highlight, the current instruments used for carbon tracking are still evolving, which often results in data that may not be fully accurate or reliable. Yan et al. (2023) and Mengis et al. (2023), note that there is a lack of robust verification mechanisms for reported carbon data. Figuring out carbon usage helps guide long-term growth. Yet hurdles still block clear tracking by firms and officials alike. Making sense of emissions remains tricky across most levels.

Questions grow around trust and reliability when it comes to carbon disclosures, especially since records lack oversight. Forests and soil, acting like natural sinks, could suffer damage from fires, clearing woods, or changes in how land is used—this might lead to quick carbon releases back into air (Brander et al. 2021; Mengis et al. 2023). Should emissions climb over set limits, known as carbon overshoot. People rely on costly, only partly tested methods to pull CO_2 from skies, yet such approaches rarely work well or grow large enough to matter (Brander et al. 2021). Currently Negative Emission Technologies remain largely untouched by investors. Their lack of traction ties back to a clear issue but the returns aren't strong enough to justify risk. As Brander and team note, current carbon tracking standards often work against these solutions. Because profits don't follow policy lines, companies stay wary of deep-scale carbon pulling efforts. Without solid backing from regulators, real progress stalls, dimming hopes for accurate carbon tracking beyond zero emissions.

6 Conclusion and Recommendations

As of now the total usage of carbon accounting is low but it is being realized that this is an important tool to move forward with sustainable development. The tool on the other hand is also evolving and is trying to develop with many details, in order to draw fine lines. An important step currently is to make connections at different levels, this includes connections between corporates, corporate practices and SDGs, connections between countries as well as regions. It is also important to give this tool more prominence in industry specific issues. Industry specific issues will help in putting specific emphasis on respective industry processes and the nitty gritty of specific industrial environments. Measuring and disclosing emission in a more transparent manner will help organizations make informed decisions, which will inculcate trust within the stakeholders. The globe will be able to create a more structured approach in the light of better disclosure of emissions in a transparent manner. In effect it acts as a tool to develop a structured approach to keep track of climate related progress for organizations and contribute towards global climate action. There are certain barriers that restrict widespread adoption of carbon accounting like formalizing standard frameworks, limited policies about data sharing, lack of universal measurement tools and lack of centralized carbon databases. The following key actions that are recommended to unlock the full potential of carbon accounting:

Establishment Global Standards that facilitate harmony and synchronization in reporting standards.

- Creation of GHGs database to promote transparent and reliable data sharing for the stakeholders.
- Need to strengthen carbon mechanisms of carbon data verification to show more accuracy and reliability.
- More investment is required for research and development on GHGs measurement tools and technology.
- Government and international bodies need to develop incentives that motivate business for more adoption of carbon accounting practices.
- Top level management or leaders should encourage employees to integrate with corporate strategy.
- Carbon accounting should not be seen as just a reporting task, it must be embedded into the heart of business strategy. When we link carbon accounting with long-term sustainability goals and operational decisions, it can lead to meaningful climate action.

Annexure

Table of SDGs and their targets aligning with carbon accounting

SDG goal	Target No.	Target
SDG 9 (industry, innovation and infrastructure)	9.4	Upgrade infrastructure and retrofit industries to be sustainable
	9.5	Enhance research and industrial innovation
SDG 12 (responsible consumption and production)	12.4	Environmentally sound management of chemicals and waste
	12.6	Encourage companies to adopt sustainable practices and reporting
SDG 13 (climate action)	13.1	Strengthen resilience to climate hazards and disasters
	13.2	Integrate climate action into policies and plans
SDG 15 (life on land)	15.2	Promote sustainable forest management and stop deforestation
	15.9	Integrate biodiversity values into development plans
SDG 17 (partnership for the goals)	17.16	Enhance global partnerships
	17.18	Boost data availability and capacity-building in developing countries

References

Abeysekera I (2022) A framework for sustainability reporting. Sustain Account Manag Policy J 13(6):1386–1409. https://doi.org/10.1108/SAMPJ08-2021-0316

Ascui F, Lovell H (2011) As frames collide: making sense of carbon accounting. Account Audit Account J 24(8):978–999. https://doi.org/10.1108/09513571111184724

Ascui F, Lovell H (2012) Carbon accounting and the construction of competence. J Clean Prod 36:48–59. https://doi.org/10.1016/j.jclepro.2011.12.015

Bowen F, Wittneben B (2011) Carbon accounting: negotiating accuracy, consistency and certainty across organisational fields. Account Audit Account J 24(8):1022–1036. https://doi.org/10.1108/09513571111184742

Brander M, Ascui F, Scott V, Tett S (2021) Carbon accounting for negative emissions technologies. Climate Policy 21(5):699–717. https://doi.org/10.1080/14693062.2021.1878009

Bui B, Fowler CJ (2019) Strategic responses to changing climate change policies: the role played by carbon accounting. Aust Account Rev 29(2):360–375. https://doi.org/10.1111/auar.12213

Butar I, Purwanto R, Putranto A, Afrianto WF (2025) Trust and the intent to use on demand charging services of electric vehicles in Indonesia: the moderating effect of perceived charging risk and environmental concern. Energy 323:135713. https://doi.org/10.1016/j.energy.2025.135713

Castka P, Balzarova MA (2008) ISO 26000 and supply chains—on the diffusion of the social responsibility standard. Int J Prod Econ 111(2):274–286. https://doi.org/10.1016/j.ijpe.2006.10.017

Chen G, Shan Y, Hu Y, Tong K, Wiedmann T, Ramaswami A, Wang Y et al (2019) Review on city-level carbon accounting. Environ Sci Technol 53(10):5545–5558. https://doi.org/10.1021/acs.est.8b07071

de Bortoli A, Bjørn A, Saunier F, Margni M (2023) Planning sustainable carbon neutrality pathways: accounting challenges experienced by organizations and solutions from industrial ecology. Int J Life Cycle Assess 28(7):746–770. https://doi.org/10.1007/s11367-023-02147-z

Dong S, Xu L, McIver R (2021) China's financial sector sustainability and "green finance" disclosures. Sustain Account Manag Policy J 12(2):353–384. https://doi.org/10.1108/SAMPJ-10-2018-0273

Drucker PF (1967) The effective executive. Harper & Row, New York

Hettler M, Graf-Vlachy L (2024) Corporate scope 3 carbon emission reporting as an enabler of supply chain decarbonization: a systematic review and comprehensive research agenda. Bus Strateg Environ 33(2):263–282. https://doi.org/10.1002/bse.3486

Hummel K, Bauernhofer K (2024) Consequences of sustainability reporting mandates: evidence from the EU taxonomy regulation. In: Accounting forum, vol 48, no 3. Routledge, pp 374–400. https://doi.org/10.1080/01559982.2024.2301854

Kaith A, Sachdeva G (2024) Bibliometric insights into green human resource management: a connection with business ethics and values. PURUSHARTHA 17(1):51–63. https://doi.org/10.21844/16202117104

King Committee on Corporate Governance. (2002) King Report on Corporate Governance for South Africa—2002 (King II). Institute of Directors in Southern Africa (IoDSA)

King Committee on Corporate Governance (2016) King IV Report on Corporate Governance for South Africa—2016 (King IV). Institute of Directors in Southern Africa (IoDSA)

Kumar K, Kumari R, Kumar R (2021) The state of corporate sustainability reporting in India: evidence from environmentally sensitive industries. Bus Soc Rev 126(4):513–538

Kumar K, Singh R, Sharma A, Lehri BPS (2024) Climate change adaptation strategies for achieving net-zero economy. In: Transition towards a sustainable future: net zero policies and environmental sustainability. Springer, Singapore, pp 163–183

Kumar K, Prakash A (2019a) Developing a framework for assessing sustainable banking performance of the Indian banking sector. Soc Responsib J 15(5):689–709

Kumar K, Prakash A (2019b) Examination of sustainability reporting practices in Indian banking sector. Asian J Sustain Soc Responsib 4(1):2

Lovell H, MacKenzie D (2011) Accounting for carbon: the role of accounting professional organisations in governing climate change. Antipode 43(3):704–730. https://doi.org/10.1111/j.1467-8330.2011.00883.x

Mengis N, Paul A, Fernández-Méndez M (2023) Counting (on) blue carbon—challenges and ways forward for carbon accounting of ecosystem-based carbon removal in marine environments. PloS Climate 2(8):e0000148. https://doi.org/10.1371/journal.pclm.0000148

Mishra A, Srivastava SK (2025) Role of trust and celebrity endorsement in shaping purchase intention towards electric vehicles. In: Implementing ESG frameworks through capacity building and skill development. IGI Global Scientific Publishing, pp 381–400. https://doi.org/10.4018/979-8-3693-6617-2.ch018

Nica I, Georgescu I (2025) The ecological impact of agricultural production on CO_2 emissions in India: pathways to sustainable agriculture. J Environ Manag 384:125548. https://doi.org/10.1016/J.JENVMAN.2025.125548

Parfitt C (2024) A foundation for 'ethical capital': the sustainability accounting standards board and integrated reporting. Crit Perspect Account 98:102477. https://doi.org/10.1016/j.cpa.2022.102477

Partnership for Carbon Accounting Financials (n.d.) About PCAF. Retrieved 6 Aug 2025, from https://carbonaccountingfinancials.com/about

Rankin M, Windsor C, Wahyuni D (2011) An investigation of voluntary corporate greenhouse gas emissions reporting in a market governance system: Australian evidence. Account Audit Account J 24(8):1037–1070. https://doi.org/10.1108/09513571111184751

Roy A, Kumar S, Rahaman M (2024) Exploring climate change impacts on rural livelihoods and adaptation strategies: reflections from marginalized communities in India. Environ Dev 49:100937

Saraswati E (2020) Carbon accounting, disclosure and measurement: a systematic literature review. Int J Account Bus Soc 28(2):17–44. https://doi.org/10.21776/ub.ijabs.2020.28.2.2

Schaltegger S, Csutora M (2012) Carbon accounting for sustainability and management. Status quo and challenges. J Clean Prod 36:1–16. https://doi.org/10.1016/j.jclepro.2012.06.024

Sharma HP, Kumar K (2021) Developing and implementing environment management practices in small and medium size manufacturing companies in India. In: IOP conference series: earth and environmental science, vol 795, no 1. IOP Publishing, p 012022

Siew RY (2015) A review of corporate sustainability reporting tools (SRTs). J Environ Manag 164:180–195. https://doi.org/10.1016/j.jenvman.2015.09.010

Singh R, Sharma A, Kumar K (2025a) Navigating carbon neutrality: strategies for sustainable future and environmental challenges. In: Climate neutrality through smart eco-innovation and environmental sustainability. Springer, Cham, pp 69–80

Singh R, Tiwary KS, Kumar K, Kumar A (2025b) Empowering a sustainable future: blockchain's role in economy, organizations, and finance towards smart circular economy. In: Blockchain technologies for smart circular economy and organisational sustainability. Springer, Cham, pp 271–283

Solikhah B, Wahyuningrum, IFS, Yulianto A, Sarwono E, Widiatami AK (2021) Carbon emission report: a review based on environmental performance, company age and corporate governance. In: IOP conference series: earth and environmental science, vol 623, no 1. IOP Publishing, p 012042. https://doi.org/10.1088/1755-1315/623/1/012042

Srivastava SK, Anshul A, Pathak P, Bansal J (2018) Responsible consumption for curbing food wastage: an exploratory enquiry. PURUSHARTHA 11(2):104–116

Srivastava SK, Singh S, Srivastava M (2021) Food waste at household and social gatherings: drivers and possible remedies. PURUSHARTHA 14(2):117–133

Srivastava SK, Sinha AK, Kumar S (2024) Workplace spirituality empowered by spiritual leadership to promote organizational effectiveness: a bibliometric analysis. PURUSHARTHA 17(1):1–22. https://doi.org/10.21844/16202117101

Stechemesser K, Guenther E (2012) Carbon accounting: a systematic literature review. J Clean Prod 36:17–38. https://doi.org/10.1016/j.jclepro.2012.02.021

Stefanescu CA (2022) Linking sustainability reporting frameworks and sustainable development goals. Account Res J 35(4):508–525. https://doi.org/10.1108/ARJ-07-2020-0196

Tosun J, Leininger J (2017) Governing the interlinkages between the sustainable development goals: approaches to attain policy integration. Global Chall 1(9):1700036. https://doi.org/10.1002/gch2.201700036

Tripathi V, Singh K, Parihar S, Srivastava SK (2022) A bibliometric analysis on supply chain collaboration. PURUSHARTHA 15(2):66–103. https://doi.org/10.21844/16202115205

Turnhout E, Skutsch MM, de Koning J (2015) Carbon accounting. In: Research handbook on climate governance. Edward Elgar Publishing, pp 366–376. https://doi.org/10.4337/9781783470600.00044

Wright LA, Kemp S, Williams I (2011) 'Carbon footprinting': towards a universally accepted definition. Carbon Manag 2(1):61–72. https://doi.org/10.4155/cmt.10.39

Yan H-S, Tan X-C, Zeng A, Kong L-S (2023) China's carbon accounting system in the context of carbon neutrality: current situation, challenges and suggestions. Adv Clim Chang Res 14(1):23–31. https://doi.org/10.1016/j.accre.2023.01.008

Navigating Carbon Accounting in Developing Nations: Insights from the Indian Perspective

Khusboo Srivastava

Abstract Carbon accounting is one of the critical tools in the global fight against climate change, allowing a country to ensure accurate measurement, reporting, and verification of GHG emissions. For developing countries, such as India, carbon accounting provides both mega challenges and strategic opportunities. This paper examines the complex landscape of carbon accounting in India by analyzing institutional, technical, financial, and policy-related barriers obstructing the accurate and transparent management of carbon data. The key challenges pertain to the shortage of technical expertise, inadequate data infrastructure, fragmented regulatory frameworks, and lack of incentives for compliance, especially in the informal and unorganized sectors. Despite all these challenges, India also offers substantial potential for leading in novel and innovative approaches to carbon accounting through the use of digital technologies, public–private partnerships, and alignment with globally accepted mechanisms such as the Paris Agreement. The study accentuates the role of national initiatives in providing a facilitative environment for enhanced carbon accounting, such as the Perform Achieve and Trade (PAT) scheme, renewable energy targets, and corporate ESG mandates. Addressing these challenges and leveraging emerging opportunities, India can not only enhance domestic environmental governance but also strengthen its position in global climate negotiations. The paper concludes by offering strategic policy recommendations for establishing a robust, transparent, and inclusive carbon accounting framework tailored to the unique needs of developing economies.

Keywords Carbon accounting · Greenhouse gas (GHG) emissions · Climate change mitigation · Environmental governance · ESG compliance

---

K. Srivastava (✉)
Centre for Distance and Education, GLA University, Mathura, Uttar Pradesh, India
e-mail: khusboo.srivastava@gla.ac.in

K. Kumar et al. (eds.), *Sustainable Value Creation: Corporate Strategies for SDGs Implementation*, World Sustainability Series,
https://doi.org/10.1007/978-3-032-24652-3_9

1 Introduction

Among all global issues, climate change is the most significant one. The IPCC (Intergovernmental Panel on Climate Change) in its Sixth Assessment Report indicates that it is highly probable for the average global temperature to rise by 1.5 °C or even more by the year 2040, if the present emission pattern continues. Consequently, various natural occurrences like extreme weather events (mainly heat waves and droughts), fast melting of ice caps, and dying out of certain animals will happen more often. The scenario will have the developing nations as the weakest links taking the largest share of the suffering due to their limited adaptive capacities in comparison to the industrialized countries. One of the important contributions to the effective climate actions is the systematic measurement, reporting, and verification (MRV) of greenhouse gas (GHG) emissions, known as carbon accounting. The proper carbon accounting not only aids countries and firms in setting reduction targets and monitoring rates but also in participating in the international climate policies such as UNFCCC and Paris Agreement (2015, 2022). For a nation like India that is still in the process of development, carbon accounting is both a matter of policy and a chance of strategy. It is the third largest GHG emitter in the world responsible for about 7% of the total emissions in 2022. The emissions trend of India has been steadily upward due to urbanization, industrial development, and rising energy demand. India has per capita emissions that are still lower than that of a large number of nations, however, its total emissions are increasing. To tackle this matter head-on, India has announced its commitment to achieving net-zero by 2070 and to generating 50% of its power from renewable sources by 2030 which will form part of its new NDCs under the Paris Agreement. This problem, in turn, has been met directly by the country with such strong measures. To, in fact, fully attain such ambitious targets, India would need to have a strong and clear carbon accounting system.

2 Carbon Accounting: Concepts and Global Standards

Carbon accounting is a systematic methodology that measures, accounts for, and manages the total amount of GHG emissions-indirect and direct, and mainly man-made-that come from the energy, industry, transportation, and land-use sectors. This activity has been one of the sources of reliable data inputs on climate policy and corporate sustainability, as it enables governments, companies, and institutions to set emission ceilings and track progress, developing plausible carbon-neutrality plans in the process. The need for carbon accounting now appears more important, with the world continuing its aim to keep global warming below 1.5 °C, a threshold defined by the Paris Agreement. The classification of emissions into three scopes has gained the most acceptance and was the most widely used method. It forms the very basis of the entire mechanism and was given a structure by the Greenhouse Gas (GHG) Protocol drawn up by the World Resources Institute (WRI) and the World Business

Council for Sustainable Development (WBCSD) (2004) as partners. The updated GHG Protocol provides more specific guidance on the nature and boundary condition applicable under Scope 1, Scope 2, and Scope 3 reporting. The 'Scope 1' category or group refers to the immediately identifiable emissions that result directly from the organization's activities. The burning of fuel or the use of company-operated vehicles are two examples of such emission sources. In contrast, the 'Scope 2' classification focuses on emissions caused by the energy that the company has consumed and for which the company has paid, such as the electricity, steam, heating, or cooling energy that the company uses and pays for. The 'Scope 3' category is an inclusive one in that it embodies the whole range of indirect emissions throughout a company's value chain. Among them, there are emissions coming from suppliers, those associated with waste disposal, employees commuting, and those generated during the use of a product.

The three scopes empower companies to have a very precise view of their emissions responsibilities, avoid the risk of two companies counting the same emission, and formulate a carbon footprint reduction strategy that permeates the entire company. To achieve consistency and comparability between geographical areas and sectors, international frameworks and standards have been introduced:

1. The GHG Protocol is the most popular of them all and is still the one that gets the most licensing and implementation. It offers powerful guidance for both corporate and project-level emissions accounting.
2. The International Organization for Standardization developed ISO 14064 as an addition to the GHG Protocol which sets out the guidelines for developing inventories and third-party verifications.
3. The Governments use The IPCC Guidelines for National Greenhouse Gas Inventories as an instrument, informing them about the levels of emissions at any given time. These methods range in tiers from the simple estimations to detailed assessments and are thus adaptable according to the availability of data and technical capacity.

On the ground, carbon accounting involves collecting and analyzing data from various sources—fuel consumption, industrial activity, transport usage, and more (Srivastava et al. 2021; Kaith and Sachdeva 2024). These inputs are paired with emission factors to estimate outputs. Increasingly, emerging technologies are transforming how this data is captured. IoT sensors, satellite imaging, blockchain, and AI-driven analytics are enabling real-time monitoring, fraud prevention, and better insights into emission patterns—particularly in hard-to-reach or informal sectors. Nonetheless, there is no uniformity in the application of these tools in various locations. The developed regions have the infrastructure, legal demands, and institutions developed enough to perform thorough carbon accounting with the help of modern tools. In contrast, the tools cannot be utilized by most of the impoverished countries that are still struggling with various issues like poor access to data, lack of skilled personnel, and inconsistent policies. The discrepancy in the abilities of different nations reveals that the world should instantly make it a priority to invest in the developing of capacities, transferring of technologies, and setting of universal standards whereby no nation is left behind in executing climate actions.

3 Institutional and Policy Landscape in India

During its carbon accounting journey, India has been affected by its situation as a responsible global player dealing with climate change and as a fast-growing economy with urgent development needs. Gradually, the country has constructed a complex system of institutions, policies, and market-based approaches to monitor and ultimately cut greenhouse gas (GHG) emissions. The current arrangement is a mirror image of India's commitment to UNFCCC, Paris Agreement, and the likes, but it also shows the country's concerns like: energy security, economic integrating, and eco-friendly development. The long-term aim of the actions is to make a single carbon trading system out of the current practices like PAT and Renewable Energy Certificates (RECs) (Mohan and Srivastava 2023). Consequently, India will be able to join the global carbon markets shortly through internationally transferable mitigation outcomes (ITMOs), as stated in Article 6 of the Paris Agreement. India's corporate governance and financial landscapes have witnessed another significant positive development. The Securities and Exchange Board of India (SEBI) has mandated the top 1000 listed companies to report their ESG performance through the Business Responsibility and Sustainability Report (BRSR) (Sarkar et al. 2022) which includes GHG emissions (Scope 1 and 2 with Scope 3 on a voluntary basis) as one of the categories of data in the required disclosures. This in turn will cause the corporate sector to almost comply with the national and international carbon accounting standards. The move by SEBI reflects the increasing acceptance that climate transparency is no longer just a compliance issue but rather a significant financial concern.

Furthermore, the voluntary activities carried out by the companies are also becoming common practice. Several large Indian companies especially from sectors like Information Technology, automobile, and energy are now providing reports on their emissions according to global standards such as the Greenhouse Gas Protocol and Science-Based Targets initiative (SBTi). The above-mentioned actions not only complement national goals but also boost the international credibility of Indian carbon data. The energy sector, which is among the largest emitters of carbon, has integrated carbon accounting through Central Electricity Authority (CEA) monitoring corresponding to the emissions intensity of power generation. Mandates such as Renewable Purchase Obligations (RPOs) and Solar Renewable Energy Certificates (SRECs) also link market performance to emissions thus, making the whole process of transition to cleaner energy more accountable. Nevertheless, the situation has not changed much and the carbon governance system of India continues to be fragmented and even cumbrous. The MoEFCC, the Ministry of Power, the Ministry of New and Renewable Energy (MNRE), and NITI Aayog are the key players in the carbon governance, and this often becomes a cause of overlapping responsibilities and lack of proper synchronization. Furthermore, there is no single national emissions inventory which can reflect real-time data for all sectors; this makes it difficult to have a thorough and updated accounting. In addition, they keep off-the-record sectors renewable and nonrenewable from the formal statistics. Integration of small-scale manufacturing, traditional agriculture, and local transport, among others, is the case here. The nature of the solution that will be required for this integration will be decentralized, flexible, and low-cost, as well as being adapted to the local conditions.

4 Key Challenges in Carbon Accounting for India

India is becoming more serious about its climate commitments, and the creation of a solid carbon accounting system is now an urgent but complicated task. There are no challenges that can be addressed only in terms of technology or regulation—the issues span across the whole of institutional coordination, financing, and socio-economic inclusion. Solving these problems is necessary in order to make sure there is no opacity, a good international rapport, and a fast changing of India into a low-carbon economy (Jain et al. 2022).

Technical Limitations and Data Gaps

The greatest challenge which has to be overcome first is the poor condition of India's Monitoring, Reporting, and Verification (MRV) system which is still not developed properly. MRV systems are necessary for, and provide the foundation of, the process of getting accurate and instant emissions data for all sectors but in India, these systems which are characterized by non-uniformity and lack of coordination, especially in the decentralized and informal sectors (Pandey et al. 2020), encounter shortage of resources. The biggest shortcoming can be seen in the case of Scope 3 emissions, which consist of indirect emissions from sources such as logistics, procurement, and waste. There are no reliable databases or interoperable platforms to give a clearer picture of these emissions, hence, they remain hidden for the most part (Garg et al. 2017). Besides, the emission factors that are presently being used for the estimation of sectoral emissions have the drawbacks of being either out-of-date or not well determined geographical-wise causing large discrepancies in the results. As pointed out by Singh and Pandey (2021), agriculture, construction, and transport are some of the industries that are in desperate need of having a common data system which can help in the preparation of large-scale and accurate nation-wide inventories that are necessary for both local policy-making and access to the world's carbon market.

Fragmented Regulatory Landscape

India boasts many climate-friendly instruments that look toward the future, the Perform, Achieve and Trade (PAT) scheme (Sharma and Banerjee 2016), Renewable Energy Certificates (RECs), Energy Savings Certificates (ESCerts), and the Indian Carbon Market (ICM), just to name a few. Nonetheless, these initiatives frequently work in isolation, and they share some responsibilities while having different rules for compliance. This scenario of non-integration and non-harmonization confuses the industry and raises the possibility of double counting the emission reductions; especially in the case of one company being counted under more than one scheme (Mandal et al. 2022). Furthermore, communication among the different government departments responsible for the policy—Ministry of Power, MoEFCC, and NITI Aayog—is still limited. This leads to less data, longer time for implementation, and a larger disparity in the application of carbon accounting standards between states and sectors.

Financial Barriers for Small and Medium Enterprises (SMEs)

Carbon accounting comes with a hefty price tag and it is a very difficult thing for small and medium-sized enterprises (SMEs) in India that represent a significant portion of the country's manufacturing and labor output. Going for MRV systems, hiring consultants, and upgrading technology to clean are usually not affordable for smaller companies that are working on low profit margins. Though the global interest in green finance is rising, the actual movement of money is still far from being sufficient. According to the Climate Policy Initiative (2021), India is going to need over $170 billion every year till 2030 to achieve its climate goals but the current investments are nowhere close to that. Small and medium enterprises (SMEs) are cut off from the capacity-building programs, which results in their lack of knowledge about the carbon market and their inability to utilize the financial instruments like green bonds or carbon credit trading (Chatterjee et al. 2020; Singh et al. 2025a, b; Prakash and Tiwari 2025). This financial imbalance limits the participation and thus the transition to a more climate friendly era is still not equal in the case of SMEs.

Weak Verification and Transparency Mechanisms

The reliability of a carbon market is based on the three main pillars of transparency, traceability and trust. In India, the voluntary carbon market has not been able to build these foundations due to the non-existence of a central registry, the small number of third-party verifications done, and the greenwashing risks-where companies make false claims or overstate their climate actions. A unified emissions tracking platform is a must to make it possible to carbon credits being accurate, verifiable, and not double-counted (World Bank 2023). Additionally, the demand for independent verifiers and auditors is still underdeveloped. Often, the standards for who can be a verifier are unclear, thus creating doubt in the entire system (Aggarwal and Sarkar 2021; Kumar et al. 2025). This uncertainty dissuades and drives away serious climate investors and puts India's potential to participate in international carbon markets as a credible player in jeopardy.

Socio-economic Exclusion and Regional Disparities

Communities that are the most at risk from climate change are frequently omitted from India's carbon accounting ecosystem. Over 90% of India's entire labor force is part of the informal sector (ILO 2020), and these sectors are usually excluded from the MRV systems because of the lack of digital access, ignorance, and unestablished incentives. This scenario is common with small farmers, local transport services and rural art makers—to mention a few; they do emit but cannot report or do not know how to report their emissions. Thus, the economy's large part is not recorded on the carbon ledger, making the national reporting less credible. Furthermore, the geographic disparities still exist. The carbon accounting tools and the support from the government, on the other hand, are more readily available in the industrially developed states, such as Gujarat and Maharashtra, while the northeastern and tribal areas are still behind in terms of basic facilities and are also blocked by socio-economic factors; hence, they are not a part of carbon accounting. The net-zero targets in India will be only effective and just if the transition is made in a fair and inclusive way. This will require the initiation of new outreach programs, grassroots training, and the active participation of underserved communities in the climate finance and reporting systems (Dubash et al. 2021; Srivastava et al. 2024).

5 Emerging Opportunities and India's Strengths

Carbon accounting system in India has been a troublesome affair, but it is also a remarkable case of opportunity. The robust and vibrant digital infrastructure of India, along with the rise of public and private partnerships and shifts in the policy trend, which are quite favourable to India, collectively and mutually enable the country to lay down a road map on the governance of carbon, which would be quite reliable and engaging (Jain et al. 2022; Joselin et al. 2025).

Leveraging Digital Innovation for Scalable MRV

India has built up enormous digital platforms like Unified Payments Interface (UPI) and CoWIN, which support the next generation systems for measuring, reporting and verifying carbon emissions. The introduction of fresh technologies such as artificial intelligence (AI), blockchain and satellite monitoring can completely alter the emissions tracking, verifying and reporting process. Blockchain would protect the records of carbon transactions to a significant extent decreasing fraud and double counting (Hua et al. 2022; Singh et al. 2025b). For the energy and carbon markets, smart contracts on the blockchain would create opportunities for decentralized trading, while at the same time, ensuring accountability and transparency throughout the process (Bhadange et al. 2022). Through IoT sensors and satellite images, it becomes possible to acquire high-quality, real-time emissions data even in informal or hard-to-reach areas; thus, monitoring of hard-to-reach operations also becomes much easier. AI-assisted analytics, for example, can facilitate the validation of self-reported data, predict emissions trends, and provide in turn more responsive and adaptive MRV systems. Put together, these technologies will operationalize carbon accounting activities much more significantly in India with respect to their accuracy, transparency, and geographical coverage.

Scaling Solutions Through Public–Private Collaboration

Recent initiatives in India demonstrate that community leadership, supported by the private sector, can effectively scale carbon accounting in a manner beneficial to society. A notable example is the carbon farming project led by NABARD in Karnataka, where 3500 mango farmers generate carbon credits through sustainable practices such as biomass management and tree planting. Although the project initially encountered challenges, including delayed payments and skepticism among some participants, it illustrates the potential of community-based models that integrate climate action with rural development. Future models may be further strengthened by the involvement of agritech startups, non-governmental organizations, private investors, and local governments, which can facilitate the establishment of community-based monitoring, reporting, and verification (MRV) systems, foster trust with farmers, and expand participation in carbon markets (Tripathi et al. 2022; Srivastava et al. 2018).

Policy Momentum and Alignment with Global Frameworks

Recently, RECs and corresponding markets are beginning to set the stage for more active engagement with carbon markets globally. The latest amendments to energy conservation regulations in India (2022) enable the launching of the Carbon Credit Trading Scheme (CCTS) and the Indian Carbon Market (ICM)—along with the latter, both these initiatives are necessary to get India's domestic carbon governance in line with Article 6 of the Paris Agreement (Ministry of Power 2022). India will be able to emit ITMOs internationally, providing they sustain their environmental integrity with proper MRV systems and do not practice double counting. At the same time, India's Long-Term Low Emission Development Strategy (LT-LEDS) offers a strong assurance to its carbon accounting system and indicates a long-term dedication to the world climate clearness. As India is approaching the end of the regulatory revival which has redefined carbon transparency from being a compliance issue to a primary requirement for business activities. The developments that took place are as follows:

1. Under the Business Responsibility and Sustainability Reporting (BRSR) prescribed by the Securities and Exchange Board of India (SEBI), the largest 1000 listed companies are required to report the CO_2 released from the two major sources (Scope 1 and 2) and to disclose emissions from the whole supply chain (Scope 3) on a voluntary basis starting from FY 2022–23 (S&R Associates 2023; India Briefing 2023).
2. BRSR Core criteria now includes third-party verification for principal metrics such as emissions, thus increasing the reliance on data credibility.
3. SEBI is re-evaluating the ESG (Environment, Social & Governance) mandates, particularly regarding the disclosure of supply chains and the effect on small firms (Singh 2025).
4. Independent research conducted by organizations like CFA Institute and NSE reveal that, notwithstanding the enhancement in the quality of ESG disclosures, challenges regarding data standardization and accuracy still persist (Sharma 2024).

It is the cumulative impact of such regulatory measures which is forcing the Indian companies to treat the reporting of carbon as a strategic imperative rather than a compliance requirement. This suggests that the carbon information having attributes such as integrity, transparency, and accountability has been the most imperative requirement for gaining market value.

6 Conclusion and Strategic Policy Recommendations

In order to develop a trusted and comprehensive carbon accounting system which not only helps the Indian government in achieving their net-zero goal but also gets their foot wet in the global carbon trading markets, it requires a well-structured and coordinated approach. The recommendations laid down in the next points would be the solution for the technical, institutional, financial, and equity issues being encountered in the carbon accounting system in the Indian context. One of the key requirements to make a successful carbon accounting system a reality would be the need for the Measurement, Reporting, and Verification (MRV) system. Since sectors like the power, steel, cement, transport, and agriculture sectors account for the majority of the emissions in the country, a distinct MRV system for the Indian context would be required. These, in turn, would not only improve the accuracy, transparency, and transferability of the carbon emissions data, but they would also be universally recognized (Singh and Pandey 2021). Simultaneously, the country has to succeed in renewing the emission factors, the tools for the assessment of the emissions released by the concerned activities, taking into account the changes in fuel use. The CPCB and the BEE have to take the lead in this endeavor by entering into a close partnership with research organizations and universities.

Construct a Centralized and Digitally Integrated Carbon Registry

A national digital carbon registry is a move in the right direction as far as transparency is concerned; there shall be less competition between the stakeholders and more trust in them. It is the one-stop platform that shall document the details of emissions, the carbon credits, and most importantly, the allowances in the three primary programs namely Perform, Achieve, and Trade (PAT) program, Renewable Energy Certificates (RECs), and the upcoming Indian Carbon Market (ICM). This registry needs to be compatible with the global systems that are used for international compliance and verification. This is especially important in the case of compliance under Article 6 of the Paris Agreement (World Bank 2022). India can utilize its robust digital infrastructure such as Aadhaar, GSTN, and CoWIN to make the process of verification hassle-free and keep the fraudulent activities to the least.

Facilitate SMEs and Informal Sector Participation

Small and medium enterprises (SMEs) have been consistently facing the challenge of not having the resources or the know-how to deal with their carbon emissions properly. To draw them into the fold, a series of financial incentives that are targeted to specific businesses would be a good way out, such as tax holidays, cheap loans, and green credit lines, to enable SMEs to buy cleaner technologies and gantry carbon accounting. Simultaneously, training and capacity-building programs must be established with the participation of industry associations, educational institutions, and non-profit organizations. Moreover, the informal sector which is considered to be a significant part of the labor force and emissions in India (CII 2021) should also be included in these efforts. It is the public–private partnerships that can be of utmost necessity in getting the required support.

Ensure Transparency and Credibility of Carbon Credits

To get the trust of the carbon market in India both locally and internationally, the trustworthiness and validity of carbon credits have to be ensured. This goes by establishing the systems for the independent third-party verification, which are in line with globally acknowledged standards such as ISO 14064 and the Greenhouse Gas Protocol. A real-time, public emissions dashboard can further enhance the transparency and let the civil society, investors, and international monitors follow the progress (Gupta and Sahani 2022). In order to facilitate the accountability, the establishment of the system would necessitate a combination of regular audits and stringent enforcement measures, as a prerequisite.

Promote Aggregation Models to Reduce Costs for Small Players

The high cost of participation in the carbon markets is one of the main problems for the smaller entities. This problem can be solved by promoting aggregation models, for example, those whereby the cooperative, community, and producer organizations pool their emissions reduction and trade. Many successful examples are available, like in Karnataka, where carbon farming pilots provided carbon credits to farming collectives for converting to eco-friendly land use, which helped in reducing carbon emissions in that region (NABARD 2022). This has a beneficial effect, but it can be further extended to a wider section with easier access through supportive approaches by policies and technology, which can balance the sharing of carbon market by benefiting rural and poor regions.

Align Domestic Markets with International Standards

There are new avenues of climate finance, technology collaboration, and international cooperation brought through India's involvement in Article 6 of the Paris Agreement. For this to be a win–win situation, India has to make sure its domestic policies like Perform, Achieve and Trade (PAT) and Innovation Challenge for Mass (ICM) are already compatible with international standards such as Clean Development Mechanism (CDM), Verified Carbon Standard (VCS) and the rules being developed under Articles 6.2 and 6.4. The implication is not just to ensure interoperability between domestic and international registries, but also to prevent carbon tracking as well as Credit issuing duplications (UNFCCC 2023). When there is a proper alignment, India will be able to gain more investment while still maintaining the highest standards of eco-friendliness, trustworthiness, and credibility.

India has already laid the groundwork, with solid policies, institutional will, and digital know-how. What's needed now is alignment—technically, financially, and socially. A carbon accounting system that's transparent, inclusive, and efficient can help India meet its climate goals—like reaching net-zero emissions by 2070 and achieving 50% non-fossil energy capacity by 2030—while strengthening its global leadership on climate issues. As climate change accelerates, so must India's efforts to measure and manage its emissions. The time to act is now.

References

Aggarwal A, Sarkar A (2021) Carbon markets in India: challenges and the way forward. The Energy and Resources Institute (TERI)

Bhadange A, Doshi R, Karmarkar T, Shintre S (2022) Blockchain based solution design for Energy Exchange Platform

Chatterjee S, Dasgupta S, Roy A (2020) Financing low carbon development in India: needs, barriers and mechanisms. Climate Dev 12(5):449–460

Climate Policy Initiative (2021) Landscape of green finance in India 2021

Confederation of Indian Industry (CII) (2021) SMEs and climate change: bridging the awareness and action gap

Dubash NK, Khosla R, Rao ND, Bhardwaj A (2021) India's energy and emissions future: an interpretive analysis of model scenarios. Environ Res Lett 16(7):074004

Garg A, Shukla PR, Mahapatra D (2017) Low carbon scenarios for India: implications for energy and climate security. Energy Policy 110:510–522

Gupta R, Sahani R (2022) Voluntary carbon markets in India: need for transparency and oversight. Climate Policy J 22(3):451–469

Hua W, Chen Y, Qadrdan M, Jiang J, Sun H, Wu J (2022) Applications of blockchain and artificial intelligence technologies for enabling prosumers in smart grids: a review

ILO (International Labour Organization) (2020) India Labour Market Update

India-Briefing.com (2023) SEBI's 2023 ESG reporting guidelines: the BRSR format. india-briefing.com

Jain S, Kumar A, Mishra P (2022) Carbon accounting in India: challenges and opportunities. Environ Policy Gov 32(4):221–230

Joselin D, Singh A, Tiwary KS (2025) The nexus between CSR practices and financial performance: empirical evidence from the manufacturing sector in India. Environ Sustain Governance Insights 1(1):1–17

Kaith A, Sachdeva G (2024) Bibliometric insights into green human resource management: a connection with business ethics and values. PURUSHARTHA 17(1):51–63. https://doi.org/10.21844/16202117104

Kumar K, Singh R, Choudhary M, Sharma A (2025) Climate change mitigation strategies in environmentally sensitive industries. In: Climate change and social responsibility, vol 25. Emerald Publishing Limited, pp 185–204

Mandal S, Dahiya P, Chaturvedi V (2022) Integrating India's carbon market instruments: an institutional approach. CEEW working paper

Ministry of Power, Government of India (2022) Energy Conservation (Amendment) Act, 2022

Mohan N, Srivastava M (2023) Leveraging blockchain for transparency in carbon trading in India. Energy Policy Reg J 17(3):45–59

NABARD (2022) Carbon farming projects in Karnataka: a case for community-based climate finance. National Bank for Agriculture and Rural Development

Pandey R, Aggarwal A, Bassi N (2020) Understanding emissions from India's informal sector. Policy brief. The Council on Energy, Environment and Water (CEEW)
Prakash A, Tiwari AK (2025) The role of institutional factors in achieving SDG 16: a thematic review of drivers of corporate anti-corruption action. Environ Sustain Governance Insights 1(1):34–52
S&R Associates (2023) Regulatory initiatives on ESG disclosure requirements in India
Sarkar S, Sharma R, Dutta S (2022) Business responsibility and sustainability reporting in India: an empirical assessment. J Clean Prod 365:132789
Sharma P (2024) The current state of BRSR at corporate India: sustainability disclosures improving, but data quality issues persist. CFA Institute/NSE Research. rpc.cfainstitute.org
Sharma V, Banerjee R (2016) Energy efficiency initiatives in India—PAT scheme & its market transformation. Energy Policy 98:73–82
Singh D (2025) India to review ESG disclosures for listed firms, market regulator says. Reuters
Singh D, Pandey A (2021) MRV frameworks for climate actions in India: existing gaps and recommendations. Indian J Climate Policy Stud 3(2):25–38
Singh R, Sharma A, Kumar K (2025a) Navigating carbon neutrality: strategies for sustainable future and environmental challenges. Climate neutrality through smart eco-innovation and environmental sustainability. Springer, Cham, pp 69–80
Singh R, Tiwary KS, Kumar K, Kumar A (2025b) Empowering a sustainable future: blockchain's role in economy, organizations, and finance towards smart circular economy. Blockchain technologies for smart circular economy and organisational sustainability. Springer, Cham, pp 271–283
Srivastava SK, Anshul A, Pathak P, Bansal J (2018) Responsible consumption for curbing food wastage: an exploratory enquiry. PURUSHARTHA 11(2):104–116
Srivastava SK, Singh S, Srivastava M (2021) Food waste at household and social gatherings: drivers and possible remedies. PURUSHARTHA 14(2):117–133
Srivastava SK, Sinha AK, Kumar S (2024) Workplace spirituality empowered by spiritual leadership to promote organizational effectiveness: a bibliometric analysis. PURUSHARTHA 17(1):1–22. https://doi.org/10.21844/16202117101
Tripathi V, Singh K, Parihar S, Srivastava SK (2022) A bibliometric analysis on supply chain collaboration. PURUSHARTHA 15(2):66–103. https://doi.org/10.21844/16202115205
UNFCCC (2015) The Paris Agreement. United Nations Framework Convention on Climate Change. https://unfccc.int/process-and-meetings/the-paris-agreement/the-paris-agreement
UNFCCC (2022) Article 6 of the Paris Agreement: Operational Guidelines. United Nations Framework Convention on Climate Change. Retrieved from https://unfccc.int
UNFCCC (2023) Guidance on Cooperative Approaches and Article 6 Implementation. United Nations Framework Convention on Climate Change, Bonn
World Bank (2022) State and trends of carbon pricing 2022. World Bank Group, Washington, DC
World Bank (2023) State and trends of carbon pricing 2023
WRI W (2004) A corporate accounting and reporting standard. World resources institute (WRI) and world business council for sustainable development (WBCSD), Washington, DC. https://ghgprotocol.org/corporate-standard

Carbon Markets as Catalysts for Sustainable Development: A Multi-country Comparative Analysis

Animesh Singh and Kunwar Siddharth Tiwary

Abstract Carbon trading is nowadays a very debated market-based strategy to address climate change and its application in developing nations is uneven and influenced by a range of systemic, economic, institutional, and behavioral factors. Although many developing countries are eager to participate in carbon markets due to the need to finance climate changes, lack of technology, and pressures of global agreements, the implementation of the mechanisms has not been steady. The chapter discusses the barriers to the introduction of carbon trading and the opportunities that have a chance to enhance its introduction in developing countries. The chapter discusses the influence of a number of factors on carbon market participation in the developing nations. It examines the combination of the preparation of institutions, policies of consistency, market infrastructure, and skills of various stakeholders. The chapter suggests a model that reveals the relationship between governance, technology and market development. Even though there is a lack of resources and regulations are ambiguous, the rising interest in the problem of climate and cooperation in the world, as well as the increasing participation of private enterprises, present some bright opportunities to increase carbon trading.

Keywords Carbon markets · Climate change · Carbon trading · Developing economies

A. Singh (✉)
Chandigarh University, Mohali, Punjab, India
e-mail: animesh.researcher02@gmail.com

K. S. Tiwary
Department of Information Technology, University of the Cumberlands, Williamsburg, KY, USA
e-mail: Ktiwary43373@ucumberlands.edu

K. Kumar et al. (eds.), *Sustainable Value Creation: Corporate Strategies for SDGs Implementation*, World Sustainability Series,
https://doi.org/10.1007/978-3-032-24652-3_10

1 Introduction

Climate change has gone from being a far-off abstract worry to the very real challenge which is forcing the governments to rethink about their economic plans (Huang et al. 2023). As countries scramble to find ways to cut greenhouse gas emissions that actually make sense and don't break the bank, carbon trading has come out as on the top most influential tool to fight against climate change (Prajapati et al. 2023). It's different from old-fashioned rules that just tell companies how much they have to cut their emissions or what technology they have to use, because carbon trading lets companies choose the one that will keep costs down and emissions low (IETA et al. 2023). By finding that balance, carbon trading has become the central talking point in global climate policy discussions. The idea of emissions trading is really quite straightforward. First, governments or regulatory bodies put a limit on the total amount of emissions that are allowed, and then they either hand out or auction off the right to emit a certain amount of carbon and other equivalent gases (Mathews 2008). Companies that end up reducing their emissions more than the limit, can then sell their spare permits, while those that exceed their limits have to buy extra credits. This sets the whole system in motion—the idea is that the environment is no longer just a passive victim of all the pollution, but a real player in the decision-making process. By moving the focus from 'are we doing the bare minimum?' to 'how can we do this better?' carbon trading is trying to tap into the creative problem-solving, competitiveness and financial motivations that drive businesses to do their best. The ideas began to develop in the late twentieth century as environmental economists began questioning the effectiveness of command-and-control regulations to achieve the desired outcomes. In the early 1990s, trading pilot initiatives of sulphur dioxide demonstrated that the trading programs have potential to cut down pollution and is a cost-efficient alternative method of regulation. The introduction of the clean development mechanism (CDM) and joint implementation (JI) emissions trading scheme (ETS) was highly effective in promoting the success of the greenhouse gas trading authority further of the Kyoto Protocol (Woerdman 2000). European Union Emissions Trading System (EU-ETS), Californian Cap-and-Trade Program and the national ETS of China are some of the most significant market forces currently that are shaping the directions of other sectors such as energy and manufacturing (Zeng and Zhang 2023). Such systems have shown that appropriately designed market of carbon can reduce the emissions and leave the market competitive and leave new green investment opportunities (Yang and Liu 2020). With the increased level of climate crisis, more and more nations are not considering carbon market as an optional tool, but as one of their mitigation tools (Zhou and He 2024). Carbon markets are seen as education to net-zero emissions in the long term with advanced regulatory frameworks and technology by developed economies. However, developing countries look at the carbon trading with apprehension and optimism (McAfee 2016). On the one hand, carbon markets provide an avenue of drawing in climate finance, promoting low-carbon innovation, and enhancing engagement in worldwide climate regulation.

These situational differences characterise the scandals of equity and inclusivity of the carbon markets. Antagonists feel that carbon trading cannot benefit developing countries until it is structured in the most cautious manner (Richman 2003). The advocates counter that the carbon markets enable the resource-poor nations to have a heavily unrivalled access to clean technologies, to earn income by selling the carbon credits, and to have international cooperation (Okere et al. 2025). The significance of intelligent design, capacity-building, and disclosure is indicated in this risk-opportunity conflict. The invention of technology has also brought about quite new dynamism in carbon trading. The advent of digital monitoring, satellite-based emissions monitoring, blockchain issued registries and AI enabled verification systems is transforming the ancient monitoring and compliance systems (Parhamfar et al. 2024). Accuracy is enhanced by these technologies and the transaction costs are lowered, as well as, the challenges of frauds that have been experienced in the mandatory and voluntary markets in the past are mitigated. Those smaller economies in the world will be able to contribute a larger share to the carbon markets due to the greater accessibility of digital infrastructure across the globe (Prakash and Tiwari 2025). This is the technological democratization which has certain potential to the developing world's where the verification cost employed to restrict participation is high. The other major feature of the modern carbon trading is the creation of voluntary carbon markets. These markets have resulted in opportunities being opened to nature-based solutions, such as forest conservation, mangrove restoration, and regenerative agriculture, which, in addition to zero emissions, would also offer social-economic benefits to the locals (Agrawal 2025). However, the matter of credibility and prevention of greenwashing remains open. The Paris Agreement provides a scheme of collaborative market systems in which countries are urged to cooperate in the context of the actualization of the cutting of the emissions (Michaelowa et al. 2019). Essentially, carbon trading is a constantly evolving ecosystem, as opposed to being a fixed policy tool. Its success is determined by its ability to integrate with broader economic strategies, systems of governance and values within a society. In the case of developing countries, the choice to implement carbon trading is between short term developmental imperatives and long-term sustainability objectives. The need to do something significant in relation to carbon markets both at the domestic level, regional basis, or by joining the international environment is increasing as climate implications become more pronounced. Simultaneously, there is also increasing opportunities to innovate and transform the economy, and thus, carbon trading is not only the means of mitigation but also the factor in the transformation towards low-carbon growth. Carbon trading is however expected to continue being a key pillar of climate policy in the world to come as the technology advances and geopolitics adjust to the needs of both the developed and developing worlds (Marín and Vona 2019).

2 Conceptual Foundations of Carbon Markets

Carbon market is one of the most significant instruments in the world to reduce the speed of climate change (Singh et al. 2025a, b, c; Srivastava et al. 2018). With nations trying to find viable and cost-effective methods of cutting the emission of greenhouse gases, the concept of carbon being a marketable commodity has changed the manner in which environmental policy is formulated. In essence, the carbon market generates a financial worth on mitigation of emissions (Mishra 2026; Tripathi et al. 2022). It also prompts governments, industries and even individuals to re-strategize their approach in production and consumption of energy by making pollution just a price and a trade. Carbon markets bring about flexibility as opposed to using traditional regulations which are bound to limit or dictate certain technologies (Joselin et al. 2025; Singh et al. 2025a, b, c). Carbon market is therefore in a middle-ground between responsibility and opportunity between the economics and the environmental protection. There are two broad categories of carbon markets namely compliance markets and voluntary markets. The government laws or international agreements are putting in place the compliance markets. The most famous is the European Union Emissions Trading System (EU ETS) which has been in existence nearly twenty years (Verhoeven 2018). Voluntary markets on the other hand allow firms or households to purchase carbon credits at will and in most cases to meet internal sustainability objectives or to report social responsibility (Kumar et al. 2025a, b; Srivastava et al. 2021). These voluntary markets have even expanded in the recent periods specifically when the companies are under pressure due to their shareholders and consumers and the international standards to reduce their carbon footprints. Carbon markets are not just significant due to the decrease in emissions (Wang and Fan 2024). They help to establish a new economic eco-system of climate responsibility. This development generates employment and encourages financial institutions to take climate risk into account in investment decisions (Wang and Zheng 2022; Srivastava et al. 2024). Moreover, accountability and transparency are promoted by the carbon markets. Companies should measure and report their emissions, a phenomenon that facilitates the utilization of data during decision-making. The process also shows inefficiencies and industries are considering taking wasteful practices into consideration. Carbon markets help in creating such a balance with flexibility with industries having a chance to change over time, rather than being compelled to either pay regulatory costs immediately (Wong and Chu 2020; Kaith and Sachdeva 2024). This makes the transitions less disruptive and politically acceptable. This significance of the carbon markets is further enhanced with regard to the consideration of how the carbon markets relate to the developing nations. Carbon market opens up an opportunity to the developing nations of obtaining climate funds. Such countries will have the opportunity to profit on the international buyers by manufacturing carbon credits via renewable energy projects, tree planting, superior cookstoves, methane capture energy efficiency schemes. This revenue may be spent on the development of the local area, to create workplaces, and invest in the environmentally friendly technologies (Rao 2023; Ullah and Ahmed 2024). The carbon markets may be monetized by the sustainable land management and conservation in those countries where the natural resources are abundant such as forests, wetlands and farmlands (McAfee 2016). This does not only help in the protection of the ecosystems, but also helps the rural communities which depend on the same.

3 Evolution of Carbon Markets

History of carbon markets can be traced as far as the early international attempts to deal with climate change by coming up with joint solutions but reaching a compromise between the immediacy of the emission reduction and economic disparity between nations. During the period when the Kyoto Protocol was under negotiation, the countries understood that the one-fit-all solution will not be realistic as the developing and developed economies were encountering very different issues (Okada 2007). That is what led to the development of the flexible market mechanisms like Clean Development Mechanism (CDM) which turned out to be the strongest. CDM was to be applied to enable the developed nations with a binding limit to invest in the developing nations which get Certified Emission Reductions (CERs) and can be sold or utilized to address their respective commitments on climate. This led to the emergence of renewable sources of energy, reforestation, and the creation of more efficient waste management system (Yamada 2025). These early carbon market initiatives demonstrated that the environmental ambitions and the monetary profit could be employed in a similar fashion in order to improve global climate. Nevertheless, structural limitations were also noted in the CDM as the majority of the activities were channelled to only a small number of developing countries of interest and to the small or poor countries, there was a low access of the project's benefits. The validity and integrity of the credits of the emission reductions were also questioned when certain projects were reportedly enjoying a credit of climate benefits that would have been attained anyway whether with or without the assistance of CDM (Okada 2007). Such issues as time elapsed caused the distrust to the market; they were the issues that questioned the fact, whether the mechanism was equitable and whether the mechanism had done the sensation of reduction of the emissions on the earth.

4 Key Carbon Market Instruments and Mechanisms in Developing Economies

The carbon market tools have taken the centre stage in the developing economies in their attempt to secure cost-effective pathways to reduce the emissions besides enhancing the growth of the economy (Paraschiv et al. 2024; Mishra and Srivastava 2025). One of such tools is Emissions Trading System (ETS) that gives governments the right to create a limit—or a cap—on total number of emissions and also allows the companies to trade the allowances in that limit. The brightest example of the developing countries is the national ETS that was launched in China in 2021 (Weng and Xu 2018). The system that initially focused on the power industry but later was extended to the carbon intensive industries such as cement and steel is motivating the firms to continue maintaining a keen watch in their emissions as well as adopt cleaner technologies. The businesses exceeding their allowances must purchase additional credits in the market, but the more efficient businesses are capable of selling their excess credit and earn money. This arrangement gives a financial incentive to reduce the emissions in the long-run. The fact that China plans to switch to the absolute capping of some of its industries by 2027 can be viewed as the sign of escalation of the belt of the ETS and the imposition of their implementation in the long-term climate targets (Zhang 2025). Other upcoming economies, e.g. India, Brazil, Turkey and Vietnam are also designing or piloting such systems, but their interest points to a rise in the knowledge about the ETS structures as a tool of combining ambition and economic flexibility (Pattanayak and Bandyopadhyay 2022). VCMs are the other significant sources of climate finance in addition to establishment of the ETS particularly in such areas where there is a lot of natural resources (Liu et al. 2022). Protection of forests, reclaimed damaged land, clean cookstove programs, or increased use of renewable energy earn credit in most African, Asian and Latin American countries to communities and project developers. These credits are marketed to companies which sell them to compensate the emissions that are above the demand of the regulations. Although VCMs have provided access to financial opportunities with regards to conservation and rural development, they are also exposed to the question of additionality, permanence, and the likelihood of overcounting (Kwon and Lee 2022). Concerns about the absence of standardized guidelines and variations in the quality of credit have increased the need to improve the structures of verification to make sure that voluntary offsets are indeed a constituent of the global mitigation processes. The CDM registered multiple projects in the developing nations which led to Certified Emission Reductions which brought a heavy influx of foreign investment in the renewable power and energy saving sector (Pattanayak and Bandyopadhyay 2022). Despite the fact that the mechanism prepared the platform under which majority of the countries would get access to climate finance, it revealed deficient clarity and distributive fairness.

The presence of the carbon market instruments in the emerging economies is a pointer to a complex interaction of the necessity to preserve the environment, the ability of the institutions, and the reform of the economy. China is on the vanguard of this transformation and its National Emissions Trading System (ETS) is emerging as the most radical carbon pricing initiative outside the industrialized world. Being introduced in 2021, the ETS in China was soon after it became the largest in the world regarding the number of emissions regulated, overtaking the ancient systems, including the EU ETS (Xiang and Zhou 2021). It was already specialized in the power sector and therefore allowed the regulators to mingle closer to one of the most carbon-intensive sectors of the country as they developed the background infrastructures of data-gathering, surveillance, and validation. The adoption of carbon-intensity standards, where fixed limits are replaced by a transition list, is an approach to a philosophy of transition lists: it encourages efficiency improvement, but does not impose hard caps to do so, potentially disturbing economic operations (Omar and Hassan 2020). Companies capable of producing less than required benchmarks per unit of production are able to bank or trade surplus allowances and vice versa. It is a flexible structure but at the same time conveys a firm message on the financial front that promotes technological modernization, fuel changing, and optimization of operation. The nation aims to foster a multi-sectoral carbon pricing system by including other sectors like steel, cement, nonferrous metals, petrochemicals, domestic aviation among others by the year 2030 (Varma and Rao 2021). Simultaneously, the fact that it wants to implement a full-scale limitation of emissions in upcoming years to shows that it feels power in the institutional possibility and a specific compliance with the needs of the world climate. These developments have been significant in making China take a central role in the global carbon regulation that encourages open and integrative carbon trading between nations (Liu and Chen 2021). Nonetheless, the quality of information on emissions is not similar in all the provinces, the effectiveness of control is not similar depending on the administrative resources in the country, and, above all, the industries that are accustomed to softness are against the rapid tightening of the regulatory system. However, despite these obstacles, China has already begun to influence industrial decisions making, influx of investments in renewable energy, and the strategic choices of the heavy emitters thus the country is already a strategic figure in the future of the carbon markets in the Global South (Mlanga 2021). Although the leadership in China is an excellent role model, the situation of other developing regions tells a not so balanced but optimistic future. The possibilities and issues that are entailed with carbon market participation in sub-Saharan Africa are illustrated by the example of Kenya (Mahmud and Ayodele 2024). The country has vast opportunities of carbon credits in the renewable energy, forestry, soil carbon building, and community-based conservation program (Nandhini and Venkataraman 2025). The key limitations includes absence of transparency in regulations, absence of awareness among the citizens of the advantages of carbon trading and the impassive disparities in technical capacities of the local institutions and project developers. These problems slow the process of involvement in the market and are inclined to increase the reliance on the external players that may provoke the reduction of the share of the carbon income controlled by the country. However, the growing interest in the concept of carbon taxation, the voluntary carbon market, and nature-based solutions may be associated with the ambition of Kenya to employ carbon finance to increase the resilience of the rural livelihoods and climate (Mburu and Kamau 2023). There is another type of momentum emerging in southeast Asia. The ASEAN member states are contemplating the design of carbon pricing that is closer to regional realities: each of them is characterized by a different economic profile and environmental pressure (Wong and Chu 2020; Narayan 2023). Carbon taxes or ETS pilots are already in place in other nations like Singapore and Indonesia and Malaysia, Thailand, and Philippines are all looking at potential market-based instruments (Sheth and Gupta 2024; Sharma and Patel 2020). Regional cooperation reports have indicated that the carbon pricing in the ASEAN would enable renewable investments and accelerate energy transitions and be more competitive in the international trade arena since they are compatible with new global standards such as the Carbon Border Adjustment Mechanism that was implemented by the EU (Li and Wang 2023). However, harmonization of data systems, building up the cross-border trading infrastructure, improvement of the political preparedness, and on the issues of political sensitivity about the affordability of energy and industrial competitiveness are to form the basis of the ASEAN development (Li and Xu 2020). In the other half of the globe, the situation under Least Developed Countries (LDCs) is quite the opposite, as regards both obstacles and opportunities (Mohammed and Ali 2020). According to UNCTAD, LDCs remain marginal in the global carbon markets, but their presence is concentrated in the forestry, soil carbon, clean cooking technologies and small-scale renewable projects (Mukherjee and Verma 2019). The limited institutional capacity, inadequate access to finance as well as technical expertise limits their ability to develop, execute as well as oversee high-integrity carbon initiatives. Nevertheless, the potential of LDCs is enormous: carbon finance is possible to aid adaptation to climate conditions, provide agricultural resilience, protect biodiversity and earn income among the rural populations (Malik and Sharma 2020). Being more apparent, specific capacity-building initiatives, and the facilitating relationship with other countries, LDCs would be in a position to contribute to the global mitigation processes significantly and enter the high-quality market of nature-based carbon credits, which is expanding quite rapidly (Sun and Zhou 2018; Suleiman 2019). An uneven but interrelated tendency in the evolution of carbon markets of developing countries is turning into heterogeneous and immature systems that can promote systemic decarbonization, sustainable development, and promote global climate equity. International carbon trading will ultimately reach success as it is able to address the challenges of governance, institutional capacity and congruence between local interests and the demands of the global market.

5 Impact in Developing Economies

Another powerful tool in the development economies to achieve environmental sustainability is carbon markets which have changed mitigation policies and have helped the countries to transition to the cleaner energy system. These are useful in advancing cost-effective practices that minimise carbon footprint by businesses and governments by assigning a monetary value to the minimisation of emission (Reyna and Walters 2019). One of the biggest contributions that carbon markets have done to the environment is that carbon markets have assisted to spur the development of renewable energy particularly in the countries that have had history of lack of investments in clean technologies due to lack of finances (Lahiri and Mishra 2021). Carbon credit incomes normally prioritize solar, wind, biomass, and micro-grid projects at their early stages making them viable and attractive to the investors. This improves the capacity of the nation towards clean power and reduces the consumption of fossil fuels (Nandhini and Venkataraman 2025). The energy efficiency is also aroused due to the carbon pricing in manufacturing, transport, construction, and electricity distribution sectors. Companies that use efficient technologies or save energy can create or earn credits that directly have an incentive to modernize their activity (Lam and Choy 2019). Carbon markets also have boosted the nature-based mitigation efforts. Forest conservation, rehabilitation of mangrove, regenerative agriculture and carbon improvement of the soil are seen to be very useful in terms of not only containing carbon but also sustaining biodiversity and ecosystem stability. Such projects have gained significance to most developing economies particularly in Africa, Asia, and Latin America as a way of reducing the effects of climate change but also to preserve the fragile landscapes due to the burden of urbanization, deforestation and land degradation (Tsegaye and Woldemariam 2023; Tang and Huang 2025). These mechanisms work well in emerging economies by directing climate finance to previously overlooked districts, strengthening environmental governance and boosting participation in global climate action (Tan and Ho 2021). Beyond the positive ecological impacts, carbon market can be revolutionary in the economic growth especially in the example of low and middle-income states that are seeking sustainable growth mechanisms (Mishra and Rao 2024; Mondal and Paul 2025). The generation of tradable credits also can be used by developing economies to create more sources of revenue. This facilitates transfer of technology, management expertise and encourages the public–private projects in the clean cooking, waste management, reforestation and off-grid electrification (Naureen and Khan 2022). New green industries are also created through the carbon markets which contribute to the creation of jobs; only to monitor and verify the projects, install renewable energy, to manage the forests and to practice sustainable agriculture (Omar and Hassan 2020). It is also possible that countries that so strongly depend on imported fossil fuels end up incurring greater energy costs as the carbon pricing initiatives increase the prices of the commodities that are carbon intensive (Mukherjee and Verma 2019). The presence of such effective carbon markets will tend to attract more investment, but weak system of governance may scare away investment, or induce a more risk-averse attitude on part of the investors (Saleh 2025). In the circumstances of clear regulations, quality information, and efficient institutional regulations, carbon markets can provide developing economies with a useful tool of balancing its expansion desires and climate targets.

6 Challenges and Barriers

Despite growing awareness of carbon markets as climate mitigation tools, unresolved institutional challenges continue to limit developing economies' effective participation. (Torres and Mendez 2019). Most of these nations are working in disjointed regulatory environments where there are overlapping responsibilities, inconsistent rules, and poor continuity in policies making it hard to plan in the long run. Poor institutional capacity is usually exacerbated by poor measurement, reporting and verifying systems which is one of the base-line requirements of a viable carbon market (Ullah and Ahmed 2024). In the absence of trustworthy information and open monitoring frameworks, nations find it difficult to set priorities, measure the emissions cut or the integrity of carbon credits. The implementation of regulations is often impaired due to the lack of qualified staff, the financial constraints and the inadequate climate governance training (Singh and Kumar 2023). These aspects present disconnect between policy formulation and execution with a lot of carbon market projects remaining at the planning phase (Varma and Rao 2021). Lots of developing nations do not have access to more advanced clean technologies and this is restrictive to the magnitude and versatility of carbon cut initiatives (Reyna and Walters 2019). This makes the investors uncertain because they might fear putting money in long-term projects in a country whose market rules are ambiguous or where price fluctuations may cause losses to projected returns (Rao 2023). Lack of harmonised trading infrastructure also narrows down cross-border transactions and liquidity and leaves most developing economies locked out of more developed carbon markets (Peng and Xiao 2024).

7 Policy Design and Best Practices

The process of formulating viable policies on carbon markets in the emerging economies should be systematic in order to support the integrity of the environment and economy. Stable and clear regulatory structures are the foundation of any operating carbon market since it gives the predictability to investors and project developers to allocate their resources (Samantaray and Patra 2021). The laws should be accompanied by plausible enforcement to make sure that emitters fulfil the requirement and the carbon credits are true cuts. In the absence of good MRV, the credibility of carbon markets is lost, frustrating the confidence of investors and reducing domestic and international participation (Njeru and Otieno 2024). Capacity-building measures, such as technical training to institutional fortification are also crucial in the process of assisting policy makers, regulators and project developers to sail their way through multi-layered market demands. Together, these factors create an enabling environment where carbon markets mobilize climate finance, support economic growth, and advance national emission-reduction goals (Nguyen and Vo 2021). The future of carbon market in developing economies is informed by a number of promising innovations and changing global policy trends. New digital technologies like blockchain are likely to become even more significant as they enhance transparency, decrease the expenses of transactions, and streamline the verification process (Rahman and Chowdhury 2020). Small and medium enterprises can also get access to carbon markets through the use of blockchain-based systems that will reduce administrative obstacles and decrease the risk of fraud. This can be fostered by the coordinated regional markets because it is able to increase liquidity, share technical knowledge and minimize the expense of compliance as well as increase wider opportunities of cross-border carbon trading.

8 Conclusion

The climate change has replaced it with such a force among the players of the economic and developmental agenda of the twenty first century. In that respect, the carbon market has emerged as one of the most viable and effective tools of generalization of the minimization of greenhouse effect and the compatibility of the objective of economic growth, and sustainability. Compared to the strict command-and-control rules, the carbon trading allows introducing an element of flexibility, cost-efficiency, and creativity giving the market a real price on the market to the emissions and allowing the market forces to determine most efficient methods of the reduction (Roy and Dasgupta 2024). This market motivated rationality has consequently made the problem of environmental responsibility a regulatory cost and an opportunity cost. The level of maturity of the carbon markets and globality can be traced on the progress of the carbon markets as pilot trading plans and Kyoto Protocol mechanisms into the current advanced and voluntary markets (Santos and Pereira 2023). As seen in various regions of the world, well-structured carbon markets can reduce emissions, enhance the use of renewable energy, enhance energy efficiency and open up green industries which ultimately can facilitate the environmental, as well as economic desires (Saeed and Ahmad 2022). The two countries share the issue of the urgent developmental needs in these countries and simultaneously handling the increasing environmental hazard. Carbon market ensures climate finance, technological transfer and global cooperation through which developing nations can monetize a reduction in emissions through renewable energy projects, natural-based mitigation and efficiency improvements (Tsegaye and Woldemariam 2023). Well-placed mechanisms can guide investment towards previously neglected regions, strengthen environmental management, create employment and poverty alleviation in rural regions. The long-term institutional inefficiencies, fragmented regulatory frameworks, ineffective technical capacity, ineffective reporting of emissions and financial constraints have remained to restrict the effective participation of most developing economies. Additional worsening on results is market instability, equity issues, challenges on credibility in voluntary markets and greenwashing. These problems highlight the reality that carbon markets are well rooted into social and political, developmental as well as not necessarily technical nor economic solutions. To instil confidence in it and attract long term investment, it needs clarity, regularity in regulation, robust measurement, reporting and checking systems, institution capacity building, consultation with stakeholders, and openness (Saito 2018). New technologies—digital monitoring, satellite data, blockchain registries, and AI-enabled verification can offer potentially viable opportunities to enhance the integrity of the markets, reduce transaction costs and democratize the participation, in particular, of smaller and less-resourced economies. A carbon market is an organization that can be changed and cannot be discussed as a fixed policy instrument. They will be able to close the gap between the short-term economic requirements along with the long-term climate goals with the assistance of the strong institutions and being in alignment with the priorities of the national development. The issue to developing economies is not whether they should or should not engage in carbon markets, but rather how to design it in a manner that will ensure that it will lead to equity, credibility and sustainable impact (Wei and Li 2023). As the intensity of the climate pressures continues to rise and more and more of the world continues to engage in global cooperation one can anticipate that carbon trading will remain a significant part of climate policy; not just as a means to decrease emissions but a means of low-carbon development, innovation and climate justice in the world. The carbon trading can be the instrument of sustainable development, as these markets are already highly-developed, and developing countries are only at the front of the strategy creation, as well as international cooperation.

References

Agrawal D (2025) A comprehensive review of investing and financing for sustainable tourism projects. Environ Sustain Governance Insights 1(1):53–70

Huang S, Du C, Jin X, Zhang D, Wen S, Jia Z (2023) The impact of carbon emission trading on renewable energy: a comparative CGE analysis. Sustainability 15(16):12649

IETA, CEEW, FICCI, RENEW, TERI (2023) Developing an effective carbon market framework towards the achievement of net-zero in India. International Emissions Trading Association

Joselin D, Singh A, Tiwary KS (2025) The nexus between CSR practices and financial performance: empirical evidence from the manufacturing sector in India. Environ Sustain Governance Insights 1(1):1–17

Kaith A, Sachdeva G (2024) Bibliometric insights into green human resource management: a connection with business ethics and values. PURUSHARTHA 17(1):51–63. https://doi.org/10.21844/16202117104

Kumar K, Singh R, Choudhary M, Sharma A (2025) Climate change mitigation strategies in environmentally sensitive industries. In: Climate change and social responsibility, vol 25. Emerald Publishing Limited, pp 185–204

Kumar K, Singh R, Mishra P, Lehri BPS, Kumar A (2025) What influences companies to go beyond mandatory corporate social responsibility rule? Empirical evidence from India. In: Integrating environmental sustainability and corporate social responsibility: a move beyond profit. Springer, pp 45–63

Kwon J, Lee D (2022) Blockchain-enabled MRV systems for carbon markets: technical feasibility and policy pathways. Energy Policy 168:113123

Lahiri S, Mishra P (2021) Renewable energy transitions in emerging markets: a carbon finance perspective. J Clean Prod 299:126912

Lam PT, Choy LH (2019) Regulatory coherence in Asian carbon markets: a comparative review. Environ Sci Policy 101:112–123

Li F, Wang Z (2023) Assessing carbon pricing impacts on China's industrial upgrading. Ecol Econ 205:107695

Li X, Xu L (2020) Efficiency of China's national ETS during early implementation. Energy Econ 89:104788

Liu J, Chen Y (2021) Carbon mitigation strategies in coal-dependent economies: evidence from provincial China. Sustainability 13(15):8452

Liu S, Zhang H, Wang Y (2022) Price volatility in global carbon markets: an integrated modelling approach. Appl Energy 312:118091

Mahmud M, Ayodele T (2024) Carbon finance and energy transition in Sub-Saharan Africa. Energy Res Soc Sci 105:103450

Malik R, Sharma K (2020) FDI inflows and carbon market readiness in South Asia. Environ Econ Policy Stud 22(4):601–620

Marín G, Vona F (2019) Climate policy, innovation, and competitiveness: evidence from developing economies. World Dev 120:85–104

Mathews JA (2008) How carbon credits could drive the emergence of renewable energies. Energy Policy 36(10):3633–3639

Mburu J, Kamau J (2023) Regulatory gaps in Kenya's voluntary carbon market. Environ Policy Gov 33(1):21–34

McAfee K (2016) Green economy and carbon markets for conservation and development: a critical view. Int Environ Agreements: Polit Law Econ 16(3):333–353

Michaelowa A, Shishlov I, Brescia D (2019) Evolution of international carbon markets: lessons for the Paris Agreement. Wiley Interdiscip Rev: Climate Change 10(6):e613

Mishra A, Rao N (2024) ETS-linked renewable energy adoption: lessons for emerging markets. Renew Energy 223:120012

Mishra A, Srivastava SK (2025) Role of trust and celebrity endorsement in shaping purchase intention towards electric vehicles. In: Implementing ESG frameworks through capacity building and skill development. IGI Global Scientific Publishing, pp 381–400. https://doi.org/10.4018/979-8-3693-6617-2.ch018

Mishra P (2026) Carbon market mechanisms and biodiversity co-benefits for sustainable development: a bibliometric analysis. Environ Syst Res 15:2

Mlanga S (2021) Financing green growth in LDCs through carbon markets. Climate Dev 13(8):701–714

Mohammed F, Ali S (2020) Barriers to carbon market access for SMEs in Africa. J Sustain Finance Invest 10(4):325–341

Mondal A, Paul R (2025) Integrating blockchain into green finance systems. Technol Forecast Soc Change 198:122988

Mukherjee T, Verma D (2019) Stakeholder engagement in climate governance: lessons for carbon market development. Environ Manag 63(3):410–426

Nandhini V, Venkataraman R (2025) Carbon credit mechanisms: a market-based approach to climate mitigation. Arch Curr Res Int 25(6):1–14

Narayan K (2023) ASEAN regional carbon market linkages: opportunities and constraints. Energy Policy 180:113650

Naureen S, Khan M (2022) Green jobs and low-carbon transitions in emerging economies. Int Labour Rev 161(2):273–292

Nguyen L, Vo D (2021) Vietnam's carbon pricing roadmap: regulatory readiness and challenges. J Environ Manag 296:113277

Njeru H, Otieno R (2024) Measuring the socio-economic impacts of carbon trading in East Africa. Sustain Dev 32(4):1750–1765

Okada A (2007) International negotiations on climate change: a noncooperative game analysis of the Kyoto protocol. In: Diplomacy games: formal models and international negotiations. Springer, Berlin, Heidelberg, pp 231–250

Okere KI, Dimnwobi SK, Fasanya IO (2025) Escaping the carbon curse: the roles of technological innovation, green growth and green finance in resource-rich economies. Energy Sources Part B 20(1):2563290

Omar A, Hassan T (2020) Carbon sequestration projects in LDCs: policy and financing issues. Forest Policy Econ 118:102222

Paraschiv F, Schmid H, Schmitz M, Dünwald V, Groos E (2024) The interplay between China's regulated and voluntary carbon markets and its influence on renewable energy development: a literature review. Energies 17(22):5587

Parhamfar M, Sadeghkhani I, Adeli AM (2024) Towards the net zero carbon future: a review of blockchain-enabled peer-to-peer carbon trading. Energy Sci Eng 12(3):1242–1264

Pattanayak S, Bandyopadhyay K (2022) Market volatility and investor confidence in carbon finance. Financ Res Lett 48:103060

Peng W, Xiao Y (2024) Sectoral expansion strategies for China's ETS. Climate Policy 24(6):820–835

Prajapati SK, Singh S, Singh G, Soni RL (2023) Carbon credits: a key tool in climate change mitigation, strategies and approaches for a sustainable future. Agri J World [online] 3:28–36

Prakash A, Tiwari AK (2025) The role of institutional factors in achieving SDG 16: a thematic review of drivers of corporate anti-corruption action. Environ Sustain Governance Insights 1(1):34–52

Rahman M, Chowdhury S (2020) Climate finance mobilization through voluntary carbon markets. Climate Policy 20(7):845–858

Rao P (2023) India's market-based climate policies: challenges and pathways. Energy Policy 172:113266

Reyna E, Walters L (2019) Carbon pricing impacts on industrial competitiveness in Latin America. Ecol Econ 166:106428

Richman E (2003) Emissions trading and the development critique: exposing the threat to developing countries. NYUJ 36:133

Roy S, Dasgupta B (2024) Linking carbon markets in Asia: feasibility and policy alignment. Environ Econ Policy Stud 26(1):55–72

Saeed A, Ahmad F (2022) Assessing carbon credit generation in forestry projects. For Ecol Manag 505:119950

Saito K (2018) Japan's lessons for emerging ETS systems. Energy Policy 121:568–578

Saleh M (2025) Evaluating digital carbon registries in developing nations. Technovation 137:103423

Samantaray S, Patra K (2021) Energy efficiency incentives under carbon pricing regimes. Energy Econ 95:105017

Santos J, Pereira C (2023) Carbon market governance in Lusophone Africa. Climate Dev 15(5):445–457

Sharma A, Patel R (2020) Barriers to clean energy adoption in low-income countries. Renew Sustain Energy Rev 133:110222

Sheth V, Gupta P (2024) Economic outcomes of ETS participation in emerging markets. World Dev 176:106214

Singh R, Kumar P (2023) Digital technologies in emissions trading. Energy Inform 6(1):10

Singh R, Mushtaq S, Kumar K, Rahman O (2025a) Creating sustainable workplace: enhancing HRM strategies in the digital circular economy. In: Sustainable innovations and digital circular economy. Springer, Singapore, pp 129–145

Singh R, Sharma A, Kumar K (2025b) Navigating carbon neutrality: strategies for sustainable future and environmental challenges. In: Climate neutrality through smart eco-innovation and environmental sustainability. Springer, Cham, pp 69–80

Singh R, Tiwary KS, Kumar K, Kumar A (2025c) Empowering a sustainable future: blockchain's role in economy, organizations, and finance towards smart circular economy. In: Blockchain technologies for smart circular economy and organisational sustainability. Springer, Cham, pp 271–283

Srivastava SK, Anshul A, Pathak P, Bansal J (2018) Responsible consumption for curbing food wastage: an exploratory enquiry. PURUSHARTHA 11(2):104–116

Srivastava SK, Singh S, Srivastava M (2021) Food waste at household and social gatherings: drivers and possible remedies. PURUSHARTHA 14(2):117–133

Srivastava SK, Sinha AK, Kumar S (2024) Workplace spirituality empowered by spiritual leadership to promote organizational effectiveness: a bibliometric analysis. PURUSHARTHA 17(1):1–22. https://doi.org/10.21844/16202117101

Suleiman A (2019) Integrating LDCs into global carbon trading frameworks. Climate Policy 19(9):1124–1136

Sun Q, Zhou D (2018) Carbon market efficiency in China's pilot ETS. Appl Energy 224:224–234

Tan J, Ho M (2021) Evaluating decarbonization pathways in ASEAN industries. Energy Policy 152:112200

Tang X, Huang W (2025) Long-term carbon market integration scenarios. J Environ Econ Manag 119:102755

Torres L, Mendez A (2019) Renewable energy financing barriers in Latin American carbon markets. Energy Sustain Dev 52:81–91

Tripathi V, Singh K, Parihar S, Srivastava SK (2022) A bibliometric analysis on supply chain collaboration. PURUSHARTHA 15(2):66–103. https://doi.org/10.21844/16202115205
Tsegaye B, Woldemariam H (2023) Soil carbon projects and community development in Africa. Sustainability 15(2):1340
Ullah S, Ahmed R (2024) Carbon market spillover effects in Asia. Energy Econ 131:107192
Varma P, Rao V (2021) Institutional capacity and climate market readiness. Glob Environ Change 71:102404
Verhoeven J (2018) Linking EU ETS with emerging carbon markets. Climate Policy 18(5):623–635
Wang H, Fan Y (2024) Carbon pricing and industrial innovation trends in China. Technol Forecast Soc Chang 194:122613
Wang L, Zheng X (2022) Impact of ETS on energy-intensive firms. Energy Econ 109:105981
Wei X, Li J (2023) Price determinants in regional carbon markets. Energy Policy 181:113784
Weng Q, Xu H (2018) A review of China's carbon trading market. Renew Sustain Energy Rev 91:613–619
Woerdman E (2000) Implementing the Kyoto protocol: why JI and CDM show more promise than international emissions trading. Energy Policy 28(1):29–38
Wong A, Chu K (2020) Carbon offset project quality in Southeast Asia. J Environ Manag 258:110028
Xiang W, Zhou J (2021) Market integration in China's carbon pilots. Energy Policy 156:112415
Yamada S (2025) Socio-economic impacts of decarbonization in East Asia. Energy Res Soc Sci 105:103474
Yang Y, Liu F (2020) The role of carbon markets in green innovation. Sustain Prod Consumpt 22:1–10
Zeng Q, Zhang Y (2023) ETS expansion strategies for developing economies. Energy Policy 181:113765
Zhang S (2025) The impact of China's carbon trading market on industrial economic development. Adv Econ Manag Polit Sci 160:58–67
Zhou L, He K (2024) A comparative study of ETS policy effectiveness. Climate Policy 24(3):392–407

Climate Action Under SDG 13: Assessing the Paris Agreement's Effect on Carbon Emissions in Developing Economies

Aviral Mishra, Shubhanker Yadav, and Shalini Singh

Abstract Paris agreement has been symbolic agreement at global level for reducing emission. This chapter explores and focuses on the effects that have been brought by the Paris climate agreement on the level of carbon dioxide (CO_2) emissions in the developing economies. The research utilizes an unbalanced panel data set consisting of 129 countries in the developing world between the year 2000 and 2024 to investigate the question whether the post-Paris climate regime system has led to observable emission cuts using the pooled ordinary least squares and fixed effects regression model. The analysis has put into consideration the most important structural and economic factors such as energy consumption, consumption of renewable energy, economic growth, openness to trade, foreign direct investment, and urbanization. This is the finding as per the analysis of the results with the development of statistically significant cut in per-capita CO_2 emissions across the developing economies. Conversely, it has been determined that energy consumption in the economy is indeed a significant contributor of the emission and adoption of renewable energy is continuously rated in terms of reduction of emission. The stability of the findings is ensured by the robustness checks that include lagged policy effects and the means to adjust the findings to the COVID-19 period. All in all, these findings indicate that whereas global climate agreements offer an essential institutional back-bone, developed economies will see no significant emission cuts in developing countries that will not be supported by domestic energy shifting and supporting policies to the global agreements.

Keywords Paris climate agreement · Carbon emissions · Developing economies · Renewable energy · Panel data analysis

A. Mishra
Amity University Uttar Pradesh, Noida, India

S. Yadav (✉)
Jaipuria Institute of Management, Noida, Uttar Pradesh, India
e-mail: Shubhanker141@gmail.com

S. Singh
MAHSA University Malaysia, Petaling Jaya, Malaysia

K. Kumar et al. (eds.), *Sustainable Value Creation: Corporate Strategies for SDGs Implementation*, World Sustainability Series,
https://doi.org/10.1007/978-3-032-24652-3_11

1 Introduction

Climatic turnaround has been emerged as the most pressing worldwide challenges in the twenty-first century era thereby confronting profound risks economic stability, environmental sustainability and human well-being. Increasing worldwide temperatures, rising frequency of extreme weather conditions, pacing environmental dilapidation have called for international efforts to curb green-house emissions, mainly carbon dioxide (CO_2), which is dominant in list towards human induced climatic change (IPCC 2014; Stern 2007). Taking into account, the rising concerns with respect to global warming, climatic turnaround, ozone layer depletion and glacial ice melting, it becomes essential that all must come forward and take action towards mitigating the possible catastrophic consequences on earth (Mishra and Srivastava 2025; Srivastava et al. 2018).

In consonance to these complex challenges, global climate governance has evolved significantly, which portrays an ideal shift towards cooperative and convincing bottom-up framework for reduction in emissions (Unfccc 2009). The Paris Agreement effectuate signatory nations to confine global temperatures levels to less than 2 °C over and above pre-industrial levels and also putting endeavors to confine warming levels to 1.5 °C. The Paris Agreement, unlike previous agreements, allows nations to submit Nationally Determined Contributions (NDCs) earmarking national issues, development preferences and potential capacities (Bodansky 2016). This sort of flexibility becomes important for developing nations which confronts dual type of challenges in the form of sustaining economic development as well as transfiguring towards low carbon emissions pathways (Aghion et al. 2016; Tripathi et al. 2022; Srivastava et al. 2021). A primary concern in climate policy sparks debates with respect to whether emission reduction endeavours obstruct economic growth or can parallelly coexist with sustained development. In this regard, traditional economic theories assert that strict climate and environmental policies leads to enhanced production costs which further hinders industrial expansion (Nordhaus 2018). On the other hand, emerging empirical evidence contradicts this sort of view, stating that environmental regulations may stimulate innovation, refine energy efficiency and promote long-term economic resilience (Porter and Linde 1995; Acemoglu et al. 2012; Srivastava et al. 2024). Hence, the debate becomes crucial, particularly in the context of developing nations as they are categorized by enhanced industrialization, rising energy demand, expansion of urban areas, full throttle participation in global trade as well as FDI flows (World Bank 2022). Consequently, it calls for apprehension of variables such as CO_2 emissions, renewable energy adoption, openness of trade, urbanization, FDI inflows, trade and GDP growth. The extent to which Paris Agreement influences these channels remains an open empirical question. Despite the growing number of climate change and related economics literatures, a standard empirical assessment of the Paris Agreement's impact on CO_2 emissions and economic growth particularly in developing nations remains scarce. Although there are existing studies contemplating developing nations, rely on aggregate global analyses or examine pre-Paris climate regimes such as Kyoto protocol (Aichele and

Felbermayr 2013; Frankel and Rose 2005). In this way, the current chapter tries to fill the gap by empirically assessing whether the Paris agreement has worked successfully towards reducing carbon emissions in developing nations and also whether such reductions are compatible with respect to sustained economic growth. This chapter contributes to the literature by providing a post-Paris Agreement panel evidence for developing economies using a fixed-effects framework with extensive controls. It further strengthens inference through lag structures and COVID-19 robustness checks, offering policy-relevant insights into why international climate agreements may have delayed effects in the Global South. Thus, this chapter empirically evaluates the impact of the Paris Climate Agreement on carbon emissions in developing economies using a panel data framework with extensive controls and robustness checks.

2 Literature Review

2.1 Paris Agreement and CO_2 Emissions

The effectiveness and efficiency of Paris climate agreements with respect to cutting down carbon emissions has been long debated in environment related economics. Previous works based on Kyoto protocol shares mixed sort of results where some studies support in favor of carbon emission reduction and some studies reflecting limited compliance as well as leakage issues (Grubb et al. 1993; Aichele and Felbermayr 2013; Kaith and Sachdeva 2024). On the contrary, the Paris Agreement is fundamentally quite different and unique as it emphasizes on worldwide participation as well as voluntary tie ups too, asking and extracting naïve questions on environmental issues and related effectiveness (Bodansky 2016). Adun et al. (2024) states a broad overview of how a decline in CO_2 emission reduction coupled with carbon dioxide removal strategies are essential in meeting up the targets related with Paris Agreement. It additionally focusses on mitigation as well as removal strategies for confining the global warming limits in consonance with Paris Agreement objectives. Jiang et al. (2025) also examines the efficacy of worldwide mitigation endeavours towards reducing CO_2 emissions under the periphery of Paris Agreement structure. It has further been witnessed that although policies have reduced the pace of emission standards, advanced economic stretch throughout economy has mainly offset such contributions leading to a state of worldwide increase in emission standards. Furthermore, the trending empirical studies states climate related agreements is somehow positively contributing towards carbon emission reductions when properly assisted with domestic policy execution, institutional quality and technological upgradations (Eskander and Fankhauser 2020). Accordingly, in case of developing economies, reductions in CO_2 emissions are proximally associated with to variations in energy consumption patterns, renewable energy application as well as structure wise economic transformation (Citaristi 2022; IRENA 2023). However, it was observed that frail enforcement as well as enactment mechanisms and financial obstructions may confine the brief and short term impact of Paris Agreement in such regions (Pizerand Zhang 2018).

2.2 GDP Growth (Economic Growth) and Climate Policy

The relationship between economic growth and environmental norms is often assessed and described through Environmental Kuznets Curve (EKC) framework which portrays an inverted-U association between income and pollution (Grossman and Krueger 1995). While some studies are in favor of EKC framework and related hypothesis but other studies argues that carbon emission reductions are more policy driven rather than income driven (Stern 2004; Dasgupta et al. 2002). Further, research works indicates that climate policies need not undermine GDP growth. Further, Lu et al. (2024) provides a quantitative relationship viewpoint on how carbon emission standards as well as GDP can be explained via novel approach i.e. Carbon Emission Indices (CEIs). Kumar et al. (2025a, b) examines a broader overview of how synthesis of interdisciplinary research works based on climatic change vitally impacts economic systems and frameworks across the globe. Further, the study also serve as a basis for determining the impact of microeconomic constructs along with global economic planning. Some additional studies also assert that investments in green energy, improvising energy efficacy and innovation altogether can counterbalance compliance expenditures and multiply productivity gains towards GDP component (Aghion et al. 2016; OECD 2021).

2.3 Energy Use and Renewable Energy Transition

Energy consumption is the pivotal determinant in case of emissions as well as growth parameter. More specifically, higher total energy distribution per capita is closely associated with increased CO_2 emissions particularly in nations that have fuel dependency (Citaristi 2022). However, the growing total percentage of renewable energy source in total energy consumption status has portrayed to reduce emission reductions without hampering the overall growth IRENA 2023; Sadorsky 2009). Awolesi et al. (2024) shares a systematic as well as structural review pinpointing how structural policy, technology innovation and adoption frames a way towards converting fossils into a renewable source. Similarly, Fan et al. (2025) states how upgradation in financial structure growth provides impetus to energy transformation. Furthermore, empirical evidence portrays that renewable energy adoption plays a pivotal role in decoupling economic growth from emissions particularly in the context of developing nations (Apergis and Payne 2014). Hence, climate agreements like Paris Agreement may enhance this transition by channelizing cross border finances and technology transfers (Popp et al. 2011; Singh et al. 2025a, b; Prakash and Tiwari 2025).

2.4 *Trade, FDI and Urbanization*

The openness in trade and FDI inflows impacts emissions standards in various ways via scale, composition and technology effects (Antweiler et al. 2001; Kumar et al. 2025a, b). During the expansion of trade, along with FDI inflows may somewhat multiply emission rate, but simultaneously offer access to cleaner technologies and more advanced manufacturing process (Frankel and Rose 2005). Further, urbanization complicates this series and association as with rise urban fraternity calls for enhanced energy requirements but on other hand it also multiplies efficiency through economies of scale (Zhang and Lin 2012; Singh et al. 2024; Agrawal 2025). Dai et al. (2025) also states the relevance how trade policies along with FDI inflows are significantly adding value towards urban growth and development, economic modification as well as sustainability aspects particularly in the context of urban universe. Sare et al. (2024) and Singh et al. (2025b) also states that Foreign Direct Investment as well as trade flow significantly impacts and marks contribution towards economic growth and development. It further channelizes the relevance of urbanization and related policy towards defining investment as well as trade related outcomes. Hence, in the context of developing nations, the intervention between climate change policy, FDI, trade and urbanization ascertain whether emission reductions are attained alongside economic/GDP growth. In this regard, the empirical findings from studies remains inconclusive calling the need for context specific analysis (Cole and Elliott 2003; Shahbaz et al. 2015; Joselin et al. 2025).

3 Data and Methodology

This study employs an unbalanced panel dataset covering 129 developing economies over the period 2000–2024. All variables are obtained from the World Bank's World Development Indicators (WDI), ensuring cross-country comparability and data consistency. Developing economies are identified based on the World Bank's income classification. The final sample size varies across model specifications due to data availability and robustness adjustments, particularly in lagged and COVID-period analyses.

3.1 Research Objective

To examine the impact of the Paris Climate Agreement on carbon emissions in developing economies.

3.2 Baseline Econometric Model

Model 1 (OLS):

$$\begin{aligned}CO_2\ \text{Emission} = {} & \alpha + \beta1 * \text{Paris Agreement} + \beta2 * \text{GDPGrowth}_{it} \\ & + \beta3 * \text{EnergyUse}_{it} + \beta4 * \text{Renewable}_{it} \\ & + \beta5 * \text{Trade}_{it} + \beta6 * \text{FDI}_{it} + \beta7 * \text{UrbanPopulation}_{it} \\ & + \mu\_i + \varepsilon\_it\end{aligned}$$

Model 2 (Panel-FE):

$$\begin{aligned}CO_2\ \text{Emission} = {} & \alpha + \beta1 * \text{Paris Agreement} + \beta2 * \text{GDPGrowth}_{it} \\ & + \beta3 * \text{EnergyUse}_{it} + \beta4 * \text{Renewable}_{it} + \beta5 * \text{Trade}_{it} \\ & + \beta6 * \text{FDI}_{it} + \beta7 * \text{UrbanPopulation}_{it} + \mu\_i + \varepsilon\_it\end{aligned}$$

Model 3 (FE-Lagged PA)

$$\begin{aligned}CO_2\ \text{Emission} = {} & \alpha + \beta1 * \text{Paris\_it} + \beta2 * \text{GDPGrowth}_{it} \\ & + \beta3 * \text{EnergyUse}_{it} + \beta4 * \text{Renewable}_{it} \\ & + \beta5 * \text{Trade}_{it} + \beta6 * \text{FDI}_{it} + \beta7 * \text{UrbanPopulation}_{it} \\ & + \mu\_i + \varepsilon\_it\end{aligned}$$

Model 4 (FE-Covid robustness)

$$\begin{aligned}CO_2\ \text{Emission} = {} & \alpha + \beta1 * \text{Paris Agreement} + \beta2 * \text{GDPGrowth}_{it} \\ & + \beta3 * \text{EnergyUse}_{it} + \beta4 * \text{Renewable}_{it} + \beta5 * \text{Trade}_{it} \\ & + \beta6 * \text{FDI}_{it} + \beta7 * \text{UrbanPopulation}_{it} + \mu\_i + \varepsilon\_it\end{aligned}$$

CO_2 denotes per capita carbon dioxide emissions in country i at time t. Paris Agreement is dummy variable for capturing pre and post agreement period. Paris_it is a lagged dummy variable capturing the post-Paris Agreement period for looking the impact on long time interval after agreement has been done. The control variables used in study includes GDP growth, energy use, renewable energy consumption,

Table 1 Variables used in study

Variable	Measurements
CO_2 per capita	CO_2 emissions (metric tons per capita)
GDP growth	GDP growth (annual %)
Energy use	Total energy supply (TES) per capita
Renewables	Renewable energy share in TFEC
Trade	Trade (% of GDP)
FDI	Foreign direct investment, net inflows (% of GDP)
Urbanization	Urban population (% of total population)

Source World Bank and IRENA Database

trade openness, foreign direct investment inflows, and urban population (Table 1). The term μ_i represents unobserved country-specific fixed effects, while ε_it is the idiosyncratic error term.

This specification controls for time-invariant country characteristics such as institutional quality, geographic conditions, and long-run energy structures that may influence emissions. Year fixed effects are not included in the baseline model, as the Paris Agreement dummy is time-specific; including year effects would absorb the policy variable and prevent identification of its impact. To account for global shocks, robustness checks incorporating alternative specifications and sub-period adjustments are subsequently employed.

4 Result and Discussions

Table 2 shows the descriptive statistics of the variables to be utilized in the empirical analysis, including 2580 observations of country years allocated to developing economies in the year 2000–2024.

There is a significant change in the carbon dioxide emitted on a per unit of capita of different countries and within different time periods. The mean of CO_2 emission is 1.94 metric tons per capita, the standard deviation is 2.38, which means that there is a lot of heterogeneity in the intensity of emissions among the developing economies. Although, various countries have minimal rate of emissions, the highest rate of 15.73 metric tons per capita indicates the existence of moderately carbon-driven economies in the sample. Paris Climate Agreement dummy is characterized by a mean of 0.45 which means that around 45 per cent of the data will be within the post 2015 range. Such a balanced distribution will provide sufficient variability to determine the effect of such agreement on carbon emissions. The average energy consumption is 990.7 kg of oil equivalent, and there is a large spread in the energy consumption of countries. These wide disparities in the energy consumption patterns, industrialization degrees and accessibility to access pipetable modern energy services among developing economies are made visible by the great margins between 9.7

Table 2 Descriptive statistics

Index	Count	Mean	Std	Min	25%	50%	75%	Max
Urban population	2580	49.4308	19.6759	9.1806	33.3588	49.3922	64.5258	92.9501
Trade	2580	76.1660	35.5321	1.9954	48.8943	71.1210	96.7877	347.7203
Renewable energy	2580	39.6589	30.3956	0.0000	11.4000	33.6000	68.0000	97.4000
Paris agreement	2580	0.4500	0.4976	0.0000	0.0000	0.0000	1.0000	1.0000
GDP growth	2580	3.6454	5.6192	− 50.3385	1.8118	4.0766	6.2429	86.8267
FDI inflow	2580	4.0288	5.8893	− 37.1727	1.1207	2.6349	5.3412	103.3374
Energy use	2580	990.7053	809.8786	9.7273	416.3835	726.7040	1357.6327	5625.6452
CO_2 emission	2580	1.9428	2.3849	0.0000	0.3498	1.0338	2.6009	15.7272

Source Authors' own compilation

and 5625.6 kg per capital. With regards to the final energy consumption, there is a 39.66% average renewable energy consumption. Nonetheless, the standard deviation of a wide magnitude (30.40) and a dynamic of 0–97.4% imply that there is a vast difference in the adoption of clean energy. Some countries are practically implying on renewable sources whereas some other nations are still heavily relying on fossil fuels.

The level of openness to trade is 76.17% (sum of exports and imports, as a percentage of GDP), which suggests that the economy is rather moderately integrated to the world economy. The highest value above 300% of GDP indicates high exposure of small and very open economies and the lower limit indicates low exposure of trade in other countries. The GDP growth, which is a measure of economic growth, is on average 3.65% in an average annual growth, however, with high levels of volatility with a range of 50.34–86.83%. This broad range encompasses recessions of the economy, and the revival of the economy following a conflict, and high growth in the economy of developing countries. The average inflows of foreign direct investment are 4.03% of GDP, which is highly skewed. The lower bound of the values show net disinvestment in some years whereas the upper limit of the values 103.34% of GDP indicates periodically high capital inflows, especially in smaller economies. The level of urbanization is 49.43% on average implying that about 50% of the people in the developing economies live in the urban areas. The large variation between 9.18 and 92.95% indicates high demographical diversity and varying levels of urban transition of countries.

Generally, this is because the descriptive statistics indicate that there is great heterogeneity in environmental, economic and structural features across developing economies. This inconsistency highlights the application of the panel data method in explaining the difference between countries in investigating the effectiveness of the Paris Climate Agreement on carbon emissions.

Table 3 presents the pairwise correlation matrix among the key variables used in the analysis, while Fig. 1 visualizes the same relationships using a correlation heatmap. The correlation results provide preliminary insights into the direction and strength of linear associations among variables and help assess potential multicollinearity concerns prior to regression estimation.

CO_2 emissions exhibit a strong positive correlation with energy use, indicating that higher per capita energy consumption in developing economies is closely associated with increased carbon emissions. This relationship is consistent with conventional energy–environment literature, which highlights fossil fuel dependence as a major driver of emissions in developing countries. In contrast, CO_2 emissions show a moderate to strong negative correlation with renewable energy consumption, suggesting that a higher share of renewable energy in the total energy mix is associated with lower emission levels. This preliminary evidence supports the role of renewable energy as a mitigation mechanism in developing economies. The correlation between CO_2 emissions and urban population is moderately positive, reflecting the environmental pressure arising from urbanization through increased energy demand, transportation, and industrial activity. Similarly, trade openness displays a mild positive association with emissions, possibly indicating scale effects of economic activity.

Table 3 Correlation analysis and multicollinearity diagnostics

Variable	VIF	1	2	3	4	5	6	7	8
CO_2 emission		1.00	0.02	0.95	0.03	0.00	− 0.59	0.13	0.45
Paris agreement	1.57		1.00	0.06	− 0.14	− 0.12	− 0.04	− 0.03	0.11
Energy use	4.12			1.00	− 0.01	0.04	− 0.57	0.23	0.51
GDP growth	1.51				1.00	0.10	0.09	0.09	− 0.10
FDI inflow	1.88					1.00	0.00	0.31	0.03
Renewable energy	2.31						1.00	− 0.27	− 0.47
Trade	6.18							1.00	0.10
Urban population	7.18								1.00

Source Authors' own compilation

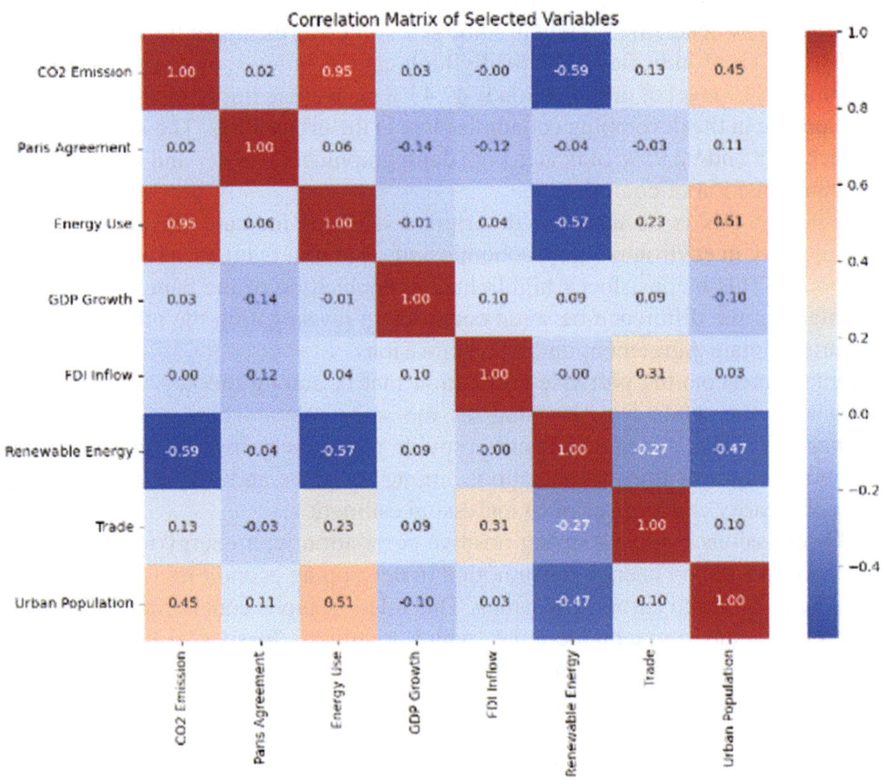

Fig. 1 Heatmap. *Source* Authors' own compilation

The Paris Climate Agreement dummy exhibits a weak and near-zero correlation with CO_2 emissions, implying that any emission-reducing impact of the agreement is not immediately observable in a simple bivariate setting and may require multivariate and dynamic modeling to capture delayed or indirect effects. The correlation matrix also shows notable associations among explanatory variables. Energy use and urban population are positively correlated, reflecting higher energy demand in more urbanized economies. Renewable energy is negatively correlated with both energy use and CO_2 emissions, indicating a substitution effect between renewable and conventional energy sources. Trade openness and FDI inflows display moderate correlations with economic growth, suggesting that external sector integration supports economic expansion in developing economies. However, none of the correlation coefficients exceed commonly accepted thresholds for severe multicollinearity. The correlation coefficients are all below the critical threshold of 0.80, and the variance inflation factor (VIF) values remain well within acceptable limits (VIF < 10). This indicates that multicollinearity is not a serious concern, validating the inclusion of all explanatory variables in the regression models.

Figure 1 support these results by visualizing the correlation heatmap which showed that there are strong positive relations between the energy use and the CO_2 emissions and strong negative relations between the renewable energy and the emissions. The fact that the correlations with the Paris Agreement dummy are relatively much less strongly shaded also confirms the fact that there are no strong contemporaneous correlations, and it is a good reason to use the fixed effects and lagged policy variables in the further econometric analysis.

Correlation analysis can be helpful to gain some initial understanding, but the analysis fails to capture unobserved heterogeneity and causal processes. As such, the following section uses panel regression methods to strictly investigate the effect of the Paris Climate Agreement on carbon emission in the developing economies (Table 4).

Country fixed effects control for time-invariant heterogeneity across developing economies.

Model 1 (Pooled OLS) provides baseline estimates without accounting for unobserved country-specific heterogeneity. The results indicate that energy use per capita significantly increases CO_2 emissions, while renewable energy consumption and trade openness reduce emissions. GDP growth and FDI inflows are positively associated with emissions, suggesting that economic expansion in developing economies remains environmentally intensive. Although the Paris Agreement coefficient is negative, it is statistically insignificant, implying no observable average reduction in emissions when cross-country heterogeneity is ignored.

Model 2 (Country Fixed Effects) controls for time-invariant country characteristics such as geography, institutional quality, and historical development paths. Once these factors are accounted for, the explanatory power of the model declines in terms of within variation, but the estimates become more credible. The Paris Agreement remains statistically insignificant, indicating that within-country changes in emissions before and after adoption are not sufficiently large in the short run. Importantly, energy consumption has strong and positive effect, while renewable energy has

Table 4 Impact of Paris climate agreement and control variables on CO_2 emissions (developing economies)

Variables	Model 1 (pooled OLS)	Model 2 (Panel-FE)	Model 3 (FE + lagged PA)	Model 4 (FE + Covid robustness)
Constant	1.3411*** (0.142)	1.8362* (0.995)	1.8534* (0.972)	1.9301* (1.076)
Paris agreement	− 0.0613 (0.040)	0.0132 (0.055)	–	0.0032 (0.053)
Paris agreement (t − 1)	–	–	0.0192 (0.049)	–
Energy use	0.0028*** (0.00005)	0.0024*** (0.0003)	0.0023*** (0.0003)	0.0024*** (0.0003)
FDI inflow	0.0132** (0.006)	− 0.0034 (0.003)	− 0.0039 (0.003)	− 0.0037 (0.003)
GDP growth	0.0090* (0.005)	0.0041** (0.002)	0.0035** (0.002)	0.0062** (0.002)
Renewable energy	− 0.0207*** (0.001)	− 0.0207*** (0.004)	− 0.0207*** (0.004)	− 0.0211*** (0.004)
Trade	− 0.0030*** (0.001)	0.0017 (0.002)	0.0021 (0.002)	0.0014 (0.003)
Urban population	− 0.0147*** (0.002)	− 0.0210 (0.018)	− 0.0207 (0.018)	− 0.0221 (0.020)
Country fixed effects	No	Yes	Yes	Yes
Year fixed effects	No	No	No	No
Observations	1432	1432	1337	1272
Number of countries	98	98	98	98
R^2 (within)	0.934	0.662	0.659	0.615
Adj. R^2	0.933	–	–	–
Durbin–Watson	0.259	–	–	–

Source Authors' own compilation
Lagged PA – Lagged Paris Agreement effect
Dependent variable: CO_2 emissions per capital
Robust standard errors in parentheses for Model (1). Clustered standard errors at the country level for Models (2 and 4)
* $p < 0.10$, ** $p < 0.05$, *** $p < 0.01$
Model 3 includes *lagged Paris Agreement* to capture delayed policy effects
Model 4 controls for *Covid-period disturbances* through sample adjustment

significant negative effect on emissions, posing to be leading variables in mitigation. The loss of significance for FDI and trade suggests that their effects in Model 1 were driven largely by cross-country differences rather than within-country dynamics.

Model 3 (Fixed Effects with Lagged Paris Agreement) adds a lagged policy variable to both the effects of climate commitments (since reductions in emissions can be delayed), and because the effects of emission reductions may be realized over a period of time. The lagged Paris Agreement variable is, however, statistically insignificant, implying that despite the time, the Agreement is not yet evident in terms of achievable emission cuts in the developing economies. Conversely, the energy intensity, renewable energy application, and economic growth are still on top of the emission performance, which substantiates the claim that policy commitments are not enough to achieve the desired outcomes without domestic transitions.

Model 4 (Fixed Effects with Covid Robustness) also examines the fixed effects more by considering the disruptions of the Covid period. The most significant results have not changed: renewable energy always lowers the emissions, and the energy consumption and GDP growth raise the emissions. The Paris Agreement still does not exhibit any statistically significant effect meaning that the null outcome is not caused by the effects of the pandemic but is as a result of the structural constraints of developing countries.

Notably, the omission of year fixed effects was done purposely to eliminate the impact of the Paris Agreement dummy that is very time-based. Though, the policy variable is not significant and it can be thought that there is no effect because of implementation gaps, lax enforcement and developmental constraints rather than an econometric artifact. The FE implications thus show that although the Paris Agreement is a significant international commitment, its effect in the reduction of emissions in developing economies is yet to be felt unless there is significant domestic structural change especially in energy systems. The Paris Agreement dummy is not statistically significant with Fixed Effects estimation; however, with a multi-country developing economy setting, the FE model is better since it accounts for the unobserved country heterogeneity, which is strongly observed to be time-invariant. Pooled OLS creates dilemma with cross-country and within-country policy effects, which may creates the irrelevant effect of countries policies. In comparison, the Fixed Effects estimator allows within-country time variation to be isolated and this offers a more conservative and methodologically appropriate measure of the effectiveness of policies.

4.1 Robustness Analysis

To ensure the reliability of the findings, multiple robustness checks are conducted.

Lagged Policy Effects

It is acknowledged that the impact of climate policies might be delayed, and a lag period of the Paris Agreement by one period is added. The lagged coefficient is not statistically significant and this shows that the missing effect of the Paris Agreement is not due to timing or slow transmission of policies. The consistency of estimates of the coefficients of energy consumption, renewable energy, and GDP growth among specifications is another indication that the core relationships are strong.

COVID-19 Shock Adjustment

Since the COVID-19 pandemic has caused unprecedented economic and environmental upheaval, another robustness check is conducted by making the adjustment to the COVID period. This study findings states that the Paris Agreement is statistically not an important event to be considered, even economies facing pandemic level shocks are considered. The key control variables have their expected signs and significance levels, which mean that the key conclusions will not be exposed to unusual world shocks.

5 Conclusion, Policy Implications and Limitations

5.1 Conclusion

Combined, the empirical data gives reason to believe that the Paris Climate Agreement does not so far translate into statistically significant carbon emission reductions in developing economies over the period of analysis. This result, however, cannot be seen as a failure of the Agreement as such but rather a symptom of structural, institutional, and developmental challenges of developing countries, such as insufficient financial ability, technological disparities, and voluntary character of nationally determined contributions (NDCs). In line with the motivation described in the introduction, the findings demonstrate that home country forces, especially, energy consumption trends and the adoption of renewable energy are a stronger determinant of the emission trends than the international commitments. The positive correlation between the economic growth and CO_2 emissions also brings out the development-environment trade-off that the developing economies experience and this supports the early-stage Environmental Kuznets Curve hypothesis where the growth rate is the cause of increased environmental pressure before the cleaner technologies and regulations gain ground.

5.2 Policy Implications

The study results indicates and suggest that global climate treaties including the Paris Agreement should be combined with robust domestic policy frameworks so as to to realize significant emission cuts in the developing economies. The priority that policy-makers should be focused on is the speed at which the shift to renewable energy ought to be done, as well as the energy efficiency and the way in which urban sprawl needs to be controlled in a way that is in accordance with the environmentally-friendly approach. Mitigation efforts can also be strengthened by harmonizing the trade policies and foreign direct investment with climate initiatives by promoting the adoption of green investments and deterring activities that are highly carbon intensive. In sum, a combined policy that responds to all three aspects of climate commitment, transition to energy, and development priorities can be the key to balancing economic growth and environmental sustainability.

5.3 Limitations

The results highlight the importance of having robust domestic policy regimes to be used in conjunction with international climate treaties like the Paris Agreement to have significant emission cuts in developing countries. The focal areas by the policymakers should be on hastening the process of shifting to renewable energy, enhancing energy efficiency, and controlling urban sprawl in an environmentally friendly fashion. By promoting green investments and discouraging carbon-intensive processes both trade policy and foreign direct investment can be used to further support mitigations by aligning the policies on climate with the objectives of climate policy. In general, to reconcile the economic and environmental sustainability, a comprehensive policy framework that fulfills climate commitments, energy transition and development priorities is necessary.

6 Scope for Future Research

The current research examines the impact of Paris agreement on carbon emissions particularly in the context of developing nations. Still, there exist healthy number of options that are opened towards future scholarly investigation. One such addition will be gone with longitudinal as well as panel data studies in order to assess the long term effectiveness of nationally determined contributions (NDCs) over and above short term emission trends. Further, the comparative region based analysis across Asia, Africa as well as America can be performed in order to extract underlying heterogeneity with respect to implementation of policies coupled with outcomes associated with carbon emissions. Since it is debatable from long ago that developing nations are not homogenous in nature; deviations in governance patterns, relying on energy sources, economic framework and technological upgradations can majorly impact the effectiveness part of climate policy. Also upcoming studies can also undertake institutional theory as well as political perspectives in order to how regulatory compliance and governance systems mediate the Paris Agreement impact. Apart from this, the study can also be conducted across different sectors such as manufacturing, healthcare, agriculture, transportation, construction and how they are specifically contributing in the direction of carbon emission reduction. The carbon intensity across such sort of industries assists in offering valuable evidence with respect to green energy adoption, smarter technologies along with compliance related behavior. Furthermore, the additional research may also accommodate mediating as well as moderating variables in the form foreign direct investment, carbon pricing frameworks, availability of green finance and renewable energy approaches via sustainable technologies. Also analyzing the role of global level climate finance organizations shall further throw light on whether financial commitments mentioned in the agreement efficiently transforms into reducing carbon emissions. Additionally, qualitative as well as mixed method studies encompassing policy makers, environmental firms and related industry stakeholders can throw light towards institution related hindrances, challenges related to implementation and policy stability and synchronization issues. In this regard, examining towards general public awareness, corporate environment responsibility and consumer behavior with respect to adopting green initiatives amongst developing nations can expand the social behavioral dimensions of global climate policy investigations.

References

Acemoglu D, Aghion P, Bursztyn L, Hemous D (2012) The environment and directed technical change. Am Econ Rev 102(1):131–166. https://doi.org/10.1257/aer.102.1.131

Adun H, Ampah JD, Bamisile O, Hu Y (2024) The synergistic role of carbon dioxide removal and emission reductions in achieving the Paris Agreement goal. Sustain Prod Consumpt 45:386–407. https://doi.org/10.1016/j.spc.2024.01.004

Aghion P, Dechezleprêtre A, Hemous D, Martin R, Van Reenen J (2016) Carbon taxes, path dependency, and directed technical change: evidence from the auto industry. J Polit Econ 124(1):1–51. https://doi.org/10.1086/684581

Agrawal D (2025) A comprehensive review of investing and financing for sustainable tourism projects. Environ Sustain Governance Insights 1(1):53–70

Aichele R, Felbermayr G (2013) The effect of the Kyoto Protocol on carbon emissions. J Policy Anal Manag 32(4):731–757. https://doi.org/10.1002/pam.21720

Antweiler W, Copeland BR, Taylor MS (2001) Is free trade good for the environment? Am Econ Rev 91(4):877–908. https://doi.org/10.1257/aer.91.4.877

Apergis N, Payne JE (2014) The causal dynamics between renewable energy, real GDP, emissions and oil prices: evidence from OECD countries. Appl Econ 46(36):4519–4525. https://doi.org/10.1080/00036846.2014.964834

Awolesi O, Salter CA, Reams M (2024) A systematic review on the path to inclusive and sustainable energy transitions. Energies 17(14):3512. https://doi.org/10.3390/en17143512

Bodansky D (2016) The Paris climate change agreement: a new hope? Am J Int Law 110(2):288–319. https://doi.org/10.5305/amerjintelaw.110.2.0288

Citaristi I (2022) International Energy Agency—IEA. In: The Europa directory of international organizations 2022. Routledge, pp 701–702

Cole MA, Elliott RJ (2003) Do environmental regulations influence trade patterns? Testing old and new trade theories. World Econ 26(8):1163–1186. https://doi.org/10.1111/1467-9701.00567

Dai S, Tang D, Li Y, Lu H (2025) Digital trade, trade openness, FDI, and green total factor productivity. Int Rev Financ Anal 97:103777. https://doi.org/10.1016/j.irfa.2024.103777

Dasgupta S, Laplante B, Wang H, Wheeler D (2002) Confronting the environmental Kuznets curve. J Econ Perspect 16(1):147–168. https://doi.org/10.1257/0895330027157

Eskander SM, Fankhauser S (2020) Reduction in greenhouse gas emissions from national climate legislation. Nat Clim Change 10(8):750–756. https://doi.org/10.5281/zenodo.3830922

Fan S, Zhao Y, Zuo S (2025) Financial development and energy transition: a literature review. Energies 18(15):4166. https://doi.org/10.3390/en18154166

Frankel JA, Rose AK (2005) Is trade good or bad for the environment? Sorting out the causality. Rev Econ Stat 87(1):85–91. https://doi.org/10.1162/0034653053327577
Grossman GM, Krueger AB (1995) Economic growth and the environment. Q J Econ 110(2):353–377. https://doi.org/10.2307/2118443
Grubb M, Edmonds J, Ten Brink P, Morrison M (1993) The costs of limiting fossil-fuel CO_2 emissions: a survey and analysis. Annu Rev Energy Environ 18(1):397–478
IPCC (2014) In: Field CB et al (eds) Managing the risks of extreme events and disasters to advance climate change adaptation. A special report of Working Groups I and II of the Intergovernmental Panel on Climate Change. Cambridge University Press
IRENA [International Renewable Energy Agency] (2023) World Energy Transitions Outlook 2023: 1.5°C pathway, vol 1. Annual report, 3rd edn. Available at: https://mc-cd8320d4-36a140ac83cc3389cdnendpoint.azureedge.net//media/Files/IRENA/Agency/Publication/2023/Jun/IRENA_World_energy_transitions_outlook_v1_2023.pdf. Accessed 26 Dec 2023
Jiang J, Shi S, Raftery AE (2025) Mitigation efforts to reduce carbon dioxide emissions and meet the Paris Agreement have been offset by economic growth. Commun Earth Environ 6(1):823. https://doi.org/10.1038/s43247-025-02743-x
Joselin D, Singh A, Tiwary KS (2025) The nexus between CSR practices and financial performance: empirical evidence from the manufacturing sector in India. Environ Sustain Governance Insights 1(1):1–17
Kaith A, Sachdeva G (2024) Bibliometric insights into green human resource management: A connection with business ethics and values. PURUSHARTHA 17(1):51–63. https://doi.org/10.21844/16202117104
Kumar K, Singh R, Choudhary M, Sharma A (2025) Climate change mitigation strategies in environmentally sensitive industries. In: Climate change and social responsibility, vol 25. Emerald Publishing Limited, pp 185–204
Kumar MS, Palanichamy NV, Shivakumar KM, Chandrakumar M, Kalpana M, Murugananthi D (2025) Impact of climate change on global economy: a comprehensive review. Reg Sustain 100274. https://doi.org/10.1016/j.regsus.2025.100274
Lu F, Ma F, Feng L (2024) Carbon dioxide emissions and economic growth: new evidence from GDP forecasting. Technol Forecast Soc Change 205:123464. https://doi.org/10.1016/j.techfore.2024.123464
Mishra A, Srivastava SK (2025) Role of trust and celebrity endorsement in shaping purchase intention towards electric vehicles. In: Implementing ESG frameworks through capacity building and skill development. IGI Global Scientific Publishing, pp 381–400. https://doi.org/10.4018/979-8-3693-6617-2.ch018
Nordhaus W (2018) Projections and uncertainties about climate change in an era of minimal climate policies. Am Econ J Econ Pol 10(3):333–360. https://doi.org/10.1257/pol.20170046
OECD (2021) Climate change mitigation and economic growth. https://www.oecd.org/climate-change/
Pizer WA, Zhang X (2018) China's new national carbon market. AEA Pap Proc 108:463–467. https://doi.org/10.1257/pandp.20181029
Popp D, Hascic I, Medhi N (2011) Technology and the diffusion of renewable energy. Energy Econ 33(4):648–662. https://doi.org/10.1016/j.eneco.2010.08.007
Porter ME, Linde CVD (1995) Toward a new conception of the environment-competitiveness relationship. J Econ Perspect 9(4):97–118. https://doi.org/10.1257/jep.9.4.97
Prakash A, Tiwari AK (2025) The role of institutional factors in achieving SDG 16: a thematic review of drivers of corporate anti-corruption action. Environ Sustain Governance Insights 1(1):34–52
Sadorsky P (2009) Renewable energy consumption, CO_2 emissions and oil prices in the G7 countries. Energy Econ 31(3):456–462. https://doi.org/10.1016/j.eneco.2008.12.010
Sare YA, Amoah JO, Bawuah B (2024) Assessing the impact of international trade and FDI in driving economic development in West Africa: the role of urbanization. SN Bus Econ 5(1):1–30. https://doi.org/10.1007/s43546-024-00765-9

Shahbaz M, Nasreen S, Abbas F, Anis O (2015) Does foreign direct investment impede environmental quality in high-, middle-, and low-income countries? Energy Econ 51:275–287. https://doi.org/10.1016/j.eneco.2015.06.014

Singh R, Kumar K, Khan S (2024) A comprehensive view of artificial intelligence (AI)-based technologies for sustainable development goals (SDGs). Artificial intelligence enabled management: an emerging economy perspective, p 183. https://doi.org/10.1515/9783111172408-012

Singh R, Sharma A, Kumar K (2025a) Navigating carbon neutrality: strategies for sustainable future and environmental challenges. Climate neutrality through smart eco-innovation and environmental sustainability. Springer, Cham, pp 69–80

Singh R, Tiwary KS, Kumar K, Kumar A (2025b) Empowering a sustainable future: blockchain's role in economy, organizations, and finance towards smart circular economy. Blockchain technologies for smart circular economy and organisational sustainability. Springer, Cham, pp 271–283

Srivastava SK, Anshul A, Pathak P, Bansal J (2018) Responsible consumption for curbing food wastage: an exploratory enquiry. PURUSHARTHA 11(2):104–116

Srivastava SK, Singh S, Srivastava M (2021) Food waste at household and social gatherings: drivers and possible remedies. PURUSHARTHA 14(2):117–133

Srivastava SK, Sinha AK, Kumar S (2024) Workplace spirituality empowered by spiritual leadership to promote organizational effectiveness: a bibliometric analysis. PURUSHARTHA 17(1), 1–22. https://doi.org/10.21844/16202117101

Stern DI (2004) The rise and fall of the environmental Kuznets curve. World Dev 32(8):1419–1439. https://doi.org/10.1016/j.worlddev.2004.03.004

Stern NH (2007) The economics of climate change: the Stern review. Cambridge University Press

Tripathi V, Singh K, Parihar S, Srivastava SK (2022) A bibliometric analysis on supply chain collaboration. PURUSHARTHA 15(2):66–103. https://doi.org/10.21844/16202115205

Unfccc UN (2009) Kyoto protocol reference manual on accounting of emissions and assigned amount. eSocialSciences

World Bank Group (2022) Global economic prospects, January 2022. World Bank Publications

Zhang C, Lin Y (2012) Panel estimation for urbanization, energy consumption and CO_2 emissions: a regional analysis in China. Energy Policy 49:488–498. https://doi.org/10.1016/j.enpol.2012.06.048

Urban Sustainability and Development Systems

Urbanization, Carbon Emissions, and SDG Aligned Strategies for Sustainable Cities

Bikram Paul Singh Lehri, Amit Kumar, and Jasreen Kaur Kakar

Abstract With the rapid growth in urbanization, carbon emission has emerged as the major challenge. Urban areas are a major contributors of global carbon emissions with transportation, factories and construction sectors contributing to the majority of greenhouse gas (GHG) emissions. Urbanisation also demands excessive energy and creates pollution. Many prior studies have reported the direct relationship between rapid urbanization and an increase in carbon emissions. This chapter aims to highlight the carbon emission related challenges to o the urbanization and actions to address those challenges. Further, this chapter also outlines various SDGs aligned strategies to reduce the carbon emissions due to urbanization and suggests policy actions to adopt green technologies that resulted in the level of carbon footprint. It provides important insights into low-carbon transportation systems, smart waste management system and green digital infrastructure to reduce the carbon footprint in the urban area and help achieve SDG7, SDG9, SDG11, and SDG 15.

Keywords Urbanization · Carbon emission · SDGs · Smart cities · Digital infrastructure

1 Introduction

The urban growth in the twenty-first century has been phenomenal. According to the United Nations, 56% of the world population resides in cities, and it will reach 70% by 2050 (United Nations report 2018). Cities contribute more than 80% of the world's

B. P. S. Lehri (✉)
University School of Business, Chandigarh University, Mohali, India
e-mail: lehribikram@gmail.com

A. Kumar
School of Business Management, Christ University, Bengaluru, India

J. K. Kakar
University Institute of Liberal Arts and Humanities, Chandigarh University, Mohali, India
e-mail: jasreen.research@gmail.com

K. Kumar et al. (eds.), *Sustainable Value Creation: Corporate Strategies for SDGs Implementation*, World Sustainability Series,
https://doi.org/10.1007/978-3-032-24652-3_12

GDP and drive economic growth, but they are also responsible for environmental degradation (WCR 2024). Urban centres are responsible for the global environmental crises, as approximately 70% of the carbon dioxide emissions are caused by them. This has happened because cities are the major hubs of construction, transportation, energy consumption, commercial activity, and industrial activity (Tan et al. 2016). Huge infrastructure, high-rise buildings, a large number of commercial vehicles, and the expansion of cities have resulted in an increased carbon footprint. The different climate risks of poor air quality, heat, and flooding affect the people living in urban centres, majorly in thickly populated and economically underdeveloped areas (Shen et al. 2012). The challenges of climate adaptation, socio-economic inequalities, and carbon footprint are increasing every day (Kumar et al. 2024; Al-Raeei 2025). Many cities in Asia and Africa have informal housing, bad public transport system, poor governance systems, and inadequate infrastructure. This leads to cities becoming less resilient to climate change and affects their ability to reduce carbon emissions.

A disciplined and systematic approach is needed to build sustainable cities, and this can be achieved by proper planning, implementation, and good governance. This involves integrating sustainability goals into urban planning and engaging in reduction of carbon footprint reduction initiatives, clean energy, and green construction (Wibowo 2024). An attempt should be made to involve citizens in governance and in the operationalization of the sustainability practices. The integration of data and technology is important to improve the effectiveness of the initiatives. Cities should be designed to counter the climate shocks, and it should be resilient enough to recover on its own (Silva et al. 2018). Despite the challenges, there are ample opportunities as well. Sustainable cities are environmentally friendly, healthier, and financially feasible (Khan et al. 2020). This chapter will explore the different strategies that can be used to reduce the levels of carbon emissions in cities in the age of fast urbanization. The best global sustainability practices, policies, and innovations will be reviewed to provide policymakers, planners, and citizens with a plan of action.

2 Urbanization and the Carbon Challenge

Urbanization is the process of changing the physical, demographic, and economic landscape of regions. Rapid urbanization is also responsible for environmental degradation and increasing carbon emissions (Luqman et al. 2023). The nature of this challenge is big, and a thorough understanding of it is important before providing a solution.

2.1 The Urban-Carbon Linkage

The most carbon-intensive urban activities fall into the following broad categories:

- **Real Estate and Construction**: All residential and commercial buildings require heating, lighting, and cooling. These requirements are fulfilled by non-renewable energy resources in many countries around the world. Most of the manufacturing industries, such as cement and steel, in cities are energy-intensive (Tan et al. 2016; Jiang et al. 2022). The construction activities increase the emissions during the construction process, and the emissions are also high during the material production stage (Yang et al. 2023).
- **Transport Systems**: Passenger and commercial vehicle mobility in urban centres is extremely high, which leads to the problem of carbon emissions, excessive fuel usage, and congestion. Public transportation systems around the world are either not developed or are overburdened (Haghshenas et al. 2015).

2.2 *Urban Sprawl and Inefficiency*

Uncontrolled development is the major reason for carbon emissions. Lateral development increases the distances and makes the public transport financially unviable and resulting in more automobile usage, carbon emissions, and huge infrastructure cost. In cities that have higher density and where land is put to mixed use, the per capita carbon emissions tend to be less in such cities. Such compact cities have a well-planned and efficient public transport system (Bamwesigye and Hlavackova 2019). The impact of the decisions made today in urban development on the carbon footprints becomes visible after decades. For example, building a city that is automobile centric will increase the dependence on fossil fuels for generations, resulting in increased carbon emissions. This problem becomes bigger in developing countries, such as India, which would require timely intervention. According to the Intergovernmental Panel on Climate Change (IPCC), urban emissions can be reduced to 90% by 2050 if properly backed by sustainable planning, governance, and efficient implementation (IPCC 2022).

2.3 *Climate Justice and Equity Concerns*

Low-income people live in areas with inadequate infrastructure, with limited green spaces, and are prone to pollution and flood risk, and are less responsible for carbon footprints. Sustainable urban planning should be environmentally friendly and socially inclusive (Bulkeley et al. 2013). The plan of action should be designed to give access to clean energy, a public transportation system, health services, and affordable housing to benefit the marginalized groups (Porter et al. 2020). The voice of the weaker communities should be adequately heard in the decision-making process.

3 Compact and Mixed-Use Urban Planning

Mixed-use development in urban areas is important in building sustainable cities that focus on less carbon emissions (Hoppenbrouwer and Louw 2005). Compact cities focus on development and infrastructure efficiency, enabling the sharing of resources, and less automobile dependency.

3.1 Understanding Compact and Mixed-Use Development

Compact cities result in higher population density on a smaller land parcel, and the focus is on vertical growth of cities. Compact cities usually have a large number of mixed-use developments where commercial, recreational, institutional, and residential are all integrated in one development, focussing on creating everything within walk able distance. Mixed-use developments are an efficient way of using the land parcel efficiently and providing multi-functional neighbourhoods. Compact cities have shops, schools, healthcare services, and workplaces near residential buildings (Coupland 2005). Green buildings are built for energy efficiency, and a public transportation system is available within walking distance (Geng et al. 2012). Such cities are not dependent on any one mode of transport for commuting. Such initiatives improves the quality of life, improves public health, and reduction in pollution.

3.2 Transit-Oriented Development (TOD)

Transit-Oriented Development (TOD) is a significant model in the development of compact cities. Commercial and residential areas are near metro stations and bus stations in a TOD (Ibraeva et al. 2020). There are many advantages of TOD, such as a rise in property value, increased tax collection, less traffic due to increased usage of public transport. Hong Kong and Portland (USA) have operationalized the TOD corridors in an effective way. In developing countries, the zoning systems separate the different zones on the basis of industrial, commercial, and residential. This leads to a situation where land is underutilized, and it also increases the travel requirements (Xia and Zhang 2022). The traditional zoning laws should pave the way for the mixed-use development to take centre stage, and it would be an important shift towards sustainability (Nasri and Zhang 2014). Urban Local Bodies (ULB) has an important role to play in this transformation. ULB should plan for mixed-use corridors in the master plan, and the high-density zones should necessarily have more green spaces. The builders should be adequately incentivized by giving them a higher Floor Area Ratio (FAR) near the transit corridor, which would boost the development of compact cities through land-use conversions (Sridhar and Mahendra 2025).

3.3 Challenges in Implementation

The compact cities have a lot to offer, but there are many challenges as well. There are many informal settlements in cities that need to be formalized by acquiring that patch of land and redeveloping it. The resistance from the inhabitants and social organizations is the biggest hurdle in this process due to security issues that come along with it. The inhabitants often feel that after redevelopment, the gentry of the place would change, which would make them unfit to live in such a society. The intervention of the government is needed in such a situation by implementing a plan that is inclusive, engaging and affordable in its entirety (Xia and Zhang 2022).

4 Low-Carbon Transport Systems

Urban transport accounts for approximately 25% of the global carbon emissions and is the main reason for urban carbon emission. With the expansion of cities, sustainable modes of mobility must become the focal point of climate strategies. The Internal Combustion Engine (ICE) is the main reason behind high levels of urban transport emissions. ICE vehicles burn different fuels, which are harmful for the environment and health, causing cardiovascular and respiratory diseases by releasing nitrogen oxide. Urban emissions are high in cities due to inadequate public transport infrastructure, and there is overdependence on private vehicles. Insufficient walking or cycling infrastructure and increased traffic congestion are also soliciting a huge investment in this area for reducing carbon emissions (Gately and Hutyra 2017; Glaeser and Kahn 2010). The suburban and light rail infrastructure should also be developed, which would facilitate the regional commuters and connect the peri-urban areas. The cost per passenger per kilometre is significantly less, and it significantly reduces the pressure on land. However, the success of public transport depends on connectivity, safety, and affordability (Tarkar 2022).

4.1 Electric Mobility (E-Mobility)

The shift from fossil fuels to Electric vehicles (EVs) is important to reduce carbon emissions. EVs can significantly reduce carbon emissions, and the effect would be more if the EVs are powered through renewable energy. In order to promote the usage of EVs, financial incentives in the form of interest-free loans, tax rebates, and subsidies should be offered (Donateo et al. 2015; Heidrich et al. 2017). Proper charging infrastructure should be developed at public places and homes. The public modes of transport should also be converted to electric mode for added benefit. Faster Adoption and Manufacturing of Hybrid and Electric Vehicles (FAME II) scheme in

India is aimed at promoting the adoption of EVs in urban areas, and it has met with decent success in Maharashtra and Delhi (states in India).

4.2 Integrated Mobility and Smart Transport Systems

The Mobility-as-a-Service (MaaS) model has proved to be a successful business model as it has integrated different modes of transport digitally. These systems give real-time data to users which are useful to them, such as waiting time and best available route (Wong et al. 2020). The MaaS model also has the provision for a single ticket window for all types of public transportation system. It tracks the vehicle on a real-time basis and provides information about the congestion on the roads. Smart parking solutions are just another benefit that such models are passing on to society. Smart mobility is not only convenient, but it also helps in reducing the carbon emissions from urban transportation.

5 Energy-Efficient Buildings and Infrastructure

The real estate sector is one of the leading contributors of urban carbon emissions, and it accounts for around 40% of energy consumption globally and energy-related CO2 emissions of more than 30%. The ever-expanding cities and their construction activities shifts the focus to having an energy efficient property, resilient, and help in the reduction of urban carbon footprints. The energy consumption of urban buildings is high due to their requirements for appliances, heating, lighting, ventilation, and cooling (Kathpal et al. 2021). In rapidly growing cities with high temperatures, the demand for fridges and ACs is high, which leads to high energy consumption (Tyler and Moench 2012; Kumar et al. 2024). The government and the regulatory bodies should intervene, otherwise the energy demand by real estate would increase exponentially. The construction materials like aluminium, steel, and cement—have a high degree of embodied carbon. The carbon emissions would increase exponentially throughout the life cycle of such construction materials (Lenzholzer et al. 2020). Sustainable built environment practices should be adopted to meet the challenges of embodied carbon.

Green Building Codes should be adopted to build energy-efficient cities. These codes lay down the standards for material usage, waste management, water conservation, and energy efficiency. Leadership in Energy and Environmental Design (LEED) is a globally renowned framework and is adopted by many companies (Gauthier and Wooldridge 2012). The Building Research Establishment Environmental Assessment Method (BREEAM) framework is extensively used in Europe and benefits society at large. Green Rating for Integrated Habitat Assessment (GRIHA) is a framework that rates green buildings in India. The municipal regulations should be tightened to ensure that all construction activities in cities meet the sustainability norms. The green building standards promote the reduction of cooling loads through the use of ventilation and natural daylight (Hightower 2013). It also promotes energy-efficient buildings and encourages rainwater harvesting. The green building standards also promote the usage of recycled steel and fly ash bricks, which are sustainable materials used in construction (Debrah et al. 2022). New green buildings should be developed, but the old and inefficient buildings need to be retrofitted for developing a sustainable city (Tarkar 2022). The various retrofitting changes that can be done to a building may include usage of sensors, smart systems, insulated roofs and walls, energy-efficient windows, and moving towards an efficient cooling and heating system. The government can introduce zero-interest loans, subsidies, and tax rebates (Liu et al. 2022).

The components of urban infrastructure—water supply, electricity network, telecommunications, roads, and drainage—consumes huge energy in building and maintenance. Sustainable infrastructure should aim at integrating carbon emissions and resilient cities in the design and execution (Ashinze et al. 2024). Sustainable infrastructure promotes the usage of low-carbon materials and centralizes the district cooling systems. Smart grids should be introduced to minimise transmission losses and build a resilient drainage system that can withstand high rainfall. The responsibility of promoting such green buildings lies on the professionals working in this sector—developers, engineers, and architects—who can appreciate the energy codes and climate-resilient designs so that the green buildings can be scaled up (Zhalmurziyeva et al. 2024). Municipal corporations can conduct awareness campaigns and give certifications to professional organizations for spreading the advantages of green buildings. They can make a mandate for all public buildings to become green buildings (Tarkar 2022). Sustainable architecture should be integrated into the curriculum for all courses in the field of the built environment sector.

6 Renewable Energy Integration in Cities

A shift towards renewable energy sources is important for achieving the target of carbon–neutral cities. Urban centres can lead this shift due to the vast resources at their disposal and which can meet the energy requirements of such centres.

6.1 Urban Energy Demand Profile

Urban population consumes nearly three-fourths of the global energy, and most of it is produced from fossil fuels. This energy meets the requirements of public transport, lighting, heating, and cooling. Most countries are still producing energy through the traditional modes of coal and oil, which leads to heavy carbon emissions and pollution. Making a shift to renewable energy will result in reducing carbon emissions, improving air quality, improving energy security, and reducing energy bills (Sharma et al. 2025). Urban population consumes nearly three-fourths of the global energy, and most of it is produced from fossil fuels (Chen 2023). This energy meets the requirements of public transport, lighting, heating, and cooling. Most countries are still producing energy through the traditional modes of coal and oil, which leads to heavy carbon emissions and pollution. Making a shift to renewable energy will result in reducing carbon emissions, improving air quality, improving energy security, and reducing energy bills (Razmjoo et al. 2022). The regulators can make it mandatory for residential, institutional, and commercial buildings to have solar rooftops, which would make energy usage cost-effective and convenient due to the falling costs of panels and metering systems. Urban local microgrids work independently or together with the main grid. Microgrids integrate batteries, wind, solar, and control systems to provide a continuous supply to communities and localities. This would help in building cities that are resilient, energy independent, and have lesser number of grid failures (Al-Thani et al. 2022).

6.2 District Energy Systems

Centralized district heating and cooling systems would power the entire district in a centralized manner, which would help in administering energy on a large scale through underground pipes. This system can be powered by renewable energy sources, such as geothermal energy and biomass (Lake et al. 2017; Bellos et al. 2022). Dubai has successfully implemented the largest cooling network in the world, which has resulted in savings of energy costs compared to the traditional method of air conditioning. Cities can benefit by building small wind turbines in urban areas and large wind farms in rural areas. The small wind turbines can be installed on the rooftops of buildings, while large wind farms in rural areas can produce wind energy on a large scale for industrial and commercial needs. Huge organic waste is produced by cities, which can be transformed into biogas using anaerobic digesters. Bioenergy plants can be powered through biomass, which is produced from organic waste. Indore and Pune in cities in India that produce electricity and compost by using food waste as raw materials (Srivastava et al. 2021, 2018). The supply of solar and wind is not regular in all seasons. Battery storage systems should be established to maintain and manage the energy supply during peak and off-peak times. Cities can focus on implementing different technologies, such as smart meters, energy storage systems, and community battery banks, which can help in dynamic pricing, preventing outages, and balancing the load (Yao et al. 2022).

7 Green Infrastructure and Nature-Based Solutions

The construction activities in urban areas are replacing the natural landscapes with roads, parking lots, and buildings, which have exacerbated the environmental problems. There are many problems that have surfaced, such as biodiversity loss and waterlogging. A paradigm shift is needed in the way a city operates, and it should be making policies that are not against nature. Green infrastructure involves planned natural spaces that deliver various ecosystem services (Štrbac et al. 2023). Green infrastructure helps in improving water and air quality, enhancing biodiversity, and mitigating climate impacts. The focus of governments has shifted to preserving, promoting, and creating green roofs, urban forests, wetlands, permeable pavements, urban parks, and river fronts (Wild et al. 2017; Tripathi et al. 2022). Green infrastructure mitigates the risk of climate change and public health. The advantages of Green infrastructure on wetlands, trees, and vegetation are significant as they absorb carbon dioxide from the environment. Green spaces can reduce the temperature by nearly 5 degrees, which would lower the demand for cooling. Effective storm water management should be implemented so that urban flooding can be reduced. Vegetation and Green infrastructure together can enhance the levels of oxygen in the environment. Usage of green spaces can lower stress levels, increase physical activity, and improve the health of people. Resilient cities can be built by making Nature-Based Projects (NBP), which can outmatch traditional cities in sustainability implementation by addressing the different challenges of construction activities. Nature-based solutions (NBS) have to be implemented to counter the challenges. Green corridors should be constructed to connect the different urban landscapes (Frantzeskaki 2019; Guo et al. 2025). Wetlands should be constructed and preserved for the treatment of wastewater and to restore the natural habitat. Rivers can be made pollution-free by constructing Riparian Buffers. Mangroves should be restored to safeguard against sea-level rise in coastal cities.

8 Waste Management and the Circular Urban Economy

Urbanisation leads to waste generation and massive resource requirements and consumption. Waste management system traditionally consists of processes that are unsustainable and carbon emissions-intensive. Circular economy models should be adopted by the cities, which would minimize the waste, recycle and convert it into a valuable resource (Möslinger et al. 2023; Singh et al. 2025a, b). Greenhouse Gas emissions are more harmful than carbon emissions. Landfills emit huge amounts of methane gas, which is hazardous to the environment. Plastic and municipal waste is burnt, and transport emissions are also harmful to the environment. The manufacturing and disposing of single-use products has also added to the carbon emissions (Agrawal et al. 2025; Daga et al. 2025).

8.1 Principles of Circular Economy in Urban Context

There are three pillars of the circular economy: Reduce, Reuse, and Recycle. It also focuses on the design and development of a system that can manage waste in an effective manner. The focus of the circular economy is on products as a service model, and it also deals with recovering materials from inorganic and organic waste (Singh et al. 2025a, b). Effective usage of waste is significant in the process, as waste of one process could become the input for the other process. Establishment of a framework is necessary, which can efficiently reuse and recycle the waste. Urban planning should embed the requirements of the circular economy in the master plan itself and accordingly implement it (Zeng et al. 2022).

8.2 Organic Waste Management

Recycling organic waste deals with transforming kitchen waste into methane to be used for cooking through biogas digesters. Various composting programs should be run for community and municipal composting plants should be set up for producing fertilizers. Despite recycling being an accepted concept, it is not the focus area for many players in the industry due to infrastructure and awareness issues (Manea et al. 2024; Hettiarachchi and Meegoda 2023). Urban recycling can be improved if the government can make policies that make it mandatory the segregate of waste at the source itself. The sorting and resale of recycled materials should be the focus area, and a digital marketplace should be created for recycled materials. Construction waste recycling plant should be set up, and policies should be made to recycle and reuse construction materials (Ferronato et al. 2023; Mishra and Srivastava 2025). A digital platform should be set-up for the trading resale of recycled industrial waste. Delhi has set-up construction waste recycling facilities that recycle materials to be used in landscaping and roadwork.

A good policy framework is required for transitioning to circular cities. Every city should target to achieve around 100% landfill conversion, and every business and household should be charged a fee based on the quantity of waste generated by them. Producers should be made responsible for the life cycle of their products (Agrawal et al. 2025). The government agencies should mandatorily purchase recycled materials to promote the usage of such materials. Plastic bags have been banned in Kigali, Rwanda, by promoting a circular economy by spreading awareness and strict legislation. The unorganised sector has a significant role to play in recycling waste in developing cities. Waste pickers should be formally organized, and they should be provided with better working conditions and social inclusion benefits. This would help in recycling more material. The government should organize training for waste workers so that they can become more skilled and they could be absorbed in recycling waste businesses (Harfadli et al. 2024; Singh et al. 2025a, b; Srivastava et al. 2024).

9 Digital Technologies and Smart Governance

The integration of urban governance and digital technologies provides many opportunities for improving sustainability in a connected world. Smart governance focuses on inclusiveness, transparency, and responsiveness, along with the reduction of carbon emissions. Cities should start using data and artificial intelligence as a strategic tool in achieving resilience (Nastjuk et al. 2022). Smart technologies have helped municipal corporations to effectively manage air quality, water usage, traffic, and energy usage. This data would provide insights and help cities in responding to problems, rather than reacting to it. Intelligent traffic systems should be installed to analyse the congestion data, which would help in reducing fuel consumption, managing waiting time at signals, and reducing carbon emissions worldwide (Kuzior et al. 2023). Urban planners can use various digital platforms, such as Building Information Modelling, to help visualise the entire building and cities on a real-time basis, which would help them understand the implications of restricting personal vehicles or improving solar energy penetration in a city. Helsinki has benefited by such an initiative, and many positive initiatives have been taken in energy management, zoning, and infrastructure development. Citizen engagement is significant for implementing smart governance in cities. Urban citizens can report their issues on digital platforms through which the government can take cognizance of such issues and take action accordingly. Citizens have greater access to data through dashboards and mobile apps, enabling them to take greener options. Smart systems have their fair share of challenges as well. Digital inequality and data privacy are major issues in cities with less connectivity and a large number of unauthorised colonies (Gautam et al. 2023). Extensive use of digital platforms for governance should not overlook the ethical aspect. Social equity and technological efficiency are the two pillars of smart governance, and the prospect of living a better urban life increases with every technological advancement (Furtado et al. 2023; Giuliodori et al. 2023).

10 Policy Frameworks and Global Best Practices

Continuous urban sustainability focuses on policy, design, and technology. Resilient and low-carbon emission cities need a framework that is robust on legal and financial parameters. Urban planning requires access to land bylaws and climate mandates for effectively implementing sustainability initiatives in cities (Karmakar 2025). The focus of many countries globally has launched initiatives to address the climate risks and infrastructure deficits. AMRUT (Atal Mission for Rejuvenation and Urban Transformation) and India's Smart Cities Mission are advocating the merging of the sustainability aspect into urban planning (Nagargoje et al. 2023). Cities are expected to prepare their climate action plans according to the climate policies of the Centre, with special focus on resilience, energy efficiency, and carbon emissions reduction.

The green policies should be implemented by closely working with the local governments in transport planning, building approvals, zoning regulations, and procurement policies. Municipal corporations have a role to play in enhancing the urban experience, without which the desired results would not be achieved. Strong green building norms, emissions caps, and renewable energy procurement from the public are strong signals of change action (Caldas et al. 2022). Cities need huge financing for investing in renewable energy and green infrastructure, and financing is a scarce resource globally. Innovative financing models are being used by cities to finance projects, such as Public Private Partnerships, International climate funds, and Green Bonds (Alhowaish 2025). Green Bonds were successfully issued in Cape Town, and the proceeds were used for the upgrading of energy infrastructure. Cities are aligning themselves with SDG goals making it safe, sustainable, inclusive and resilient. There are various international forums, such as the C40 Cities Network, that are enabling the sharing of information and knowledge with each other, through which cities can learn from each other's experiences. Effective monitoring systems should be installed by cities for better accountability and tracking progress. The main performance indicators should be regularly tracked, such as public transport usage, passenger vehicle usage, renewable energy share, and carbon emissions (Kalfas et al. 2023). Civic participation and trust are fostered if the data and dashboards are freely available, which promotes transparency (Hyder and Haque 2022). Policies should be practical and enforceable as the urban landscape is rapidly changing. An iterative solution is the need of the hour, which allows real-time course correction, adopting innovative technologies, and achieving resilience in long-run (Kathpal et al. 2021).

11 Conclusion

Urbanization is growing at a rapid pace globally, and the sustainability of cities is an area of concern. Urban areas are accelerating economic growth rapidly and also form the nucleus of pollution, carbon emissions, and resource consumption. Having sustainable cities has become a need today and is no longer a choice. This chapter has explained a multifaceted view of urban sustainability to growing urbanization and carbon emissions. Compact cities and walk able neighbourhoods reduces the dependence on vehicles and also reduces infrastructure costs. Green buildings and the use of renewable energy will reduce the dependence on fossil fuels. Green infrastructure enhances the resilience of a city, and the circular economy model converts waste into an asset. Smart governance and digital transformation create an environment where financial instruments and policies act as enablers which help cities in tackling different challenges. The policies and the technologies used should be resilient and inclusive. A single strategy cannot deliver the expected outcome. A huge impact can be created when there is proper synergy between digital transformation, organizational capacity, and community participation. Cities should act as a laboratory in which the testing and monitoring of sustainability interventions are done continuously, and the suggestions are also implemented in the next round of testing. The future is climate conscious, and the cities have a big role to play in it. This is possible by community engagement and political will. Urban cities can become viable, liveable, sustainable, and regenerative.

References

Agrawal D, Agarwal H, Singh A, Kumar K (2025) From excess to essential: embracing minimalism for resource sustainability in the circular economy. Rethinking resources: strategies, policies, and partnerships for a circular future. Springer, Singapore, pp 127–144

Alhowaish AK (2025) Green municipal bonds and sustainable urbanism in Saudi Arabian cities: toward a conceptual framework. Sustainability 17(9):3950

Al-Raeei M (2025) The smart future for sustainable development: artificial intelligence solutions for sustainable urbanization. Sustain Dev 33(1):508–517

Al-Thani H, Koç M, Isaifan RJ, Bicer Y (2022) A review of the integrated renewable energy systems for sustainable urban mobility. Sustainability 14(17):10517

Ashinze UK, Edeigba BA, Umoh AA, Biu PW, Daraojimba AI (2024) Urban green infrastructure and its role in sustainable cities: a comprehensive review. World J Adv Res Rev 21(2):928–936

Bamwesigye D, Hlavackova P (2019) Analysis of sustainable transport for smart cities. Sustainability 11(7):2140

Bellos E, Iliadis P, Papalexis C, Rotas R, Nikolopoulos N, Kosmatopoulos E, Halmdienst C (2022) Dynamic investigation of centralized and decentralized storage systems for a district heating network. J Energy Storage 56:106072

Bulkeley H, Carmin J, Broto VC, Edwards GA, Fuller S (2013) Climate justice and global cities: mapping the emerging discourses. Glob Environ Change 23(5):914–925

Caldas P, Dollery B, Marques RC (2022) Measuring what matters in local government: a Municipality Sustainability Index. Policy Stud 43(4):738–758

Chen P (2023) Urban planning policy and clean energy development Harmony-evidence from smart city pilot policy in China. Renew Energy 210:251–257

Coupland A (2005) An introduction to mixed use development. In: Reclaiming the city. Routledge, pp 1–30

Daga S, Kumar K, Yadav K, Singh P (2025) Financing and investment in a circular economy: paving the way for sustainable growth. Blending approaches towards sustainability focus. Springer, Cham, pp 17–31

Debrah C, Chan APC, Darko A (2022) Green finance gap in green buildings: a scoping review and future research needs. Build Environ 207:108443

Donateo T, Licci F, D'elia A, Colangelo G, Laforgia D, Ciancarelli F (2015) Evaluation of emissions of CO_2 and air pollutants from electric vehicles in Italian cities. Appl Energy 157:675–687

Ferronato N, Fuentes Sirpa RC, Guisbert Lizarazu EG, Conti F, Torretta V (2023) Construction and demolition waste recycling in developing cities: management and cost analysis. Environ Sci Pollut Res 30(9):24377–24397

Frantzeskaki N (2019) Seven lessons for planning nature-based solutions in cities. Environ Sci Policy 93:101–111

Furtado LS, da Silva TLC, Ferreira MGF, de Macedo JAF, Cavalcanti JKDML (2023) A framework for digital transformation towards smart governance: using big data tools to target SDGs in Ceará, Brazil. J Urban Manag 12(1):74–87

Gately CK, Hutyra LR (2017) Large uncertainties in urban-scale carbon emissions. J Geophys Res: Atmos 122(20):11–242

Gautam D, Prakash A, Kumar K (2023) Shaping sustainable smart cities as a model for urban development in India. In: 2ND international conference on futuristic and sustainable aspects in engineering and technology: FSAET-2021, vol 2721, no 1. AIP Publishing LLC, p 060007

Gauthier J, Wooldridge B (2012) Influences on sustainable innovation adoption: evidence from leadership in energy and environmental design. Bus Strateg Environ 21(2):98–110

Geng Y, Dong H, Xue B, Fu J (2012) An overview of Chinese green building standards. Sustain Dev 20(3):211–221

Giuliodori A, Berrone P, Ricart JE (2023) Where smart meets sustainability: the role of Smart Governance in achieving the Sustainable Development Goals in cities. BRQ Bus Res Q 26(1):27–44

Glaeser EL, Kahn ME (2010) The greenness of cities: carbon dioxide emissions and urban development. J Urban Econ 67(3):404–418

Guo Y, Liu S, Dong Y, Wu G, Liu J, Wang W, Wang J (2025) Research progress and prospects of nature-based solutions in green infrastructure: a bibliometric analysis. Ecol Front 45(3):561–571

Haghshenas H, Vaziri M, Gholamialam A (2015) Evaluation of sustainable policy in urban transportation using system dynamics and world cities data: a case study in Isfahan. Cities 45:104–115

Harfadli MMA, Ramadan BS, Rachman I, Matsumoto T (2024) Challenges and characteristics of the informal waste sector in developing countries: an overview. J Mater Cycles Waste Manag 26(3):1294–1309

Heidrich O, Hill GA, Neaimeh M, Huebner Y, Blythe PT, Dawson RJ (2017) How do cities support electric vehicles and what difference does it make? Technol Forecast Soc Change 123:17–23

Hettiarachchi H, Meegoda JN (2023) Microplastic pollution prevention: the need for robust policy interventions to close the loopholes in current waste management practices. Int J Environ Res Public Health 20(14):6434

Hightower R (2013) Investigating the green Leadership in Energy and Environmental Design (LEED) servicescape scale in Brazil. Constr Innov 13(3):242–265

Hoppenbrouwer E, Louw E (2005) Mixed-use development: theory and practice in Amsterdam's Eastern Docklands. Eur Plan Stud 13(7):967–983

Hyder MB, Haque TZ (2022) Understanding the linkages and importance of urban greenspaces for achieving sustainable development goals 2030. J Sustain Dev 15:144

Ibraeva A, de Almeida Correia GH, Silva C, Antunes AP (2020) Transit-oriented development: a review of research achievements and challenges. Transport Res Part a: Policy Pract 132:110–130

IPCC (2022) Climate change 2022: Mitigation of climate change. https://www.ipcc.ch/report/ar6/wg3/downloads/report/IPCC_AR6_WGIII_FullReport.pdf

Jiang J, Zhu S, Wang W, Li Y, Li N (2022) Coupling coordination between new urbanisation and carbon emissions in China. Sci Total Environ 850:158076

Kalfas D, Kalogiannidis S, Chatzitheodoridis F, Toska E (2023) Urbanization and land use planning for achieving the sustainable development goals (SDGs): a case study of Greece. Urban Sci 7(2):43

Karmakar J (2025) Revitalizing urban green spaces: insights from West Bengal's Atal Mission for rejuvenation and urban transformation. Available at SSRN 5511559

Kathpal S, Kumar K, Kumar P, Mishra S (2021) Building a low carbon Indian economy. In: IOP conference series: earth and environmental science. IOP Publishing, vol 795, no 1, p 012023

Khan HH, Malik MN, Zafar R, Goni FA, Chofreh AG, Klemeš JJ, Alotaibi Y (2020) Challenges for sustainable smart city development: a conceptual framework. Sustain Dev 28(5):1507–1518

Kumar K, Singh R, Sharma A, Lehri BPS (2024) Climate change adaptation strategies for achieving net-zero economy. Transition towards a sustainable future: net zero policies and environmental sustainability. Springer, Singapore, pp 163–183

Kuzior A, Pakhnenko O, Tiutiunyk I, Lyeonov S (2023) E-governance in smart cities: global trends and key enablers. Smart Cities 6(4):1663–1689

Lake A, Rezaie B, Beyerlein S (2017) Review of district heating and cooling systems for a sustainable future. Renew Sustain Energy Rev 67:417–425

Lenzholzer S, Carsjens GJ, Brown RD, Tavares S, Vanos J, Kim Y, Lee K (2020) Awareness of urban climate adaptation strategies—an international overview. Urban Climate 34:100705

Liu T, Chen L, Yang M, Sandanayake M, Miao P, Shi Y, Yap PS (2022) Sustainability considerations of green buildings: a detailed overview on current advancements and future considerations. Sustainability 14(21):14393

Luqman M, Rayner PJ, Gurney KR (2023) On the impact of urbanisation on CO_2 emissions. NPJ Urban Sustain 3(1):6

Manea EE, Bumbac C, Dinu LR, Bumbac M, Nicolescu CM (2024) Composting as a sustainable solution for organic solid waste management: current practices and potential improvements. Sustainability 16(15):6329

Mishra A, Srivastava SK (2025) Role of trust and celebrity endorsement in shaping purchase intention towards electric vehicles. In Implementing ESG frameworks through capacity building and skill development. IGI Global Scientific Publishing, pp 381–400. https://doi.org/10.4018/979-8-3693-6617-2.ch018

Möslinger M, Ulpiani G, Vetters N (2023) Circular economy and waste management to empower a climate-neutral urban future. J Clean Prod 421:138454

Nagargoje SV, Dubey DD, Hampannaver PR, Patil SG (2024) Overview of sustainable development initiatives in India check for updates. In: Sustainable built environment: select proceedings of ICSBE 2023, vol 451, p 127

Nasri A, Zhang L (2014) The analysis of transit-oriented development (TOD) in Washington, DC and Baltimore metropolitan areas. Transp Policy 32:172–179

Nastjuk I, Trang S, Papageorgiou EI (2022) Smart cities and smart governance models for future cities: current research and future directions. Electron Mark 32(4):1917–1924

Porter L, Rickards L, Verlie B, Bosomworth K, Moloney S, Lay B, Pellow D (2020) Climate justice in a climate changed world. Plan Theory Pract 21(2):293–321

Razmjoo A, Gandomi AH, Pazhoohesh M, Mirjalili S, Rezaei M (2022) The key role of clean energy and technology in smart cities development. Energy Strateg Rev 44:100943

Sharma A, Kumar K, Dwesar R (2025) "Keenly aware of environmental issues, yet wavering in conviction": understanding the factors influencing the consumers' willingness to pay for green energy. Sustain Futures 9:100427

Shen L, Peng Y, Zhang X, Wu Y (2012) An alternative model for evaluating sustainable urbanization. Cities 29(1):32–39

Silva BN, Khan M, Han K (2018) Towards sustainable smart cities: a review of trends, architectures, components, and open challenges in smart cities. Sustain Cities Soc 38:697–713

Singh R, Sharma A, Kumar K (2025a) Navigating carbon neutrality: strategies for sustainable future and environmental challenges. Climate neutrality through smart eco-innovation and environmental sustainability. Springer, Cham, pp 69–80

Singh R, Tiwary KS, Kumar K, Kumar A (2025b) Empowering a sustainable future: blockchain's role in economy, organizations, and finance towards smart circular economy. Blockchain technologies for smart circular economy and organisational sustainability. Springer, Cham, pp 271–283

Sridhar KS, Mahendra A (2025) Economic benefits of the metro and relaxed floor area ratio: evidence from Bengaluru, India. Case Stud Transp Policy 19:101395

Srivastava SK, Anshul A, Pathak P, Bansal J (2018) Responsible consumption for curbing food wastage: an exploratory enquiry. PURUSHARTHA 11(2):104–116

Srivastava SK, Singh S, Srivastava M (2021) Food waste at household and social gatherings: drivers and possible remedies. PURUSHARTHA 14(2):117–133

Srivastava SK, Sinha AK, Kumar S (2024) Workplace spirituality empowered by spiritual leadership to promote organizational effectiveness: a bibliometric analysis. PURUSHARTHA 17(1):1–22. https://doi.org/10.21844/16202117101

Štrbac S, Kašanin-Grubin M, Pezo L, Stojić N, Lončar B, Ćurčić L, Pucarević M (2023) Green infrastructure designed through nature-based solutions for sustainable urban development. Int J Environ Res Public Health 20(2):1102

Tan Y, Xu H, Zhang X (2016) Sustainable urbanization in China: a comprehensive literature review. Cities 55:82–93

Tarkar P (2022) Role of green hospitals in sustainable construction: benefits, rating systems and constraints. Mater Today: Proc 60:247–252

Tripathi V, Singh K, Parihar S, Srivastava SK (2022) A bibliometric analysis on supply chain collaboration. PURUSHARTHA 15(2):66–103. https://doi.org/10.21844/16202115205

Tyler S, Moench M (2012) A framework for urban climate resilience. Climate Dev 4(4):311–326

United Nations (2018) Retrieved 24 Feb 2026 from https://www.un.org/uk/desa/68-world-population-projected-live-urban-areas-2050-says-un

Wibowo MA (2024) Green construction and local wisdom integration for sustainability: a systematic literature review. Civil Eng J 10(11)

Wild TC, Henneberry J, Gill L (2017) Comprehending the multiple 'values' of green infrastructure—valuing nature-based solutions for urban water management from multiple perspectives. Environ Res 158:179–187

Wong YZ, Hensher DA, Mulley C (2020) Mobility as a service (MaaS): charting a future context. Transport Res Part a: Policy Pract 131:5–19

World Cities Report (2024) Cities and climate action. https://unhabitat.org/sites/default/files/2024/11/wcr2024_-_full_report.pdf

Xia J, Zhang Y (2022) Where are potential areas for transit-oriented development (TOD)—exploring the demands for built environment for TOD planning. Sustainability 14(14):8364

Yang L, Liang Z, Yao W, Zhu H, Zeng L, Zhao Z (2023) What are the impacts of urbanisation on carbon emissions efficiency? Evidence from Western China. Land 12(9):1707

Yao X, Ma SC, Fan Y, Zhu L, Su B (2022) An investigation of battery storage operating strategies in the context of smart cities. Ind Manag Data Syst 122(10):2393–2415

Zeng X, Yu Y, Yang S, Lv Y, Sarker MNI (2022) Urban resilience for urban sustainability: concepts, dimensions, and perspectives. Sustainability 14(5):2481

Zhalmurziyeva K, Tokbolat S, Durdyev S, Mustafa MY, Karaca F (2024) Assessment of sustainability indicators for urban water infrastructure in a developing country. Int J Build Pathol Adapt 42(3):337–351

Urban Governance for SDGs Achievement Through Sustainable Innovation in Developing Economies

Shubhra Mishra

Abstract Urbanization in developing economies brings opportunities for economic and social development. However, it also creates a high risk of unsustainable growth, informality, and environmental instability. This chapter discusses the dynamic nature of urban sustainability governance using a comparative lens based on theoretical and documented case studies and best practices. By analysing key governance models such as metropolitan, community-based, and city-to-city collaboration, the chapter shows how multi-actor engagement and key policy tools enable urban transformation. The chapter also highlights some of the universal best practices. It also explains the significance of having participatory governance, cross-level collaboration, flexible regulatory frameworks, and goal alignment with global sustainability agendas. In conclusion, it provides suggestions for the design of responsive and innovative governance arrangements for sustainable urban futures to policy-makers, city managers and, development partners.

Keywords Urban sustainability · Developing economies · Participatory governance · Green infrastructure · SDG · Sustainable innovation

1 Introduction

In the developing world, the concept of urbanization is significant as the magnitude and rate of urbanization are unparalleled and reshaping society. More than half of the earth's human population currently lives in cities with over 95% of the urban expansion of sub-Saharan Africa, Asia, and Latin America projected in the future (United Nations 2018). The rapid shift from rural to urban is motivated by the search for improved lifestyles and economic prospects (Kumari et al. 2025). This growth also leads to significant sustainability challenges from environmental degradation,

S. Mishra (✉)
Indian Institute of Technology Jodhpur, Jheepasani, India
e-mail: reachshubhra11@gmail.com

K. Kumar et al. (eds.), *Sustainable Value Creation: Corporate Strategies for SDGs Implementation*, World Sustainability Series,
https://doi.org/10.1007/978-3-032-24652-3_13

to housing insufficiency, and high social inequality. The problems between urbanization and sustainability are clearly seen in developing countries. Here, these challenges include informal settlements with poor amenities, services and infrastructure as well as pollution (Bashir et al. 2023). Carbon emissions, and rising exposure to climate hazards such as flooding and heat extremes can be seen as environmental impacts arising from these urbanization challenges (Shen et al. 2016). Therefore, the need for strong urban governance becomes important. The success of urban government structures determines whether cities can be the engines of national and global sustainability agendas, or will enhance existing developmental pressures (Shen et al. 2016). Governance here is not limited to formal government structures but includes a range of actors and processes such as public bodies, private corporations, civil society and global networks which work within porous and frequently fragmented institutional terrains (United Nations 2018). There are a variety of reasons for privileging urban areas as key sites within sustainability agendas. One, cities are where the economic action takes place. More than 70 percent of global GDP and a large proportion of resource consumption come from cities (World Bank 2020). Second, urban areas are not only the origin and the place of many sustainability problems but they also represent the greatest potential to put in place large-scale solutions whether in the area of energy, transportation, housing and resources (Kumari et al. 2025; Shen et al. 2016). Third, they are living laboratories for policy innovation, creating paths of resilient, inclusive, and low-carbon development that are increasingly acknowledged by international agendas such as the SDGs and Paris Agreement (Amorim et al. 2022; United Nations 2018). In developing economies, the governance of urban sustainability is shaped by a distinct set of pressures and possibilities. Budgetary limitations, rapid demographic transitions, and the prevailing informality stand in the way. Yet, they also stimulate creative responses including participatory budgeting, green infrastructure, and community-based adaptation that have much to teach on urban transformation worldwide (Zeng et al. 2016).

The objective of this chapter is to develop a comparative framework of urban sustainability governance in developing cities, pursued along three interrelated aims. First, it adds to our theoretical understanding in the literature on urban governance and sustainability, outlining how governance architectures influence urban outcomes. Second, it categorically presents each of the most critical barriers such as financial, institutional, political, and socio-environmental to the pursuit of sustainable urban development. Third, it also focuses on innovative policies and best practices, based on secondary literature and documented experiences in several regions, to contribute to policy and academic discussions.

2 Conceptual Underpinnings: Urban Governance and Sustainability

Urban governance is a complex structure of approaches, actors, mechanisms and decision-making processes affecting how cities are designed, planned, developed and managed (Baud et al. 2021). It refers to systems in place to enforce policy and guide collective decision-making among different government and nongovernment agencies (UN-Habitat 2016). No less than three key dimensions are at stake: multi-level governance (coordination of governance among local, regional, national institutions), polycentric governance (many centres of authority interacting in multiple, overlapping jurisdictions) and participatory governance (inclusion of civil society, business and other constituents) (Ostrom 2017). Municipal governments and urban agencies, as well as non-state actors (businesses, community organizations, and international donors) bring unique capabilities and viewpoints to the urban policy arena (Pierre 2011). Negotiation, coordination and mediation across interests and scales is the process of effective urban governance (Bulkeley 2013). It includes policy making, resource allocation, implementation, monitoring, and mechanisms of accountability (Broto 2017). These processes have positive effects on collaboration, transparency and agility in addressing the problems of cities. These qualities are well suited to the dynamic settings of emerging markets.

Sustainability in the urban context can be discussed as the ability to maintain a balance quality of life that maintains environmental integrity and fairness between current and future generations (Portney 2013; Singh et al. 2025a, b; Prakash and Tiwari 2025). It encompasses the environmental dimension, which deals with resource efficiency, pollution control and climate change resilience (UN-Habitat 2020). The foundation of social sustainability comprises equitable distribution of resources among residents, social inclusion and accessible basic services and participatory opportunities for all, especially marginalized groups (Evans and Thomas 2023). Economic sustainability focuses on productive places of work, the stimulation of innovation and the long-term prosperity of the economic bases of cities (Portney 2013). These pillars form the backbone of many international agendas and the United Nations' Sustainable Development Goals (SDGs) in particular, related to the aim to "make cities and human settlements inclusive, safe, resilient and sustainable" (United Nations 2015) and global climate adaptation and mitigation initiatives. The association between urban governance and sustainability is critical: governance structures (inclusiveness, power and coordination) condition cities' ability to create sustainable outcomes (Bulkeley 2013; Kumar et al. 2025). Cities that have strong participatory structures and interlinked policies are also likely to be able to enact sound environmental and social policies. On the other hand, fragmented or top-down governance tends to produce policy incoherence, exclusion of critical stakeholders, and unplanned urban sprawl (UN-Habitat 2020). There are several conceptual debates that structure the discussion on urban governance for sustainability. The centralization-decentralization debate centres on whether governance is more responsive and efficient when power is pooled at higher-level authorities or pushed down to the grassroots level. On the other hand, decentralization processes may be expected to spur innovation and the development of contextually sensitive decisions (Smoke 2015; Singh et al. 2024; Agrawal 2025; Joselin et al. 2025). A second central debate places focus on formal versus informal governance, the arrangement between official institutions and informal actors and practices that are more common to developing cities. Although informal governance provides a form of flexibility and some community-based solutions, it can also be a means to reproduce inequality and dismiss accountability (Auerbach and Thachil 2018; Singh et al. 2025b). The governance of informality: how states make formal and informal settlements and economies 'fit' or repress them, is therefore a key dimension of the shaping of urban sustainability pathways (Parnell and Simon 2014).

3 Governance Structures in Developing Economy Cities

The nature of urban governance in developing economies is dynamic and multi-layered, involving a variety of actors, institutions, and decision-making logics. The challenges of service delivery, infrastructure provision, inclusion and sustainability that come with increasing population growth and socio-economic demands stretch city governance arrangements in all world regions. The urban governance space is characterized by three main models: metropolitan governance, both large and small city networks, community-led governance and participatory budgeting, and the role of regional/national authorities. They interact across a variety of environments, which are frequently characterized by institutional pluralism and fragmentation, and key coordination challenges at multiple scales (e.g. local–global). This section outlines the main urban governance types, delves into challenges with coordination, and highlights key theoretical perspectives such as polycentric and collaborative governance.

3.1 Typology of Models

(a) **Metropolitan Governance and City Networks**

Metropolitan governance, defined as the coordination of policymaking, planning, and services across large urban regions often across multiple municipalities, peri-urban regions, and surrounding jurisdictions, is an important aspect of the local governance system. Metropolitan models emerge based on the interdependence of economic, social and environmental challenges that occur across administrative borders, including transportation, housing, pollution, and water (Bahl and Linn 2014). In a multitude of developing contexts, metropolitan governance exists as formal metropolitan authorities or informal collaboration among local governments. City networks, on the other hand, reflect systematic collaborations of cities both domestically and internationally. Initiatives like the C40 Cities Climate Leadership Group and Global Covenant of Mayors dominate their governance in coherence with sustainability in the developing economies (Acuto and Rayner 2016). Metro governance such as with transport coordination in Bogota or regional planning in Johannesburg, are examples that show that this form of governance also offers opportunities for integrated urban sustainability (UN-Habitat 2020).

(b) **Community-led Governance and Participatory Budgeting**

Community-based governance places an emphasis on the role of local communities, community organizations, and community actors in determining urban priorities and developing solutions for local development (Coaffee and Deas 2008). In poor, resource-scarce, and largely informal Global South cities, such grassroots initiatives attend to fundamental needs, enriching social capital, and contribute to greater resilience in the face of shock. The governance of the informal settlements, as well as collective access to water and sanitation and community based disaster risk management is, indeed, often initiated by these organized groups at the local level (Fox et al. 2023). Participatory budgeting, in which citizens vote directly on which portions of the public budget should be used for what purposes, has been given great impetus in Latin American cities including Porto Alegre and Belo Horizonte and is now being taken up elsewhere in Africa and Asia (Cabannes 2015). Participatory budgeting has shown effects in terms of transparency, accountability, and social inclusion, particularly when empowered by enabling institutional and legal frameworks (Wampler and Hartz-Karp 2012). It is emblematic of a new generation of more democratic and deliberative urban governance, which gives low-income communities and communities of colour long alienated from traditional, top-down planning processes, a share of the power.

3.2 Regional and National Authorities' Role

Regional and national governments largely determine the legal, financial and policy framework in which city-level governance is enshrined (Corfee-Morlot et al. 2009). Cities are highly dependent on higher levels of government for fiscal transfers, regulatory approval and large scale infrastructure investment in most developing countries (Smoke 2015; Srivastava et al. 2024; Tripathi et al. 2022). The roles and responsibilities of national, regional and local governments are sometimes not clear or disputed, resulting in overlapping mandates or lack of authority (UN-Habitat 2016). It is national urban policies- land-use codes, housing rules, climate action strategy that establish the boundaries within which local innovation and accountability can and should take place. For instance, the interventions of India's Smart Cities Mission or the Integrated Urban Development Framework in South Africa demonstrate strong central leadership but also the difficulty that comes with aligning the national-level goals with local capabilities (Denis et al. 2017; Pieterse and Parnell 2014). Intermediary or regional authorities (state/provincial government) are the focal point for coordination in some areas, but also add another layer of red tape or politics. Cities in developing economies tend to harbour institutional pluralism, meaning that multiple public organizations, NGOs, private sector and informal institutions coexist sometimes working together, frequently in competition (Batley and McLoughlin 2015; Srivastava et al. 2021). This pluralism carries potential for innovation, diversity and responsiveness but also risks of fragmentation and inefficiency. The lack of transparent assignment of accountabilities or the breakdown of communication between urban and regional agencies often leads to duplication, contradictions or service holes (Lindell 2008). Institutional coordination issues are particularly apparent between city, region and national levels. Fiscal decentralization, for one, is frequently incomplete or ad hoc, undermining the autonomy of cities to fully implement their mandates (Bahl 1999). Furthermore, the pace of growth of urban populations in informal settlements largely unfolds outside the formal policy arena, compounding coordination challenges related to land use, infrastructure and risk reduction (Parnell and Robinson 2012; Srivastava et al. 2018). In the end, the organization of urban governance in developing economies may be seen as a continuing experiment with forms and arrangements optimized to strike a balance between efficiency, responsiveness, inclusion, and the long-term sustainability of urban products and processes. Although the models of organization vary in their institutional configuration and their efficacy, success often depends upon the ability to coordinate diverse actors and levels.

4 Policy Instruments for Sustainable Urban Development

Sustainable urbanization in emerging economies is largely determined by how the main policy tools and measures are implemented and coordinated. These strategic priorities, based on good governance models, responding to particularities at local level are green infrastructure, sustainable mobility, integrated planning, eco-budget with green finance and good harmonisation with international networks and standards (Table 1).

The subsequent analysis draws on secondary literature as well as documented best practice to present a diffuse picture of the topic:

Green infrastructure (GI, e.g. parks, urban forests and green roofs) provides essential ecosystem services such as temperature reduction, air-quality improvement, stormwater management and community well-being (Herath and Bai 2024). In these fast-urbanizing and frequently informal settings GI has the potential to address challenges that could not be handled by the traditional infrastructure (Fang et al. 2023; Kaith and Sachdeva 2024). Community-led intervention and inter-sectoral collaboration are vital components for a successful GI intervention, such as introduced to new low-income urban settlements are decentralized wastewater treatment with constructed wetlands (Prescott et al. 2021). The efficacy of GI is enhanced when incorporated into urban networks, enabling connections between biodiversity, recreation and economic services (Fang et al. 2023; Prescott et al. 2021). Nevertheless,

Table 1 A comparative representation of policy tools and instruments for urban sustainability

Policy tool/ instrument	Key functions	Best practices/ examples	Challenges	Enabling conditions
Green infrastructure	Urban cooling, air quality, stormwater management, recreation, social wellbeing	Parks, city gardens, constructed wetlands in informal areas	Competition for land, level of technology, management responsibility	Community engagement, cross-sector collaboration, governance alignment
Sustainable mobility and TOD (transit oriented development)	Sustainable transport, integration with land use, and accessibility for the marginalised populations	Bogota TransMilenio (BRT), bike lanes, pedestrian areas	Informal area exclusion, affordability and urban sprawl	Political leadership, coordination among agencies, financing and development
Integrated urban planning and land use	Area Upgrading of Informal Settlements, disaster risk reduction, zoning reforms	Pukou district (China) informal regularization (UN-Habitat)	Resistance to reform, spatial fragmentation	Secure tenure, participatory planning, flexible legal frameworks
Eco-budgeting and green finance	Integrating sustainability in the budgets of the city; unlocking investment through green bonds/funds	Asian city municipal green bonds, city climate funds	Poor credit rating, regulatory vacuum, investor apathy	Transparency, financial independence, sustainable reporting mechanisms
International standards and partnerships	Benchmarking, capacity, aspiration (SDGs, Paris Agreement), financing opportunities	C40 Cities, GCoM, UN-Habitat	Disconnect between local and global structures, uneven take-up	Robust city diplomacy, data-based reporting, network engagement

there are issues to be resolved, including land competition, technical capability and maintenance costs, and the effects are highly influenced by governance and cultural fit.

Transit-oriented development (TOD) and investments in sustainable mobility bus rapid transit (BRT), cycling infrastructure, walkable urban design serve as an access point for low-carbon, inclusive cities (World Bank 2019). Positive examples of this in Latin America (for instance, Bogota's TransMilenio) have demonstrated reductions in emissions and gains in access for disadvantaged populations by integrating land use and transport policy (Fang et al. 2023). Political leadership, inter-agency cooperation, and financing models are key to successful implementation. However, there continue to be challenges in how to add informal settlements into formal transit plans to ensure equitable access at affordable costs, especially in areas where urbanization is sprawling and market forces determine land use (World Bank 2019).

Comprehensive urban planning, such as zoning changes, regularization of informal settlements, and management of land readjustments are all key to guiding sustainable city growth (Wei et al. 2018). Unstructured growth results in fragmented green spaces and susceptibility to hazards; spatial planning models tested at a higher level, for example, the Pukou district in China, maintain an arrangement of urbanization for green features (Wei et al. 2018). Recent best practice strongly supports the upgrading and regularization of informal settlements where security of tenure and access to basic services is ensured rather than moving thousands of people to another location (UN-Habitat 2020). This is the agenda of equity, social inclusion and urban resilience (Batley and McLoughlin 2015).

Urban fiscal-policy innovation has swung the spotlight onto eco-budgeting and green financing instruments including municipal green bonds and city climate funds (Jena 2021). Eco-budgeting mainstreams environmental targets across the municipal planning and expenditure cycles, to allow systematic reporting against sustainability objectives. Green bonds and funds facilitate access to innovative sources of capital for investments in energy-efficiency measures, green infrastructure, and climate adaptation visible in early initiatives throughout cities in Asia and Africa (Jena 2021). Despite this promise, barriers remain. Green finance has little chance of being effective without clear regulation, with cities that are creditworthy, and with investors that are interested, all of which are scarce in the developing world (Prescott et al. 2021). Alignment with global frameworks and city networks including the C40 Cities and the Global Covenant of Mayors (GCoM) which enable shared knowledge and capacity building, and access to finance (Acuto and Rayner 2016). Through peers and technical support, these partnerships provide benchmarking and technical support to improve the chances of localizing international targets (SDG 11, the Paris Agreement, etc.) (UN-Habitat 2020). Long-term progress requires sustained effort, such as continued political commitment, investment in developing technical and managerial capacity and governance structures that can learn and adapt (Acuto and Rayner 2016). Empirically, cities that strategically implement green infrastructure, build in support of sustainable mobility, and focus on participation and community-led planning generate significant positive economic, social, and environmental returns (Wei et al. 2018; Prescott et al. 2021). This kind of urban sustainability requires local ownership and cooperation among cities.

5 Major Challenges and Barriers to Urban Sustainability Governance

Challenges for urban sustainability in developing economies include fiscal, human, institutional, and socio-political dimensions. These factors also determine the capabilities of cities to deliver sustainable, equitable, and resilient urban initiatives.

Restrictions on accessing financial capital for municipalities are very common barrier among all. Cities often lack strong sources of revenue and are reliant on intergovernmental transfers, which are inadequate for sustainability goals and objectives (Smoke 2015). In addition, lack of investment into crucial infrastructure, such as water, transport, green spaces, and waste management, is also an important barrier to look at (Jena 2021). These are financial constraints which lead to dependence on donor finance or international development agencies with resultant patchwork, project-based interventions rather than integrated, city-driven strategies (Batley and McLoughlin 2015; UN-Habitat 2020). Moreover, major technical, managerial, and institutional capacity gaps also undercut urban governance in the Global South. For instance, many city administrations suffer from a severe lack of well-trained planners, engineers, financial managers, and data analysts responsible for developing and executing increasingly complex sustainability interventions (UN-Habitat 2016). There is, all too often, insufficient or insufficiently integrated knowledge of urban development, environmental management and participation across the agencies. Some cities continue to face challenges surrounding training and staff retention, particularly in rapidly growing cities or weak institutional cultures (Acuto and Rayner 2016). These capacity deficits limit cities in their ability to plan, implement, monitor and adjust sustainability policies and to engage in meaningful ways with stakeholders.

One more major structural barrier to integrated urban governance is fragmentation. Confusion and gaps in execution of related policies often result from overlaps between multi-level governance- city, regional, and national mandates (Bahl and Linn 2014). In reality, jurisdictional fragmentation between agencies can exist in infrastructure, housing, land use, and environmental regulation with little coordination or vision between them (Lindell 2008). A lack of inter-agency coordination and competing regulatory frameworks delays approvals, undermines accountability, and too often results in the duplication or omission of important matters (An et al. 2021). Wherever the metropolitan or municipal level is loath to genuinely devolve authority and resources, this just becomes another front of political rivalry, if not outright conflict (Smoke 2015). In addition, socio-political dynamics are a strong hindrance to sustainable urban development. Due to the widespread nature of informality within both settlements and economies a large portion of the urban population resides and earns their livelihood in informal governance and policy space (Auerbach and Thachil 2018). Informal settlements have little to no legal standing and their residents tend to lack access to basic services, as well as being excluded from official planning processes. The exclusion perpetuates cycles of poverty, insecurity, and environmental degradation. It also reflects governance failures. The needs of marginalized communities especially, such as women, the poor, migrants, and ethnic minorities tend to be invisible from urban decision-making and allocation of public resources as well as urban interventions (Evans and Thomas 2023; UN-Habitat 2020). Top-down approaches create a greater deal of mistrust, and tokenistic participation does not change anything. In the absence of inclusive governance, informal and grassroots governance structures may develop, but they too can replicate social hierarchies and remain devoid of formal accountability (Parnell and Pieterse 2014). Numerous examples have pointed to regulatory and institutional "lock-ins" i.e., the entrenched rules, norms, and practices that bind cities to past investment decisions which may hinder urban sustainability (Bulkeley 2013). The need for sustainable development remains hampered by issues such as legacy infrastructure, stuck zoning, and continued heavy reliance on fossil fuels, as well as land tenure laws that make it practically impossible to upgrade in-place or enter into joint ventures. Yet, path dependency can also be found in planning cultures that are more averse to change, where previous decisions fix certain alternatives and reduce the ability to adapt to new technologies or participatory techniques (Ostrom 2017; Healey 2020). Even when evidence amply demonstrates the need to change, institutional inertia creates barriers to reallocating resources, incorporating new actors, and testing new governance structures. Politics, often driven by rent-seeking behaviour of landowners, service providers or bureaucratic elites, may thwart reform processes (Bulkeley 2013).

5.1 Critical Reflections: Real Cases

Taken together, above-discussed challenges are starkly reflected in the urban resilience gap of small and medium-sized cities with limited, sometimes no resources, resource endowments and policy space; unlike the capital cities or megacities (UN-Habitat 2020). For instance, the fact that many African and South Asian secondary cities are not being made more resilient to climate shocks, flooding, extreme heat, disease outbreaks. They do not just reflect financial and technical shortfalls but also fragmented governance, exclusion of informality, and regulatory rigidity (Parnell and Robinson 2012; Acuto and Rayner 2016). In Latin America, while participatory budgeting practices in cities, such as Porto Alegre, are now celebrated as world best practice, attempts to copy them elsewhere failed, in the absence of enabling institutions and political leadership, therefore implying the need for systemic, context sensitive strategies (Cabannes 2015). Incomplete mandates among Urban Local Bodies, Parastatal Agencies, and state governments in India have resulted in stalled projects for infrastructure upgradation and also insecurity in service delivery still persists (Aijaz 2016). These examples highlight the need for unified, multi-scalar approaches that can consider across constraints. The goal is not just to find a few band-aids, but to design new policies that could accelerate efforts to overcome financial bottlenecks and capacity constraints.

6 Comparative Policy Innovations in City Governments:

Urban governance innovations in the Global South shed light on the multiple pathways. Many cities are following these innovations to pursue sustainability. Although the challenges are significant, significant examples from Latin America, South Asia, and Africa show that effective governance structures, can make a real contribution to better urban sustainability and social inclusion (UN-Habitat 2020; Acuto and Rayner 2016).

- **Case 1: Green Mobility Reforms in Bogota**

An important example in the green mobility revolution is Bogota, Colombia. Bogota led a radical restructuring of its urban transport network in the late 1990s and early 2000s through implementation of the TransMilenio bus rapid transit (BRT) system. This project provided a low-cost mass transport alternative to costlier metro networks. This also helped to significantly decrease traffic congestion, travel times and vehicle pollution. It also led to increasing accessibility for low and middle-income users (Gilbert 2008). Additionally, the city funded a large amount of cycling infrastructure including more than 300 kms of "Cicloruta" bicycle lanes and events like Sunday Ciclovia that closed major roads to cars in order for community recreation (Cervero et al. 2009). These included changes in commuting patterns, air pollution reduction, improved road safety and quality of life benefits. Yet, maintaining these benefits has necessitated continued political will, adaptation to growing urbanisation, and the need to tackle persistent fiscal and social inequities (Hidalgo and Huizenga 2013).

- **Case 2: Pune's Integrated Waste Management**

Another example is Pune, India where integrated sustainable solid waste management developed through a collaborative model with informal sector workers. By the early 2000s, Pune was experiencing increasing waste generation, overflowing landfills, and informality in waste collection. The city then formed a decentralized, door-to-door waste collection system with the help of "SWACH". It is a cooperative run by women waste pickers which formalized the employment of thousands of informal workers (Chikarmane and Narayan 2023). Not only did this partnership innovate traditional strategies to improve waste segregation and recycling rates, it also used this project to directly increase labour rights, safety and social status for a marginalised population (Baud and De Wit 2008). Formal negotiations between the municipal corporation, local NGOs, and associations of waste pickers established governance arrangements with contracts for payment, capacity-building programs, and regular stakeholder platforms to facilitate dialogue and monitor progress. This led to an increase in household source segregation in Pune, reduced reliance on landfills and improved livelihoods and dignity at the workplace for thousands of women (Chikarmane and Narayan 2023). Citizen awareness campaigns, combined with integrative feedback loops, helped keep the movement alive while the system adapted to the urban growth and the new neighbourhoods (Baud and De Wit 2008). Pune's journey mapping echoes the untapped potential of co-production, the processes through which city and community together shape the services they need to realize the twin goals of social inclusion and environmental sustainability through the heed given to systemic integration of expert local knowhow (Chikarmane and Narayan 2023).

- **Case 3: The Informal Settlement Upgrading Program of African Cities**

More broadly, the challenge of informal settlements across African cities has led to emerging models of participatory upgrading and land tenure reform. Responses to informality have historically leaned toward forced eviction and top-down planning that has continued to increase social and environmental vulnerability (Roy 2005). More recently, approaches have focused on in-situ upgrading: assuring secure tenure, improving infrastructure incrementally, and facilitating planning by the residents (Lines and Makau 2018). An instructive example of this co-design in practice comes from the Mukuru Special Planning Area in Nairobi, where partnerships of city authorities, grassroots organizations, NGOs and international donors have collaborated on phased roll-out of essential services including clean water and sanitation, roads and electricity alongside local resident participation in governance bodies (Lines and Makau 2018). It institutionalised participatory principles such as community enumeration, flexibility in standards, and layered financing to promote tailored solutions as opposed to blanket approaches (Satterthwaite et al. 2020) in the design of a National Upgrading Support Programme (NUSP) in South Africa. Enabling regulatory schemes, acknowledgment of informal leadership, sustained investments by government and civil society, and a focus on improving social cohesion along with hard infrastructure have been the cornerstones of these successes. These examples show that, despite challenges in scaling up and maintaining such models with fiscal and political constraints, significant upgrading can be achieved if policy instruments are context-specific and multi-actor coalitions are mobilized (UN-Habitat 2020).

Combining these regional cases, there are some cross-cutting governance features that can be identified as drivers of successful transitions toward sustainability in resource-constraint cities.

- One, is the importance of participatory governance – participatory processes ensure that interventions are locally relevant, legitimate and sustainable (Baud and De Wit 2008; Lines and Makau 2018).
- Secondly, adaptive institutional arrangements whether city-waste picker partnerships, cross-sector alliances or place-based upgrading compacts drive innovation and learning (Chikarmane and Narayan 2023).
- Third, inspired leadership and effective public communication can override resistance and activate broad-based advocacy towards reform (Cervero et al. 2013).
- And last, flexible forms of regulation that exercise informality and stimulate multi-level integration also enable cities to innovate within constraints. Together, these experiences also show that while technological and financial fixes are necessary, the governance architecture based on participation, partnership, and flexibility is still the basis for realising urban sustainability in the Global South (UN-Habitat 2020; Acuto and Rayner, 2016).

7 Recommendations for Urban Sustainability Governance

With well-integrated, inclusive, and well-coordinated governance frameworks, urban sustainability in developing economies can be ensured. Building multi-level and cross-sectoral cooperation is fundamental. Both vertical (local, regional, national) and horizontal (city-to-city) partnerships facilitate knowledge transfer, policy alignment and resource mobilization (Acuto and Rayner 2016). It means that active engagement is fundamental to work with government agencies, civil society, and the private sector to be able to tackle urban complexity over time (Batley and McLoughlin 2015). One of the top priorities needs to be developing local capacity. Numerous cities have a lack of technical, managerial, and institutional skills, restricting their capability to introduce and maintain creative sustainability policies. Specialized training and peer networks like structured mechanisms help the city administration to design and implement transformative initiatives (Evans and Thomas 2023). Moreover, financing innovation is also crucial. Cities should leverage climate finance, launch municipal green bonds, and create blended finance instruments to mobilize resources for sustainability (Jena 2021). Lastly, due to the risks associated with conflicting actions by different sectors, achieving policy consistency for housing, climate, transport, and economic development is crucial (UN-Habitat 2020). All these actions, with strong local leadership and global partnership, can bring us resilient, sustainable urban futures.

8 Conclusion

This chapter discusses the urbanization challenges focusing on developing economies. It highlights that barriers like fiscal constraints, social inequalities, path dependency in regulatory systems and unstructured institutions could hamper urbanization in developing economies for sustainable development. It also explains through theory and regional case-studies that these challenges can be avoided and economies could have stronger governance structures that are more capable of encouraging innovations like green infrastructure, waste management, sustainable transport systems and upgrading informal settlements. This will lead to benefits for overall health and economic outcomes. However, these practices need follow-up and continuity. Investment in training, knowledge-sharing and brainstorming are essential for adaptive and resilient urban governance. In conclusion, urban governance that is participatory, technical, and also in sync with regional realities forms the foundation of sustainable urban futures in developing economies. Policy-makers and practitioners can develop new solutions to ensure urban sustainability remains relevant for both today and for the future.

References

Acuto M, Rayner S (2016) City networks: breaking gridlocks or forging (new) lock-ins? Int Affairs 92(5):1147–1166

Agrawal D (2025) A comprehensive review of investing and financing for sustainable tourism projects. Environ Sustain Govern Insights 1(1):53–70

Aijaz R (2016) Challenge of making smart cities in India. Asie Visions 87. IFRI, Paris
Amorim EER, Menezes M, Fernandes KVG (2022) Urban living labs and critical infrastructure resilience: a global match? Sustainability 14(16):9826
An B, Tang S-Y, Leach WD (2021) Managing environmental change through inter-agency collaboration: protective governance in mandated sustainability planning. Environ Sci Policy 125:146–156
Ansell C, Gash A (2008) Collaborative governance in theory and practice. J Publ Admin Res Theory 18(4):543–571
Auerbach AM, Thachil T (2018) How clients select Brokers: competition and choice in India's slums. Am Polit Sci Rev 112(4):775–791
Bahl RW, Linn JF (2014) Governing and financing cities in the developing world. Lincoln Institute of Land Policy, Cambridge
Bahl R (1999) Implementation rules for fiscal decentralization. In: International studies program working paper 30. Georgia State University, Atlanta
Bashir S, Sarker T, Talib MNA, Akram U (2023) Financing options for green and affordable housing: evidence from South Asia. Sustainability 15(15):11645
Batley R, McLoughlin C (2015) The politics of public services: a service characteristics approach. World Dev 74:275–285
Baud ISA, de Wit J (2008) New forms of urban governance in India: shifts, models, networks and contestations. SAGE, New Delhi
Baud I, Jameson S, Peyroux E, Scott D (2021) The urban governance configuration: a conceptual framework for sustainability transitions. Geogr Compass 15(5):e12562
Broto VC (2017) Urban governance and the politics of climate change. World Dev 93:1–15
Bulkeley H (2013) Cities and climate change. Routledge, London
Cabannes Y (2015) The impact of participatory budgeting on basic services. Environ Urb 27(1):257–284
Cervero R, Guerra E, Al S (2013) Beyond mobility: planning cities for people and places. Island Press, Washington
Cervero R, Sarmiento OL, Jacoby E, Gomez LF, Neiman A (2009) Influences of built environments on walking and cycling: lessons from Bogotá. Int J Sustain Transp 3(4):203–226
Chikarmane P, Narayan L (2023) Pune waste Pickers' innovative efforts to extend decent work and social protection. WIEGO, Manchester
Coaffee J, Deas I (2008) The search for policy innovation in urban governance. Public Policy Admin 23(2):167–187
Corfee-Morlot J, Kamal-Chaoui L, Donovan MG, Cochran I, Robert A, Teasdale PJ (2009) Cities, climate change and multilevel governance. OECD, Paris
Denis E, Zérah M-H, Mukhopadhyay P (2017) Subaltern urbanisation in India. Springer, New Delhi
Evans J, Thomas C (2023) Environmental governance. Routledge, London
Fang X, Li J, Ma Q (2023) Integrating green infrastructure and nature-based solutions for urban sustainability. Sustain Cities Soc 98:104843
Fox A, Ziervogel G, Scheba S (2023) Strengthening community-based adaptation in informal settlements. Local Environ 28(7):837–851
Gilbert A (2008) Bus rapid transit: is TransMilenio a miracle cure? Transp Rev 28(4):439–467
Healey P (2020) Collaborative planning: shaping places in fragmented societies. Bloomsbury, London
Herath P, Bai X (2024) Benefits and co-benefits of urban green infrastructure. Sustain Sci 19(3):1039–1063
Hidalgo D, Huizenga C (2013) Implementation of sustainable urban transport in Latin America. Res Transp Econ 40(1):66–77
Jena LP (2021) Financing climate-resilient infrastructure in cities. Emerg Econ Stud 7(2):107–114
Joselin D, Singh A, Tiwary KS (2025) The Nexus between CSR practices and financial performance: empirical evidence from the manufacturing sector in India. Environ Sustain Governan Insights 1(1):1–17

Kaith A, Sachdeva G (2024) Bibliometric insights into green human resource management: a connection with business ethics and values. PURUSHARTHA J Manage Ethics Spiritual 17(1):51–63. https://doi.org/10.21844/16202117104
Kumar K, Singh R, Choudhary M, Sharma A (2025) Climate change mitigation strategies in environmentally sensitive industries. In: Climate change and social responsibility. Emerald Publishing Limited, vol 25, pp 185–204
Kumari D, Shashwat S, Verma PK, Giri AK (2025) Urbanization, energy use and environmental sustainability in India. Int J Energy Sector Manage 19(3):731–746
Lindell I (2008) The multiple sites of urban governance. Urban Stud 45(9):1879–1901
Lines K, Makau J (2018) Taking the long view. Environ Urban 30(2):407–424
Mitlin D (2008) With and beyond the state: co-production as a route to political influence. Environ Urban 20(2):339–360
Ostrom E (2017) Polycentric systems for coping with collective action and global environmental change. In: Global justice. Routledge, London, pp 423–430
Parnell S, Pieterse E (2014) Africa's urban revolution. Zed Books, London
Parnell S, Robinson J (2012) (Re)theorizing cities from the global south. Urban Geogr 33(4):593–617
Pierre J (2011) The politics of urban governance. Bloomsbury, London
Portney KE (2013) Taking sustainable cities seriously. MIT Press, Cambridge, MA
Prakash A, Tiwari AK (2025) The role of institutional factors in achieving SDG 16: a thematic review of drivers of corporate anti-corruption action. Environ Sustain Govern Insights 1(1):34–52
Prescott MF, Dobbie MF, Ramirez-Lovering D (2021) Green infrastructure for sanitation in the global south. Sustainability 13(4):2071
Roy A (2005) Urban informality: toward an epistemology of planning. J Am Plan Assoc 71(2):147–158
Satterthwaite D, Archer D, Patel S et al (2020) Building resilience to climate change in informal settlements. One Earth 2(2):143–156
Shen L, Shuai C, Jiao L, Tan Y, Song X (2016) A global perspective on the sustainable performance of urbanization. Sustainability 8(8):783
Singh R, Kumar K, Khan S (2024) A comprehensive view of artificial intelligence (AI)-based technologies for sustainable development goals (SDGs). Artif Intell Enabled Manage Emerg Econ Perspect 183:9783111172408–012
Singh R, Sharma A, Kumar K (2025a) Navigating carbon neutrality: strategies for sustainable future and environmental challenges. In: Climate neutrality through smart eco-innovation and environmental sustainability. Springer Nature Switzerland, Cham, pp 69–80
Singh R, Tiwary KS, Kumar K, Kumar A (2025b) Empowering a sustainable future: Blockchain's role in economy, organizations, and finance towards smart circular economy. In: Blockchain technologies for smart circular economy and organisational sustainability. Springer Nature Switzerland, Cham, pp 271–283
Smoke P (2015) Rethinking decentralization. Public Admin Dev 35(2):97–112
Srivastava SK, Anshul A, Pathak P, Bansal J (2018) Responsible consumption for curbing food wastage: an exploratory enquiry. PURUSHARTHA J Manage Ethics Spiritual 11(2):104–116
Srivastava SK, Singh S, Srivastava M (2021) Food waste at household and social gatherings: drivers and possible remedies. PURUSHARTHA J Manage Ethics Spiritual A J Manage Ethics Spiritual 14(2):117–133
Srivastava SK, Sinha AK, Kumar S (2024) Workplace spirituality empowered by spiritual leadership to promote organizational effectiveness: a bibliometric analysis. PURUSHARTHA J Manage Ethics Spiritual 17(1):1–22. https://doi.org/10.21844/16202117101
Tripathi V, Singh K, Parihar S, Srivastava SK (2022) A bibliometric analysis on supply chain collaboration. PURUSHARTHA J Manage Ethics Spirit 15(2):66–103. https://doi.org/10.21844/16202115205
UN-Habitat (2016) World cities report 2016: urbanization and development. UN-Habitat, Nairobi

UN-Habitat (2020) World cities report 2020: the value of sustainable urbanization. UN-Habitat, Nairobi

United Nations (2015) Transforming our world: the 2030 agenda for sustainable development. United Nations, New York

United Nations (2018) World urbanization prospects: the 2018 revision. United Nations, New York

World Bank (2019) Transit-oriented development: a sustainable urban growth model. World Bank, Washington, DC

World Bank (2020) Urban development overview. World Bank, Washington, DC

Zeng C, Deng X, Dong J, Peng H (2016) Urbanization and sustainability in BIC countries. Sustainability 8(4):400

Measuring What Matters: Optimising Procurement Metrics for Sustainable Economic Development in Developing Economies

David Chikwere, Justine Mbudaya, and Dumisani Mawonde

Abstract Public procurement represents a substantial economic lever within emerging economies, offering immense potential to drive green economic development through strategic market influence. However, conventional procurement metrics predominantly prioritise immediate cost savings and operational efficiency failing to adequately capture the crucial environmental and broader socio-economic contributions essential for sustainable transformation. This misalignment hinders effective mobilisation of public spending towards achieving national green growth programs and international sustainability commitments like the United Nations Sustainable Development Goals (UNSDGs). This chapter aims to identify and propose procurement Key Performance Indicators (KPIs) that effectively measure and promote green economic outcomes within the unique contexts of emerging economies. To achieve this, the chapter pursues three key objectives: (1) critically assess the effectiveness of existing public procurement performance metrics in promoting environmental sustainability in emerging economies, (2) identify key challenges and gaps in designing and implementing environmentally focused procurement metrics and (3) synthesise international best practices and emerging innovations in green procurement measurement relevant to developing contexts. The research identifies critical gaps by employing a systematic literature review (SLR) of Scopus, Web of Science, and Google Scholar academic databases supplemented by grey literature (UN, governments publications) from 2012 to 2025. Key findings reveal that current metrics undervalue long-term environmental benefits, lack standardisation for green outcomes, offer insufficient strictness for monitoring progress, and often neglect social co-benefits integral to green development. Consequently, the chapter recommends the strategic integration of outcome-based environmental

D. Chikwere (✉)
Leeds Business School, Leeds Beckett University, Leeds, UK
e-mail: D.chikwere@leedsbeckett.ac.uk

J. Mbudaya
BA ISAGO University, Francistown, Botswana

D. Mawonde
Faculty of Business Management, Nelson Mandela University, Gqeberha, South Africa

K. Kumar et al. (eds.), *Sustainable Value Creation: Corporate Strategies for SDGs Implementation*, World Sustainability Series,
https://doi.org/10.1007/978-3-032-24652-3_14

KPIs (lifecycle carbon footprint, percentage spend on certified sustainable goods/services), socio-economic indicators (job creation in green sectors, local sustainable supplier development), alongside robust data collection systems and capacity building for procurement entities. By redefining value through optimised metrics, this chapter provides a practical roadmap for policymakers and procurement practitioners to transform procurement into a powerful, measurable catalyst for inclusive and sustainable economic development in developing economies.

Keywords Emerging economies · Environmental sustainability · Green economy · Key performance indicators · Sustainable procurement

1 Introduction

Public procurement is considered an important economic tool in developing countries because of its potential to shape market behaviour and impact the direction of economic development (Ambe 2019; Chikwere et al. 2023). This is critical in terms of poverty reduction through sustainable development. These countries' governments are the biggest buyers in their respective markets and dominate spending that determine manufacturing and allocation of resources (Wade 2018; Ambe 2019). That spending power is no passive transaction, but it can be used to incentivise ecologically innovative intensification, stimulate social growth that is inclusive, and empower green economic development. Manta et al. (2022) contend that nowhere can this be seen more relevant than in developing countries which face a dilemma on how to balance fast pace of development and environmental sustainability with these competing pressures. Therefore, public procurement provides an institutional mechanism through which governments can influence the supply chain and market provision to deliver the sustainable development agenda.

The greening of public procurement as also termed as Sustainable Public Procurement (SPP), can be a driver for a green economy if all three dimensions (economics, environment and social) are considered instead of only considering cost efficiency (Manta et al. 2022; Nsiah-Sarfo et al. 2023). This strategic leverage is more important in view of the developmental pathways and the ecological insecurities of most emerging economies. Wise use of procurement can also drive forward the uptake of greener technologies, increased use of resources and have beneficial socio-economic effects such as job creation in sustainable industries (Yosef et al. 2023). This alignment with the sustainable development frameworks such as the UNSDGs, highlights procurement's growing role beyond the conventional concept of purchase to that of an instrument of change.

Recognising the wide role that procurement has played, previous studies have underscored the need for a focused analysis and enhancement of procurement metrics that are able to capture its role in green economic development (Nsiah-Sarfo et al. 2023). Performance measures such as KPIs are not only a useful in making procurement decision but may also contribute as the foundation for the policy formulation or the judgment of the effectiveness of the policy (Cayolla-Trindade et al. 2017). Without proper measurement frameworks, the considerable purchasing power of public procurement can also be misdirected or underused. Moreover, developing countries have their own institutional, economic and social conditions, which demand customised procurement strategies and measures for them to successfully achieve their sustainable development goals (Hafsa et al. 2021; Nsiah-Sarfo et al. 2023).

Conventional procurement Key Performance Indicators (KPIs) have typically been on short term cost savings and operational efficiency (Cayolla-Trindade et al. 2017). This focus has been inspired by the fact that procurement has historically been seen as a cost-saving machinery in public administration, and value for money (Chikwere et al. 2023). While these issues remain relevant, their exclusive consideration severely limits the potential for procurement to contribute towards the goal of sustainable development. Yosef et al. (2023) argue that the traditional metrics fail to include the more intense environmental and social measures associated with procurement such as emissions, resources consumption and quality of life of the communities.

The limited approach in procurement metrics can cause a disconnect between the current trend of development for sustainability and green growth, particularly in the context of developing countries where vulnerabilities to the environment and social inequalities remain high. The short-term financial focus appears to contradict the principles of sustainable procurement and long-termism benefits (such as ecosystem protection, social equity development and economic viability) (Cayolla-Trindade et al. 2017). Furthermore, such procurement mechanisms lack the necessary flexibility and precision for the tracking of environmental achievements even to judge impact of progress towards national environmental goals or fulfilment of international obligations like SDGs (Manta et al. 2022; Waqar et al. 2025). It is worth mentioning that when the procurement metrics poorly reflect the environmental-social outcomes, national investment could be directed to subside unsustainable development and thus undermine national green growth strategies (Kumar 2022). Furthermore, the lack of standard published outcome-based KPIs results in poor transparency and accountability, making it a challenge for policymakers and procuring authorities to elicit areas for improvement or obtain any value achieved from an investment in an effective or efficient way. It is these gaps that contribute to constraining the redeployment of public expenditure towards sustainability transitions and maintaining a procurement culture dominated by a focus on cost, rather than ecological soundness and social inclusivity (Manta et al. 2022; Waqar et al. 2025).

In the developing world, the problems are exacerbated by institutional barriers, limited capacity, and multiple market dynamics, making it even more difficult for sustainability to be incorporated into procurement metrics in a meaningful way (Delmonico et al. 2018). Therefore, there is an urgent need to critically review the current procurement metrics, and to examine their limitations especially regarding an environmental sustainability focus on procurement, while consolidating what constitute expanded dimensions other than financial cost and transaction efficiency (Malolitneva and Dzhabrailov 2019; Nsiah-Sarfo et al. 2023).

To achieve this aim, the chapter sets up three primary goals. It evaluates, firstly, the extent to which current public procurement performance measures have been successful in driving environmental sustainability in developing countries. It evaluates the extent to which existing indicators are suitable, applicable and relevant and the deficiencies in the metrics design and implementation as a barrier to green growth. Second the chapter outlines the crucial challenges and gaps in environmental procurement metrics, and third, the chapter reviews the international best practices of green procurement measurement, assesses their applicability and transferability to developing country contexts. In drawing from a range of geographies and policy contexts, the analysis offers recommendations that are tailored to the institutional context and developmental imperatives of developing economies.

2 Literature Review

2.1 Theoretical Framework

The theoretical foundation of this study is embedded in the theory of Sustainable Supply Chain Management (SSCM) advanced by Carter and Rogers (2008). It offers a solid theoretical basis by ushering a new way of thinking compared to the traditional concepts on how value and performance in procurement must be defined and measured (Seuring and Müller 2008). SSCM embraces the dimensions of environment and society, as well as those of economics popularly known as the Triple Bottom Line (TBL) as integrated components of the management, measurement, and optimisation of supply chain opportunities (Yosef et al. 2023). This is firmly beyond old-school, cost/speed/quality focused, linear production and supply chain doctrines, realising that creating value is about environmental protection and social fairness across the span of the product lifecycle.

In this perspective public procurement is cast as a governance node within complex, sometimes global supply networks. SSCM theory does not consider procurement as a mere purchasing transaction but conceptualised as a strategic lever that contributes to upstream and downstream activities (Johnsen et al. 2018). It obliges consideration of environmental (such as resource extraction impacts, production emissions, logistics footprint, end-of-life management) and social factors (worker standards, community impacts, supplier diversity) encapsulated in the goods and services purchased. Such an approach requires a systems focus where procurement decisions reverberate through the supply chain to affect sustainability outcomes well beyond the immediate contract (Johnsen et al. 2018; Yosef et al. 2023).

SSCM theory highlights the need for performance measurement that reflects its integrated TBL view. It inherently requires KPIs that reflect the environmental efficiency (such as life cycle carbon footprint, water usage, recycled content) and social responsibility (such as supplier compliance with labour standards, local economic development) as well as the traditional business metrics (Seuring and Müller 2008).

These measures should also capture the interdependency of the supply chain and go beyond mere input or output measures to evaluate the effects and consequences of procurement decisions in relation to natural resource conservation, pollution control and socio-economic welfare (Yosef et al. 2023). The theory holds that transparency and traceability are crucial for measurement and accountability that matters.

If the promotion of green business is targeted as part of the development of a green economy in emerging economies, then emphasis would be placed on SSCM theory. It forms the theoretical foundation to develop purchasing KPIs aimed at stimulating domestic green industry development, circular economy principles, and inclusive development by building supplier capacity (Cayolla Trindade et al. 2017). Embedding SSCM principles in performance metrics can in turn help to steer procurement systematically towards the encouragement of sustainable supply chains which not only provide goods or services but also deliver tangible ecological benefits and socio-economic welfare, in line with wider national and international sustainability objectives.

2.2 *Existing Public Procurement Metrics and Their Effectiveness*

The terrain of public procurement performance measurement is currently dominated by conventional measures such as cost savings, process efficiency, compliance, and timeliness (Chikwere et al. 2023). These traditional measures tend to focus on quantifiable outputs of the contracting experience, such as award rates, cycle time (time elapsed from start to finish of the procurement event), and short-term cost savings, which reflect the traditional emphasis on financial efficiency, waste avoidance, and stewardship of public funds (Kumar 2022). However, they fall short when measured against in terms of environmental sustainability.

Empirical research and policy analysis concludes that existing procurement metrics do not adequately capture environmental performance or socio-economic outcomes of procurement decisions (Kumar 2022; Yosef et al. 2023). For example, the treatment of environmental sustainability in monitoring frameworks of procurement is limited and focused mainly on procedural, rather than on substantive, compliance results. This complicates the evaluation of the real sustainability impacts of procurement activities and restricts the ability of governments to guide the markets in the direction of greener production and consumption (Waqar et al. 2025).

Furthermore, most available frameworks do not fully capture the complete life cycle environmental effects of purchased goods and services, for example, carbon emission at the production stage or resource depletion along supply chains (Yosef et al. 2023). This leaves long-term environmental benefits underpriced and fails to align procurement with circular economy models (Kumar 2022). Similarly, social aspects such as jobs creation in green industries and building- capacity of local, sustainable suppliers, are seldom not explicitly considered by conventional procurement metrics.

Literature (Cayolla et al. 2017; Hafsa et al. 2021; Yosef et al. 2023), suggests an increasing consensus about moving from input and process-oriented KPIs to outcome-based indicators that capture the true environmental and socio-economic progress effectively. Delmonico et al. (2018) opine that there are still difficulties in operationalising these performance-based measures, especially in developing countries with data constraints, limited institutional capacity, and a fragmented regulatory structure. In general, both theoretical criticism and empirical analyses lead to the conclusion that performance measurements systems must go beyond traditional efficiency measures and systematically incorporate sustainability dimension (Yosef et al. 2023; Waqar et al. 2025). For such an aim, it is necessary to integrate lifecycle assessments, social impact indicators, and objective performance targets in all procurement assessments underpinning any policy objectives that are truly related to a green economy development (Renukappa et al. 2015; Cayolla et al. 2017; Hafsa et al. 2021).

2.3 Environmental and Socio-economic Dimensions in Procurement

Environmental and socio-economic aspects are the two cornerstones for sustainable public procurement (Kumar 2022). Recent studies (Cayolla et al. 2017; Hafsa et al. 2021; Yosef et al. 2023), recommend lifecycle-based environmental performance measures that cover the entire environmental footprint of purchased products and services. These comprise lifecycle carbon footprint, material resource efficiency, energy use and waste generated. Thus, the procurement departments can consider and mitigate the wider ranging consequences beyond the actual procurement activities. The implementation of these environmental performance metrics into procurement approaches is necessary to reflect impacts that are in line with objectives of green growth and climate adaptation (Waqar et al. 2025). For example, lifecycle analysis forms the scientific foundation for calculating the greenhouse gas load of a product in raw material acquisition, production, use and disposal (Yosef et al. 2023). Such holistic measurement allows fairer trade-offs and better-informed procurement decisions that prefer lower carbon options or recycled materials, for public spending to better reflect these environmental sustainability imperatives.

Equal importance also lies in the expansion of the public procurement to embrace the socio-economic factors that can contribute to an inclusive growth. This includes green job creation, local sustainable supplier development, labour standards training, and small and medium enterprise (SME) development. Procurement can provide social co-benefits such as poverty alleviation and community adaptation as well as environmental sustainability. Nonetheless, the evaluation of these is challenging but for indirect effects, no standard criteria and limited long-term data (Delmonico et al.

2018). Tracking faces even more challenges due to lack of information and capacity. Multi-dimensional measures that include social and environmental dimensions are therefore necessary. Success relies on effective monitoring, sound data collection and institutional capacity to implement findings (Liu et al. 2024; Van Opstal et al. 2024).

2.4 Innovations and International Best Practices in Green Procurement

Although traditional methods for the measurement of environmental performance are considered limited, in the last couple of years there have arisen new greening procurement metrication created to overcome these gaps. An extensive life-cycle assessment based on the so-called "cradle-to-grave" approach traces a product or process from the stage of initial resource extraction and selection of suppliers (cradle) to its ultimate disposal (grave) (ISO 2006; Hauschild 2019). To eliminate wastes and costs, this perspective helps users in identifying environmental impacts all the way through the entire life cycle. This approach also includes collecting, comparing, and interpreting information based on life-cycle assessment (ISO 2006; Hauschild 2019). The cradle-to-gate analysis meanwhile focuses on the stages between the raw material extraction (cradle), through to material processing (gate) just before the product is transported and delivered to customers (Patel et al. 2025).

A best practice has been shifted towards green KPIs which helps to assess performances along with its environmental aspects and not merely compliance to the process (Yosef et al. 2023). For instance, percentage spend-based procurement systems associated with spending on certified sustainable goods or services have proven more effective in enforcing green goals by assuring the quality and environmental authenticity of procurements. These changes also involve the implementation of circular economy into the procurement metrics via evaluation of product recyclability, manufacturability, and material cycling rate (of a given product), (Waqar et al. 2025). These practices are more than linear use patterns that migrate toward resource efficiency and shrinking of waste in line with sustainability (Yosef et al. 2023; Waqar et al. 2025).

Life-cycle costing (LCC), social life-cycle assessment (S-LCA) and life-cycle assessment (LCA) can be implemented based on International Organisation for Standardization (ISO) 14,044 standard, to promote the performance of sustainable procurement (Patel et al. 2025). This requires, a strategically embedded measurement, for example, through the embedding of environmental and socio-economic KPIs within procurement policy, evaluation templates and reporting mechanisms of government and public bodies. According to ISO 14040:2006 and ISO 14044:2006 (ISO 2006) adopted by ISO, harmonisation of methods of conducting a life-cycle assessment (LCA) is important covering the LCA conceptualisation, goal and scope definition, inventory analysis, impact assessment, and interpretation (Hauschild

2019). There is a need for augmenting stakeholder awareness around environmental matters, improved strategic sustainability outcomes, and stronger market pull for green products and services. Such measures are important to low-income countries seeking to have tailored procurement indicators linked to their institutional and market realities.

Nonetheless, the transfer of best practice must be mindful of the specific context in which it is implemented, considering that there are unique institutional challenges and developmental priorities within emerging economies. Metrics need to be flexible, pragmatic and simple enough to implement with current resources but strong enough to contribute to meaningful progress in terms of sustainability (Patel et al. 2025). Capacity building, knowledge generation to raise awareness, and enabling policy environments are essential complements to metric uptake (Moraga et al. 2019; Swensson et al. 2021; Al-Sinan and Bubshait 2022).

3 Materials and Methods

To satisfy the objectives of this study, a literature review was conducted. From a methodological perspective, literature reviews can be understood as content analysis, which combines quantitative and qualitative elements to evaluate both content and structural (descriptive) criteria (Fink 2019). Reviews of the literature typically have two goals: first, they provide an overview of the body of research by highlighting trends, themes, and problems. Second, this can aid in the construction of theories and aid in identifying the field's conceptual content (Meredith 1993). The dilemma is that it is not feasible to read everything and that complete reviews might only be feasible for topics that are newly developing or have limited definitions (Seuring and Müller 2008).

The chapter recognises that there are some key areas of consensus on the use of procurement metrics towards fostering the sustainable economic development of developing countries and related challenges that led to the contextualisation of the research questions. That was informed by researchers' ability to select a sample of studies which provided clear idea of the questions of study and guided them in formulating an appropriate procedure of search strategy (Leonidou et al. 2020; Del Vecchio et al. 2022). Thus, the research questions below were formulated.

RQ1: How effective are public procurement metrics in supporting sustainability objectives in developing economies?

RQ2: What challenges are being faced in adopting environmental procurement metrics?

RQ3: What are strategies in integrating environmentally focused procurement metrics?

3.1 The Search Strategy

The research design followed a procedure for reviewing literature related to investigating optimising of procurement metrics towards fostering the sustainable economic development of developing countries. The first step was to decide upon specific inclusion and exclusion criteria which guarantee that relevance, quality and focus are maintained. Refereed journal articles, conference papers, books and the grey literature published from the year 2012 to 2025 in English language were included yielding 97 publications. Selected papers needed to explicitly relate to any of the areas of public procurement key performance metrics, sustainability measures, or economic development within the framework of developing countries.

Articles not on procurement and sustainability, those published outside the required period and those that do not focus on emerging economies which totalled 50 documents were excluded from the review. This rigorous process was designed to guarantee academic in relation to the topic relevance and thematic specificity to restrict the impact of tangential or non-academic sources. Academic databases including Scopus, Web of Science, and Google scholar were selected for their broad coverage in economics, management, environmental science and policy studies. The search terms included combination of general and specific terms such as "sustainable procurement," "economic development," "developing countries," "performance measures." Boolean operators and filters for "subject area" and "publication type" were used to narrow the search resulting in 25 documents being excluded. This search yielded 22 documents that met the inclusion criteria across all three specified combinations in the final sample. This design of research made it possible to provide a broad, focal, academic solid understanding of the current context and discussion within sustainable procurement practices (Fig. 1).

4 Study Findings

4.1 Number of Published Documents Per year

Figure 2 shows the annual number of publications between 2012 and 2025. Between 2012 and 2019, the output of publications was consistent, with no more than one article reported intermittently over an almost decade period indicating little academic interest. A gradual upward trend begins in 2020, signalling a growing academic interest likely driven by the global shift toward sustainability and the increased emphasis on resilient supply chains in the aftermath of COVID-19. A sudden surge is identified from 2022 to 2024, and six articles peak in 2024 and 2025. Such an explosive increase speaks to the boom of the academic research possibly resulting from the relationship of sustainable development to efficient public procurement in developing countries. The increasing trend emphasises the emerging importance of

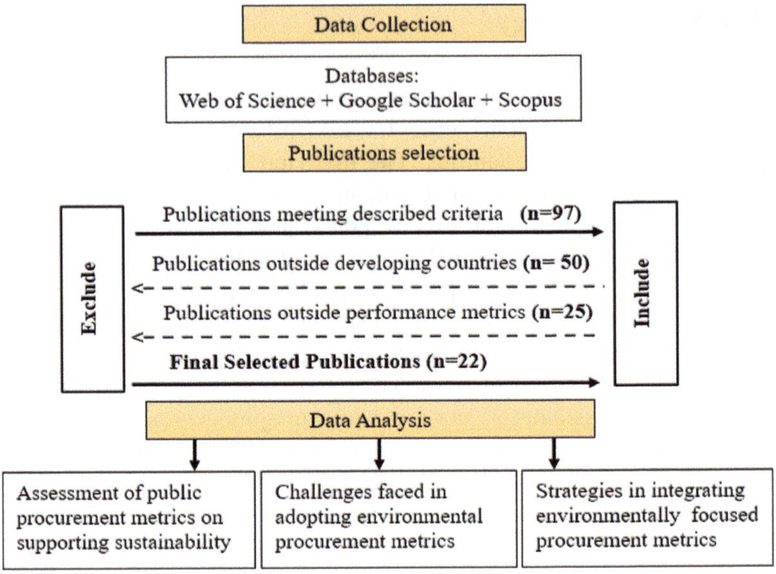

Fig. 1 Research design. *Source* Literature analysis

the topic and the development of the topic as a major research area with respect to the global development priorities.

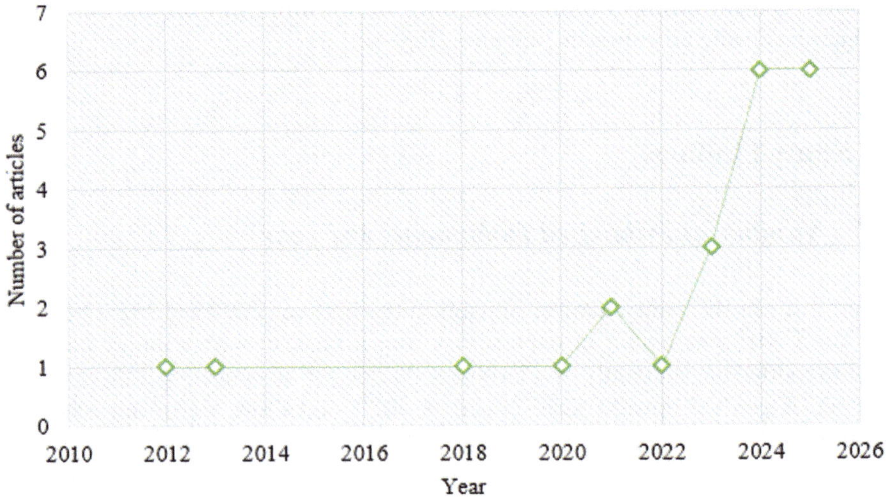

Fig. 2 Number of published documents per year. *Source* Literature analysis

4.2 Published Documents by Subject Area

Figure 3 shows that Business Management, and Accounting is the first with 12 publications, reflecting the influence of procurement on economic and managerial disciplines. Social Sciences (7) and Engineering (6) rank as the second and third most prevalent disciplinary categories respectively, reflecting conscientiousness to the socio-technical complexity of procurement systems within a development environment. Chemistry, Engineering and Business are moderately represented, alluding to cross-disciplinary interest for sustainability concerns. A lower frequency of contributions from Energy, Medicine and Decision Sciences can be considered evidence of yet-nascent interest. Disciplines such as Agricultural Sciences, Arts and Humanities, and Materials Science have few activity levels suggesting potential research gaps. The cumulative percentage curve shows that only a few disciplines account for most of the scholarly literature with more than 80% of publications focused on top five fields. This bias highlights the importance of wider interdisciplinary engagement for procurement metrics to be fully optimised in delivering sustainable development.

5 Summary of Findings

See Table 1.

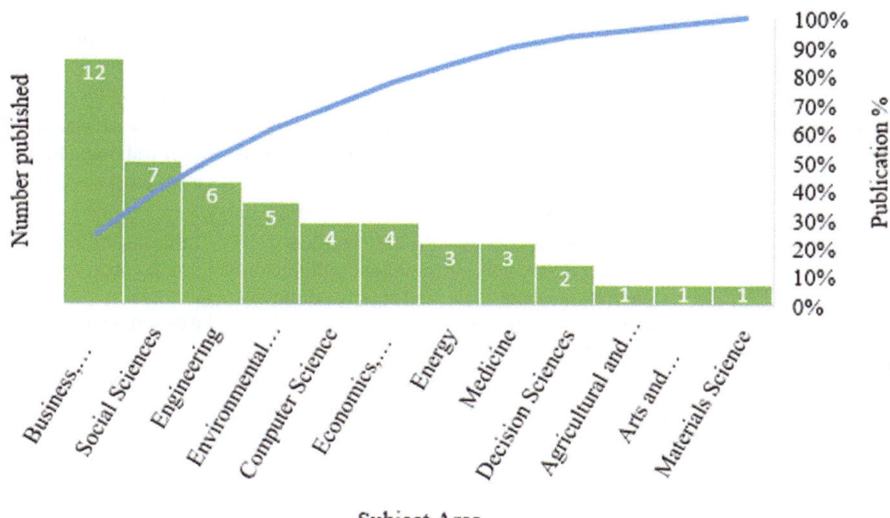

Fig. 3 Published documents by subject area. *Source* Literature analysis

Table 1 Summary of findings

Authors	Public procurement metrics effectiveness in supporting sustainability objectives	Challenges faced in adopting environmental procurement qualities	Strategies in integrating environmentally focused procurement
Puspitasari et al. (2025)	SPP aligns with SDGs by considering economic, social, and environmental factors	Limited availability of sustainable suppliers, high costs, lack of sustainable human resource (HR)	Use of Relative Importance Index (RII) to prioritise drivers such as sustainable policies, HR, and supplier availability
Terman and Smith (2018)	Green procurement in local governments reflects genuine sustainability commitment	Disparity between policy adoption and actual environmental impact	Adoption of permissive and mandated green procurement policies
Haffar and Özceylan (2024)	Blockchain ensures transparency and sustainability in supplier selection	Risk of fraud, inefficiency, human error in traditional systems	Use of blockchain-based systems for supplier evaluation (economic and sustainability criteria)
Ball et al. (2023)	Responsible sourcing improves stakeholder value, supply chain performance	High costs, industry constraints, lack of awareness	Development of ethical sourcing frameworks, awareness and stakeholder collaboration
Dudić et al. (2024)	Digital procurement enhances efficiency and environmental responsibility	Transition risks, lack of training, management complexity	Risk-aware planning, staff training, strong supplier partnerships
Zhang et al. (2025)	Procurement decisions (such as organic tea sourcing) promote sustainability, student awareness	Poverty-stricken suppliers' vulnerability and supply continuity	Ethical sourcing and linking procurement to social development goals
Nsiah-Sarfo et al. (2023)	Sustainable procurement in Ghana is enabled by isomorphism and leadership	Institutional voids and inconsistent leadership culture	Coercive/mimetic pressures and strong sustainable leadership drive change
Ali et al. (2023)	Green procurement in construction sector boosts viability	Industry-specific constraints, limited generalisability	Adoption of practices like sustainable materials, local sourcing, reverse logistics

(continued)

Table 1 (continued)

Authors	Public procurement metrics effectiveness in supporting sustainability objectives	Challenges faced in adopting environmental procurement qualities	Strategies in integrating environmentally focused procurement
Smale et al. (2025)	Green Intensive Care Unit (ICU) approach links procurement and sustainability in critical care	Tension between sustainability and clinical priorities	Co-creation, local adaptation, and cross-functional collaboration (procurement and clinicians)
Zaman, Ali and Khan (2024)	Pharmaceutical green procurement needs consumer involvement	Barriers such as lack of understanding and supportable product information	Visualisation of barriers/enablers via grey-DEMATEL; consumer education
Prosperi et al. (2024)	Sustainability embedded via supply chain indicators in food/agriculture	Smallholder farmers' integration into complex procurement systems	Identifying key metrics in green public procurement and short supply chains
Hoejmose, and Adrien-Kirby (2012)	Literature indicates socially and environmentally responsible procurement (SERP) is conceptually strong but practically fragmented	Lack of cohesion, limited practitioner-oriented research	Establish a unified research agenda, stronger empirical-practice linkage

Source Literature analysis

5.1 Discussion of Results

Theme 1: Public Procurement Metrics Effectiveness in Supporting Sustainability Objectives

Public procurement is increasingly used as a strategic instrument to pursue sustainability. As shown by Puspitasari et al. (2025), SPP is in line with the SDGs by considering economic, social and environmental aspects. Procurement is no longer be merely an operational practice, but a tool for integrating (institutionalising) sustainability into public systems. For instance, Zhang et al. (2025) emphasise how university procurement decisions (such as sourcing of organic tea) advance sustainability learning and community development. Likewise, the work by Smale et al. (2025) on Green ICUs illustrates the ways in which procurement measures can manifest themselves in measurable environmental impacts in sensitive healthcare environments. Further, systems based on the blockchain (Haffar and Özceylan 2024) offer a transparent and standardised solution to incorporate environmental and social aspects in the evaluation of suppliers which enhances the efficiency of procurement in sustainability governance. These studies also reinforce the growing significance of public procurement as a measurable stimulant for wider sustainability successes.

Theme 2: Challenges Faced in Adopting Environmental Procurement Qualities

Even though sustainable procurement has brought a host of promising developments, there are still main constraints to making green features a standard requirement for public procurement. One of the challenges is a lack of sustainable suppliers and higher price of environmentally friendly products as reported by Puspitasari et al. (2025). In Ghana, Nsiah-Sarfo et al. (2023) discovered institutional voids and weak leadership to be challenges to the practical application. Additionally, Terman and Smith (2018) find that while many municipal governments pass sustainability-oriented policies, policy is not always synonymous with concrete environmental actions that necessarily reflect substantive responses to significant environmental problems. Other limitations are industry specific, such as context limited generalisability and logistics restrictions (Ali et al. 2023). Furthermore, Ball et al. (2023) highlight the absence of knowledge and high economic costs as barriers of responsible sourcing in the building design profession. These challenges are perpetuated by other systemic issues including fragmented information bases (Hoejmose and Adrien-Kirby 2012) and information asymmetry in pharmaceutical procurement (Zaman et al. 2024), so it is hard to have universal regulation.

Theme 3: Strategies in Integrating Environmentally Focused Procurement

Many integration methods have been developed to pass the obstacles in adopting environmentally focused procurement. One approach to prioritising sustainability drivers is to apply the analytic tools such as the Relative Importance Index (RII) (Puspitasari et al. 2025). Meanwhile, Dudić et al. (2024) highlight risk-conscious planning, skills development and successful supplier relationships as the main drivers of digital procurement transformation. Smale et al. (2025) calls for co-creation and cross-functional collaboration especially between those involved in clinical practice and procurement to directly support clinicians and buyers in making environmental purchasing decisions in high-sensitive sectors such as healthcare. Zhang et al. (2025) propose ethical sourcing frameworks that integrate procurement and broader social development processes, whereas Ball et al. (2023) demand for more stakeholder engagement and the anchoring of ethical sourcing structures. At a systems level, Nsiah-Sarfo et al. (2023) and emphasise how isomorphic pressures (coercive and mimetic) and sustainability leadership, contribute to stimulating compliance and change of behaviour. Integrating consumer education, as recommended by Zaman et al. (2024), and coherent research agendas (Hoejmose and Adrien-Kirby

2012), are crucial to infuse sustainability deep into procurement culture. Combined, these approaches suggest that a holistic multi-stakeholder effort is needed to catalyse meaningful transformation.

6 Study Implications

6.1 *Implications to Theory*

This chapter adds to the connection between procurement metrics and sustainable economic development in the context of emerging economies. It raises important weaknesses and gaps in traditional measurement of procurement which has been mainly based on cost-effectiveness, thus advancing procurement theory to include environment and social/economic dimensions as part of the paradigm.

Furthermore, the study highlights the importance of SSCM theory in understanding and adoption of sustainable procurement. It serves as the theoretical underpinning for the creation of purchasing KPIs that, by enhancing supplier capacity, promote the growth of local green industries, the circular economy, and inclusive development. In keeping with broader national and international sustainability goals, integrating SSCM principles into performance metrics can also aid in methodically guiding procurement towards the promotion of sustainable supply chains, which not only offer goods or services but also provide observable ecological benefits and socioeconomic welfare.

By synthesising these theoretical backgrounds, the chapter provides a conceptual step forward in situating procurement metrics beyond mere technical instruments into institutionalised artefacts influenced by policy environment, organisational culture and developmental objectives. This synthesis provides the basis for further theoretical investigation in the dynamic interaction between measurement systems, sustainable procurement and economic development effects.

6.2 *Implications to Practice*

The chapter offers a practical basis for policy and procurement professionals who want to transform public procurement KPIs to be more aligned with green economic goals. It argues for the strategic inclusion of environmental performance indicators (such as lifecycle carbon footprints, proportion of spend on certified sustainable products) that reflect the outcomes sought in an environmental context, as well as (socio-economic) performance indicators that cover job creation and local supplier development.

Additionally, the study brings to light the importance of strong data collection systems and capacity building in the procurement agencies to both measure correctly and sustainably implement these advanced metrics. It recognises the practical challenges of developing countries and proposes flexible approaches leading to incremental metrics adoption according to institutions' capabilities and resource limitations. The principles help procurement leaders move from legacy cost-based models to more holistic approaches to assessment that improve transparency, accountability and policy coherence. It also implies that embedding the sustainability criteria in procurement process, and training of the procurement officers and use of technology for data management would drive the transformation of the procurement into a sustainable and impactful tool economic development.

7 Areas for Further Research

7.1 Development and Validation of Outcome-Based Procurement KPIs

There is a critical need for empirical research in the development and validation of KPIs for outcome-based procurement evidence-based long-term environmental and socio-economic impacts. This includes conceptual propositions and several pilot studies but, systematic testing in different developing countries context is scarce in the existing literature. In this respect, there is a scope for future research to develop indicators that are empirically developed and tested for their applicability, reliability and validity in various procurement contexts, and are modified accordingly. This includes the resolution of methodological challenges concerning data gathering, templating, and the attribution of outcomes to decisions in procurement. Furthermore, environmental adjustment efforts are required to adapt metrics according to different institutional settings, economic foundations and phases of development typical of an emerging nation. It will drive the practical applicability and applicability of sustainable procurement metrics.

7.2 Integration of Circular Economy Principles in Procurement Metrics

With the shifting focus on circular economy (CE) and its principles, there is a need for in-depth studies of how circularity indicators can be incorporated into procurement measurement frameworks. Areas for research could include the measurement of KPIs for assessing the efficacy of circular procurement such as product life extension, reuse rates and resource recovery. Further, there is a need for studies that investigate the supply chain and the engagement of local suppliers when the supply chain is run in circular economy configurations to understand how procurement indicators can create systemic circularisation. The situational challenges in developing countries-power capacity limitations and market readiness demand critical analysis for successful execution of CE metric deployment.

8 Limitations of the Study

8.1 Data Limitations and Scope Constraints

The research is based upon the available academic literature and grey literature to which access is limited, and this may have been insufficient regarding data coverage and bias. The vast differences and the heterogeneity in emerging economies call for caution in generalising results and policy recommendations across countries. Data limitations relating to environmental and socio-economic outcome measurement in developing countries also limit the scope of this review. These factors may restrict the granularity and specificity of potential procurement metrics and the extent to which they would work in all emerging market contexts.

8.2 Metric Standardisation and Measurement Challenges

It may continue to be a challenge to obtain uniform and standardised KPIs across various jurisdictions. The complexity of these measurements, in addition to the difficulty of quantifying complex socioeconomic benefits and monitoring significant long-term effects on the environment, complicates the standardisation issue. In addition, differences in institutional capabilities, legal limits, and market contexts between emerging economies hinder the standardisation of procurement performance indicators. These issues require a more adaptive, yet scientifically rigorous framework for indicator development and use.

9 Conclusion

The integration of the concept of sustainability in public procurement is a very important shift in the way institutional procurement can be utilised to accomplish environmental and socio-economic objectives. Evidence is emerging from sectors as diverse as health care, construction and local government that knowledge-based procurement can be a powerful lever for SDG progress provided there are robust metrics and inclusive policies to back it up. However, barriers such as supplier capacity, cost of implementation and institutional challenges continue to limit widespread adoption of environmental procurement expectations.

However, the complexity to address barriers is by no means just a regional one. Analytic tools such as RII, digital tools such as block chain and capacity development approaches are relevant to introduce a more transparent, accountable and performant system. Policy approaches that include ethical sourcing, buy-in from leadership, and collaborating with stakeholders at all stages of the procurement process are of interest. In addition, cocreation models as well as collaborations across sectors enable that context sensitive solutions may be cast with the idea of sustainability into reality. Ultimately, the key to the success of environmentally friendly purchasing is not only the availability of green products, but institutions and leadership that can harmonise it. As the literature indicates, adopting a more holistic, metric-oriented procurement practice with a keen social consciousness is key to achieving environmentally friendly outcomes. However, this requires institutional investment in monitoring, education and innovation to translate sustainability ambitions into measurable procurement results.

References

Ali HH, Kalf HAI, Shaikhan MH, Al Lami GK, Shatawi HH, Abdulaali HS, Hanoon TM (2023) Investigating the integration of environmentally friendly practices and green materials throughout the construction supply chain. Int J Constr Supply Chain Manage 13(1):276–288

Al-Sinan MA, Bubshait AA (2022) The procurement agenda for the transition to a circular economy. Sustainability 14(18):11528

Ambe IM (2019) The role of public procurement to socio-economic development. Int J Procure Manage 12(6):652–668

Ball S, Booth CA, Prabhakaran A, Mahamadu AM, Glass J (2023) A systematic review of responsible sourcing in the architecture, engineering, and construction sectors of the UK. Buildings 13(4):889

Carter CR, Rogers DS (2008) A framework of sustainable supply chain management: moving toward new theory. Int J Phys Distrib Logist Manag 38(5):360–387

Cayolla Trindade P, Antunes P, Partidário P (2017) SPP toolbox: Supporting sustainable public procurement in the context of socio-technical transitions. Sustainability 10(1):67

Chikwere D, Chikazhe L, Tukuta M (2023) Value for money in public procurement: experience from Zimbabwe's rural district councils. Cogent Soc Sci 9(2):2244746

Del Vecchio P, Mele G, Siachou E, Schito G (2022) A structured literature review on Big Data for customer relationship management (CRM): toward a future agenda in international marketing. Int Mark Rev 39(5):1069–1092

Delmonico D, Jabbour CJC, Pereira SCF, de Sousa Jabbour ABL, Renwick DWS, Thomé AMT (2018) Unveiling barriers to sustainable public procurement in emerging economies: evidence from a leading sustainable supply chain initiative in Latin America. Resour Conserv Recycl 134:70–79

Dudić Ž, Vrhovac V, Vulanović S, Dakić D, Erdeji I, Perović V (2024) A risk-aware approach to digital procurement transformation. Sustainability 16(3):1283

Fink A (2019) Conducting research literature reviews: from the internet to paper. Sage Publications

Haffar S, Özceylan E (2024) Blockchain-based system for supplier selection in sustainable and leagile supply chains. IEEE Access

Hafsa F, Darnall N, Bretschneider S (2021) Estimating the true size of public procurement to assess sustainability impact. Sustainability 13(3):1448

Hauschild MZ (2019) Life cycle impact assessment. In: CIRP encyclopedia of production engineering. Springer, Berlin, Heidelberg, pp 1053–1057

ISO 14040; Environmental Management (2006) Life cycle assessment, principles and framework ISO 14040:2006. International Organization for Standardization, Geneva, vol 3, p 20

Johnsen T, Howard M, Miemczyk J (2018) Purchasing and supply chain management: a sustainability perspective. Routledge

Kumar S (2022) Emergence of sustainable public procurement as a strategic instrument. Understanding sustainable public procurement: reflections from India and the world. Springer International Publishing, Cham, pp 17–85

Leonidou E, Christofi M, Vrontis D, Thrassou A (2020) An integrative framework of stakeholder engagement for innovation management and entrepreneurship development. J Bus Res 119:245–258

Liu W, Cao Y, Hou J, Cheng Y, Chan HK, Tang O (2024) Green procurement or green supply? A meta-analysis of their impacts on firm sustainability performance. Int J Logist Res Appl 1–35

Malolitneva V, Dzhabrailov R (2019) Strategic public procurement: facilitating sustainable development in Ukraine. Eur J Sustain Dev 8(2):91–91

Manta O, Panait M, Hysa E, Rusu E, Cojocaru M (2022) Public procurement, a tool for achieving the goals of sustainable development. Amfiteatru Econ 24(61):861–876

Meredith J (1993) Theory building through conceptual methods. Int J Oper Prod Manag 13(5):3–11

Moraga G, Huysveld S, Mathieux F, Blengini GA, Alaerts L, Van Acker K, Dewulf J et al (2019) Circular economy indicators: what do they measure? Resour Conserv Recycl 146:452–461

Nsiah-Sarfo DJ, Ofori D, Agyapong D (2023) Sustainable procurement implementation among public sector organisations in Ghana: the role of institutional isomorphism and sustainable leadership. Clean Logist Supply Chain 8:100118

Patel K, Feng H, Ruparathna R (2025) Rethinking sustainable procurement: investigating the impact of system boundary on building proposal selection. Sustainability 17(4):1408

Prosperi P, Soltanpour Y, Kaliji SA, Ouahi L, Hou MA, Achillas C, Camanzi L et al (2024) Agricultural and food business dynamics in the Mediterranean region: identifying key indicators for sustainable supply chain systems originated by small-scale farming production. New Medit 3:103–127

Puspitasari NB, Devytasari AKA, Susanty A (2025) What drivers factors for the implementation of sustainable public procurement in Indonesia? Int J Ind Eng Prod Res 36(1):87–98

Renukappa S, Akintoye A, Egbu C, Suresh S (2015) Sustainable procurement strategies for competitive advantage: an empirical study. Proc Inst Civil Eng Manage Procur Law 169(1):17–25

Seuring S, Müller M (2008) From a literature review to a conceptual framework for sustainable supply chain management. J Clean Prod 16(15):1699–1710

Smale E, Baid H, Balan M, McGain F, McAlistar S, de Waele JJ, Hunfeld N et al (2025) The green ICU: how to interpret green? Crit Care 29(1):80

Hoejmose SU, Adrien-Kirby AJ (2012) Socially and environmentally responsible procurement: a literature review and future research agenda of a managerial issue in the 21st century. J Purch Supply Manag 18(4):232–242

Swensson LF, Hunter D, Schneider S, Tartanac F (2021) Public food procurement as a game changer for food system transformation. Lancet Planet Health 5(8):e495–e496

Terman J, Smith C (2018) Putting your money where your mouth is: green procurement as a form of sustainability. J Public Procur 18(3):202–216

Van Opstal W, Borms L, Brusselaers J, Bocken N, Pals E, Dams Y (2024) Towards sustainable growth paths for work integration social enterprises in the circular economy. J Clean Prod 470:143296

Wade R (2018) Governing the market: economic theory and the role of government in East Asian industrialization

Waqar A, Othman I, Shafiq N (2025) The role of sustainable procurement in attaining sustainable development goals: insights from sustainable logistics and sustainable metamorphosis practices. J Clean Prod 500:145285

Yosef FA, Jum'a L, Alatoom M (2023) Identifying and categorizing sustainable supply chain practices based on triple bottom line dimensions: evaluation of practice implementation in the cement industry. Sustainability 15(9):7323

Zaman SI, Ali MR, Khan SA (2024) Exploring interrelationships among barriers and enablers of green procurement for a sustainable supply chain. Int J Procur Manage 19(2):226–252

Zhang RY, Al Muneem A, Byrd K (2025) Sustainability in food procurement: tea sourcing at a university. J Hosp Tourism Cases 13(2):78–84

Sustainable Business Models, Reporting, and Market Transformation

Sustainability Reporting in the Era of Circular Economy: A Bibliometric Network Visualization

Mohit Narang, Neha Puri, and Suman Lodh

Abstract This Chapter explores the relationship between Sustainability Reporting (SR) and Circular Economy (CE) through the lens of bibliometric review. By performing a bibliometric review, where *research articles* were sourced from *Scopus* repository, spanning the years *2000 to 2024*, this study tries to map global research trends and collaborations, identify most cited publications and authors, and explore emerging themes and clusters that gives future direction. Our methodology involves examining a total of *1604* research articles which were published in *English* language during the year *2000 to 2024*. The methodological framework utilizes advanced bibliometric tools such as VOSviewer and Biblioshiny, to create pictorial representations through co-citation and co-occurrence analysis. Our analysis reveals a significant overlap between SR and CE. The findings indicate that while research contribution span from *82 countries*, with USA, China and UK emerging as the top contributors, substantial challenges still persist in integrating the sustainability reporting with circular economy principles. As a result, our research provides a strategic roadmap for future scholars in the field of sustainability and sustainability-based reporting, identifying influential contributors and depicting emerging themes indicates a shift towards integrated frameworks a collaboration among industry, academicians, and policymakers so that the gap is bridged between theoretical and practical approach.

Keywords Sustainability reporting · Circular economy · Bibliometric analysis · Sustainable development goals · Non-financial reporting

M. Narang (✉) · N. Puri
ACCF, Amity University, Noida, Uttar Pradesh, India
e-mail: mohit.narang@s.amity.edu; mohitnarangmcr@gmail.com

N. Puri
e-mail: npuri@amity.edu

S. Lodh
Kingston University, London, UK
e-mail: S.Lodh@kingston.ac.uk

K. Kumar et al. (eds.), *Sustainable Value Creation: Corporate Strategies for SDGs Implementation*, World Sustainability Series,
https://doi.org/10.1007/978-3-032-24652-3_15

1 Introduction

1.1 Background

In recent times, there has been an unprecedented rise in resource utilisation all over the globe, and by 2050, it is expected that the same will become threefold. This is mainly because of developing countries contributing to their development (Jaeger and Upadhyay 2020). The steep rise in consumption of resources reflects the need for adopting sustainable practices, which aim to *'fulfill the needs of current generation, while considering simultaneously the potential needs of future generations'* (WCED 1987). In line with the same objective, Sustainable Development Goals (the SDGs) were set and issued in 2015 by the United Nations Organization (United Nations, 2015). These goals were issued under the United Nations Environment Programme (UNEP), with 193 member countries committing to the same (Adams 2017). These goals and the underlying project set the target for 2030 to make the world better and more sustainable (Adams 2017; Sachs 2015). The period 2000 to 2024 was selected to cover full evolutionary cycle of non-financial disclosure. The cycle began in 1997 with the establishment of GRI and the last two decades saw the institutionalization of non-financial reporting into formal reporting frameworks. Sustainable development, since then has become a readymade key to solve problems connected to the economy and environment (Hahn and Kühnen 2013).

From the conceptual standpoint, *Sustainability Reporting* (SR) is a key component to achieve sustainability in the society. SR allows businesses to monitor their efforts in environmental, economic and social areas (Bocken et al. 2016). One major step that the corporate entities are taking is to simply move from only following sustainable practices to in fact, actively measuring and reporting the same in their performance. Also, with the rising significance of Circular Economy, importance of sustainability reporting has been strengthened. In contrast with the linear models of '*take-make-dispose*', the circular economy promotes a more regenerative model (Geissdoerfer et al. 2017). The focus has now shifted to minimising waste, reusing materials, and increasing the durability of resources, which now closely aligns with the SDGs, particularly the *'SDG 12: Responsible Consumption and Production'*. This ensures that there is sustainable pattern for production and consumption both. The objective is to shield the environment, support economic stability, and conserve natural resources.

Additionally, after the introduction of SDGs, the reporting has gone beyond just financial performance. Now it underscores the contribution of an organisation towards circularity and sustainability principles. Many initiatives have come up with an aim to improve transparency in financial and non-financial disclosures. One of the initiatives that have come up and gained importance is the launch of '*Global Reporting Initiative (GRI)*', which is aimed at guiding the companies to report about their qualitative efforts such as reducing waste, conserving resources, and value creation through a product's usual lifecycle (Kolk 2010). Incorporating these *circular economy* principles into traditional reporting and mapping them with SR helps businesses in showcasing their adoption of a smarter and sustainable approach to resource utilization. Efforts like these promote environmental upkeep, social equality, and economic strength. In line these global initiatives, different countries have started developing frameworks based on their internal dynamics. For instance, SEBI (Securities and Exchange Board) in India, has launched the *'Business Responsibility and Sustainability Report (BRSR)'* framework.[1] It is applied mandatorily to the top 1000 listed firms in India (Singh et al. 2024). The goal underlying such initiatives is to standardise ESG disclosures and encourage firms to report on sustainability issues. This in turn, reinforces a country's dedication to sustainable development. By accepting to report on sustainability, corporate firms can align their operations with internationally recognised frameworks like the SDGs, GRIs, and even CDP *(Carbon Disclosure Project).* This simultaneously shows their commitment to responsible and comprehensive development (Global Reporting Initiative 2016).

Additionally, the continuous integration of sustainability reporting with the principles of circular economy shows how operations within the businesses are moulding towards circularity. Also, businesses are starting to adopt sustainability and circularity principles into their everyday operations, showcasing a forward-looking approach towards the changing global scenario (Singh, K., & Misra, M. (2022). Eventually, steps like these strengthen the 'pillars of sustainable development'; *the first pillar—economic pillar*, which promotes inclusive growth, efficient resources, and sustainable business models; *the second pillar—social pillar*, which emphasises equity in society and focuses on basic human rights and access to essential services viz. education, healthcare, and decent work; *the third pillar—environmental pillar* which focuses on the preservation of natural resources and biodiversity and seeks to control environment degradation and impact on climate.

1.2 Rationale and Justification

The last few decades have seen a decent growth of scholarly interest around SR and CE. Still, there are limited contributions asking for more deeper studies and interconnectedness of CE and SR (Murray et al. 2017). This study aims to fill this gap by employing a bibliometric analysis and the underlying tools & techniques. The study also tries to present an all-inclusive analysis of the existing literature available in this area. In addition, the research promotes the knowledge of global undercurrents that influence SR and CE (De Pascale et al. 2021). It also answers calls for a more synchronised and inclusive review in this field. Ultimately, the research establishes a strong foundation for future research by highlighting key contributions in the research area, the collaborative efforts, and the emerging themes.

---

[1] Singh, Harjit, Neha Puri, and Nandita Mishra, eds. *Sustainability Reporting and Blockchain Technology*. London: Routledge, 2024. https://doi.org/10.4324/9781003378341.

1.3 Research Objectives and Questions

The objective of this study is to analytically discover the interconnectedness of SR and CE principles. This is done by applying a rigorous bibliometric approach which involves tracing trends in article publications, identifying articles which are cited the most, exploring their influence on the research, pinpointing collaborations, and recognising leading journals, authors, and institutions. The research also helps in depicting emerging themes that can help future researchers in the field. To guide further into the analysis, the research proposes to find answer to the following research questions:

RQ 1: Who are the most prolific *authors* contributing to this research field and which articles have been highly cited in the research field?

RQ 2: What are the leading *sources* (*Journals*), publishing most of the work in this research field?

RQ 3: Which *countries* are leading the research in this field? What are the *collaboration patterns* among the countries in this research field?

RQ 4: What are some *emerging themes* and trends related to this field?

RQ 5: What are some *research gaps* present in the field, and *what path should future studies follow* to fill such gaps?

While finding answers to the above questions, our study aims to identify key contributors, geographic centres of research, and some highly influential publications. The results of the research also highlight the core themes and unexplored areas in the existing body of literature.

2 Review of Literature

2.1 Evolution of Sustainability Reporting

Sustainability Reporting (SR) has been evolving since the *1980s*. In the beginning, it was a recommendary policy for disclosures, which now has gradually become a vital component of the corporate reporting and strategy (Drexhage and Murphy 2010). Earlier, the reporting was mostly done under regulatory pressures but the establishment of Global Reporting Initiative (GRI) in 1997 proved to be a turning point in this approach (Kolk, 2004). *GRI* provided standardized guidelines and has aimed to widen the reporting framework for entities. Since the establishment, GRI has been a prominent step in guiding organisations address ESG related concerns. But despite its role in transparency, GRI is often criticized for its lack of rigor, which can lead to 'greenwashing' (Marquis et al. 2016). Firms selectively report positive metrics while skipping systematic loopholes in their models (Adams 2017). Similar to GRI, there have been other initiatives such as the *'Carbon Disclosure Project (CDP)'* and *'Sustainability Accounting Standard Board' (SASB)* which were formed globally. The *'Business Responsibility and Sustainability Report' (BRSR)* in India is also one of such initiatives. Similar steps have been taken from time to time to guide corporate firms in bringing more transparency to improve stakeholder engagement (Singh et al. 2024).

2.2 *Emergence of the Circular Economy*

Concept of Circular Economy (CE) came into limelight in late *twentieth century.* It gained attention as the standardized framework for resource efficiency, waste minimization, and value creation as noted by Ghisellini et al. (2016). The scholarly interest in CE grew sharply due to its alignment with sustainability principles. CE started to emphasize changes in production, consumption, and disposals and to deal with threats like climate change and lack of resources (Ellen MacArthur Foundation, 2013).

2.3 *Convergence of Circular Economy (CE) and Sustainability Reporting (SR)*

The need for convergence of circular economy and SR is evident in most of the existing literature. This is an interdisciplinary field that focuses on circular principles to obtain stakeholder-driven reporting (Enric et al. 2020). While *sustainability* focuses on resource consumption, the circular economy talks about maintaining efficiency in resources, waste generation and management (Korhonen et al. 2018). However, their convergence is limited, with a handful of studies covering both the concepts (Murray et al. 2017).

2.4 *Existing Gaps and Future Research Avenues*

The recent scholarly contributions and regulatory efforts by countries highlight a clear alignment towards the convergence of two concepts: SR and CE. For instance, GRI standards such as *GRI 306 on 'Waste' and GRI 301 on 'Materials'* guide organisations to keep track of their progression in Sustainability while keeping in mind they follow the circularity principles (Rockström et al. 2009). The consistency of this progression also differs significantly across sectors. This indicates that the adoption of circularity principles is uneven in various sectors (Kirchherr et al. 2018).

In addition to this, there is not enough data available about how blending CE practices into reporting affects issues like the company's success, investors' perception about the company, or policies impact. There is also a visible shortage of empirical research on SR and CE, particularly, examining the impact of SR and CE's integration on these aspects (Upadhyay et al. 2021). Bibliometric studies in this field have indicated that the research is mostly centred on concepts like *waste management, life cycle assessment, and green technologies*. However, there is limited exploration into how *SR* can function as an imperitive tool to include *circular economy practices* within the organisation. Thus, our aim is to create an intellectual landscape around the field that addresses these limitations, reflecting patterns of collaboration, and highlighting key thematic trends for future research.

3 Methodology

3.1 Data Retrieval

The data for this research is taken from *'Scopus'* repository. Out of several databases available that provide valuable information about the existing literature, Elsevier's '*Scopus'* (www.scopus.com) and Clarivate Analytics' '*Web of Science*' (www.webofscience.com/wos/) are the most popular among scholars. Out of these, *'Scopus'* was chosen as the database repository for this study's data extraction due to comprehensive coverage over social science topics unlike '*Web of Science*' that majorly includes science domain research (Falagas et al. 2008). The initial search had keywords such as 'circular economy', 'sustainability reporting,' and 'sustainable development' with the 'AND' operator that might appear in research titles, given abstracts, and some keywords. This resulted in obtaining *3190 articles* which were then refined using different filters such as the language, document type, subject area, and period of study. After these filters, the result was a number of 1604 documents as shown in the Flowchart (Fig. 1).

3.2 Data Filtering and Metrics

After pre-processing of the data set, *1604 articles* were obtained, onto which bibliometric analysis was performed. The dataset spanned from year 2000 to 2024, coming from 551 sources, predominantly journals, with 1592 articles and 12 conference papers. The annual growth rate of publications is 32.07%, indicating a staggering annual growth rate over the period, while the average document age is 3.82 years, reflecting a relatively recent body of literature. The dataset includes 125,128 references, with the average citations of 25.46 per document, underscoring the field's impact. It contains 3774 Keywords Plus and 4464 author-defined keywords,

Fig. 1 Flowchart showing document filters. *Source* Authors compilation

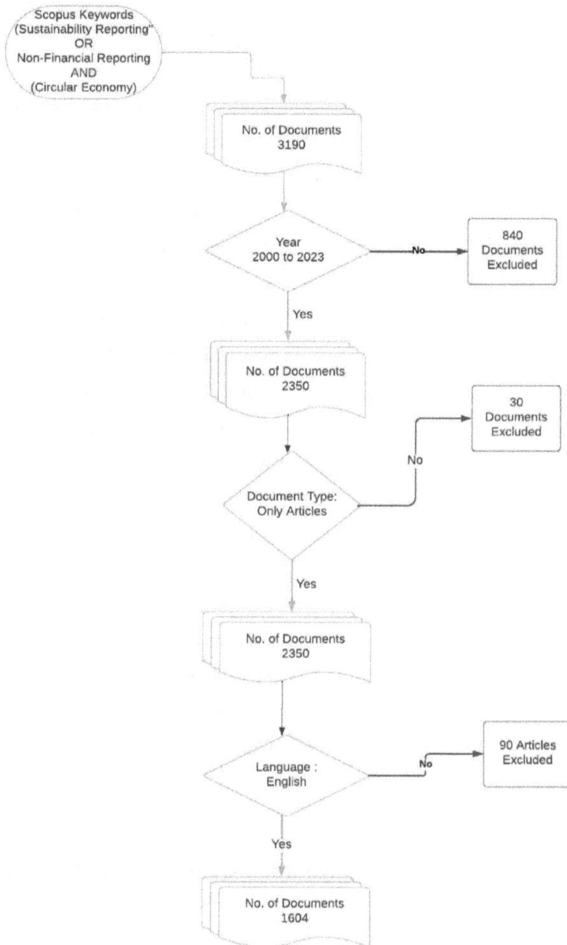

providing a rich set of terms for thematic analysis. The author base comprises 4494 authors, with 148 contributing to single-authored documents and 161 single-authored documents in total. On an average, every document in the dataset has 3.34 authors, and 32.61% of the documents contain international co-authorships, highlighting a collaborative research effort in SR and the circular economy (Table 1).

The bibliometric analysis was conducted using tools such as Biblioshiny (Aria and Cuccurullo 2017) and VOSviewer. These are the two widely used tools for mapping and visualising scientific literature. Biblioshiny, part of the Bibliometrix package *(through R software—version 4.5.2)* was employed to perform quantitative analyses such *as* author productivity, citation analysis, and thematic evolution, and to generate most visual representations, including thematic maps, three-field plots, and keyword co-occurrence maps. VOSviewer, on the other hand was used specifically for its

Table 1 Data description

Description	Results
Timespan	2000:2024
Source (Journals, books and more)	551
Documents	1604
Annual growth rate %	32.07%
Document average age	3.82
Average citations per doc	25.46
References	125,128
Document contents	
Keywords Plus (ID)	3774
Author's Keywords (DE)	4464
Authors	
Authors	4494
Authors of single-authored docs	148
Authors collaboration	
Single-authored docs	161
Co-authors per doc	3.34
International co-authorships %	32.61
Document types	
Article	1592
Conference paper	12

Source Biblioshiny

ability to depict visualisations, such as co-authorships and keyword co-occurrence networks.

4 Results of Bibliometric Analysis

This section provides a comprehensive picture of the intellectual evolution within the SR and CE research domains. By analysing the dataset of *1604 articles*, retrieved from *'Scopus'*, and following a deeper analytical approach, the following results have been obtained. The results obtained through different tools and visual representations are shown below.

4.1 Leading Authors and Sources

The study of total 1604 research articles from the Scopus database during *2000 to 2024*, uncovered contributors in the field integrating CE and SR. Rather than providing a simple tally of articles, the bibliometric analysis evaluates the intellectual value of these publications, highlighting most prolific authors. The figure indicates most prolific authors in the research field.

The above Fig. 2 helps in identifying the intellectual pillars of the SR-CE research. Authors who have published the *highest number of documents* can be seen in figure. As per figure, *Lozano R., Ramos T.B., and Zhang Y.* are the primary contributors, each contributing with nine publications. Apart from production of articles, their productivity is also analysed *with Lozano R's 692 citations* serving as foundational author. This is complemented by *Ramos T.B* with *590 citations* whose work in the field bridges the gap between theoretical management frameworks and practical compliance. Apart from them, *Zhang Y* has got *433 citations* whose work majorly explores how CE principles are structurally integrated into systematic reporting.

The above authors have established the field's robust foundations, the conceptual work is also rooted in the work of authors like *Camilleri M.A., Kumar A., and D'Adamo I.* With more than 1100 citations, these authors have provided the primary lens through which CE is integrated into Corporate Sustainability Reporting Practices (Fig. 3).

The dramatic expansion of this research field is evidenced by the emergence of some high impact journals. These journals have transitioned the SR-CE research from a conceptual term to a mainstream practical initiative. These intellectual hubs, specifically journals like *'Journal of Cleaner Production' and 'Sustainability,'* have been imperative to institutionalize the circularity principles within the corporate sustainability practices. This is explained by the figure above with 244 publications by

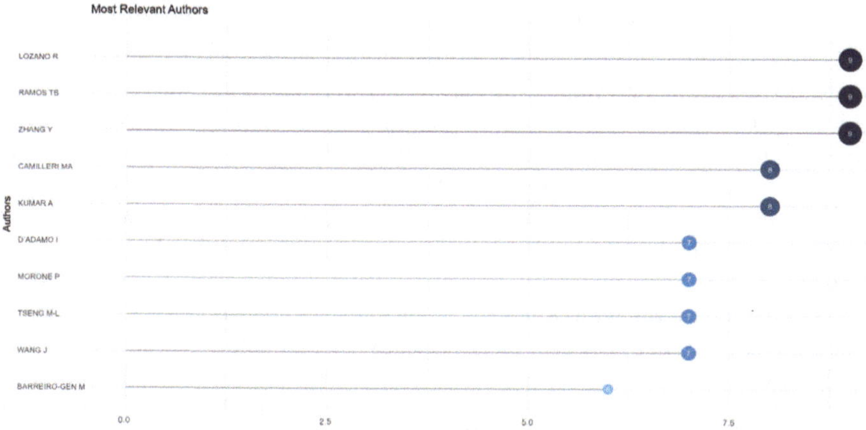

Fig. 2 Most prolific authors. *Source* Biblioshiny

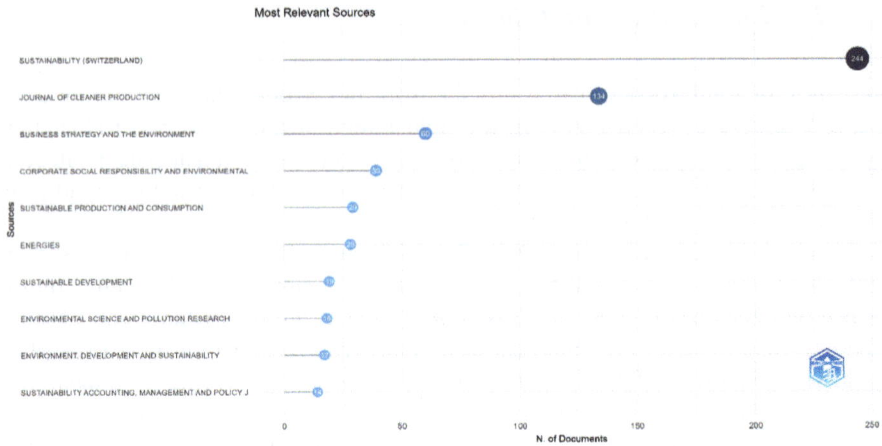

Fig. 3 Most relevant sources. *Source* Biblioshiny

'*Sustainability'* journal, predominantly becoming the knowledge hub for scholars, followed by *'Journal of Cleaner Productions'* with 134 publications and also the '*Business Strategy and the Environment*' journal with *60 articles.* Beyond these primary hubs, journals such as '*Corporate Social Responsibility and Environmental Management*' with *39 publications* and *'Sustainable Production and Consumption*' with *29 publications* also provide sector specific sustainability transitions. The consistent growth of output in these journals indicate a broadening scholarly consensus that the integration of CE principles in SR has evolved from a peripheral concern to a central pillar of ESG strategies.

4.2 Global Collaboration and Country-Wise Contributions

To suggest avenues for future exploration, a three-field plot was created using Biblioshiny to map the connections among authors, keywords, and journals. This visualisation was based on a dataset of 1604 articles retrieved from the Scopus database, spanning the years 2002 to 2022. Figure 4 illustrates these interconnections, with the left field representing the most prolific authors (e.g., Lozano R, Ramos TB, Zhang Y), the middle field displaying frequently occurring keywords (e.g., "circular economy," "sustainability reporting," "stakeholder engagement"), and the right field listing prominent journals (e.g., *'Business Strategy and the Environment,' 'International Journal of Production Economics'*). The thickness of the connecting lines indicates the strength of relationships, reflecting the extent to which specific authors contribute to certain themes and published journals. The bibliometric analysis exposes several intellectual blind spots within the current intersection of SR and CE research. A striking disconnect exists between the field's leading contributors and the keywords "social impact" and "emerging economies," which points toward a systemic under-theorization of the social pillar of circularity, particularly concerning its implementation in developing regions. Furthermore, while foundational journals such as *'Business Strategy and the Environment'* highlight the core discourse on circularity, there is also a lack of engagement within small and medium enterprises (SMEs) for digital transformation. This also represents a significant missed opportunity for future inquiry into how smaller firms can institutionalize circular models through standardized reporting, as well as the potential for digital enablers to refine ESG disclosures. These observations directly corroborate the thematic evolution discussed in Sect. 4.3, which indicates a diminishing scholarly focus on the specific challenges and nuances of the SME sector over time. Future research should explore several key directions to strengthen the evolving link between CE principles and corporate sustainability frameworks. *Firstly*, researchers should prioritise the development of standardised metrics for assessing circularity in SR, with particular attention to quantifying social impacts. *Secondly*, a deeper investigation is needed into SR practices in emerging economies, particularly where the implimentation of circular economy models may be affected due to diverse institutional, political, and economic scenarios. *Thirdly*, scholars should examine how digital technologies, such as blockchain and the Internet of Things (IOT) can facilitate the adoption and implimentation of circular economy principles.

SR could yield valuable insights, especially for the small and medium-sized enterprises that strive to face the challenge of strengthening their sustainability performance. The country collaboration map (Fig. 5) highlights the leading countries contributing and their collaborative research in the fields of Circular Economy and SR. Based on the map above, the *United Kingdom* stands out as a central node, indicating its dominance in publishing a significant number of documents and fostering collaborations across countries.

Additionally, the *United States, China, and Italy* are the key players in the growing body of research within this field, representing significant engagement with others. Some other countries, such as *Spain, Germany, and India,* also stand out for their active participation in building ties with other countries. The map's colour gradient (Fig. 5) reflects the timeline of publications, with darker tones representing older works and the lighter ones pointing to more recent ones. Particularly, *China and India*, and most *Asian* countries appear to have picked up their research output in recent years, as evidenced by their lighter-colored nodes. These patterns highlight the interdisciplinary and global nature of the research, with meaningful contributions coming from both developed and developing economies. Overall, *European* institutions, especially those based in the *United Kingdom and the Netherlands* demonstrate strong collaborative networks.

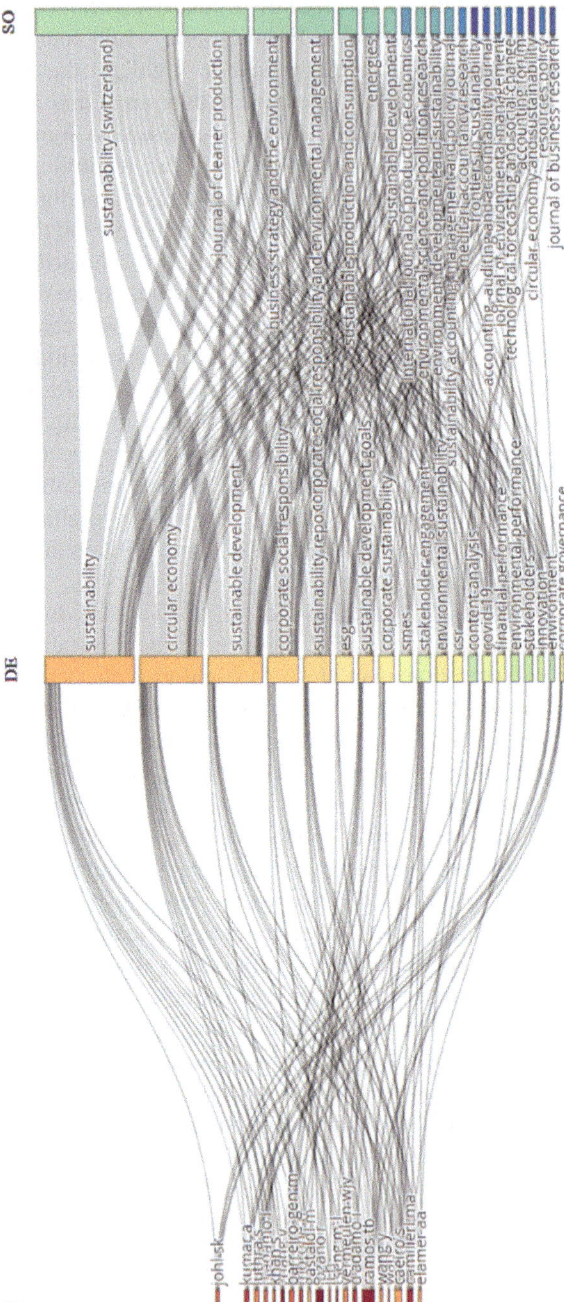

Fig. 4 Three field plot. *Source* Biblioshiny

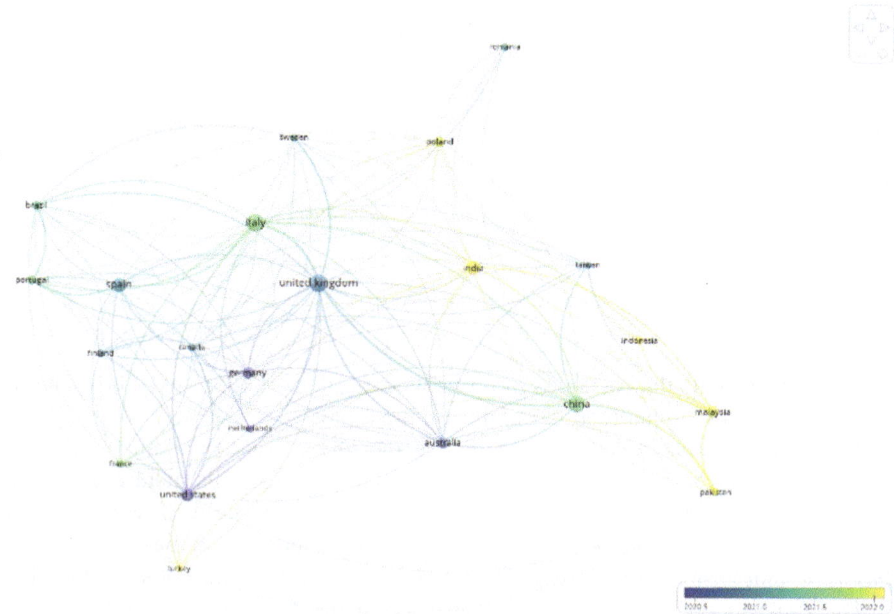

Fig. 5 Countries collaboration network. *Source* VOSviewer

4.3 Thematic Analysis and Keyword Insights

Discovering the themes related to Circular Economy (CE) and SR through the word cloud shown in Fig. 6 gives us a clear picture of how the research area is developing. The most frequently used terms in the word cloud are 'sustainability' and 'sustainable development,' underscoring their significant role in addressing global issues. Inclusion of words like "circular economy" highlights the increasing significance of principles of circularity in the reporting frameworks. Additional terms like 'decision-making,' 'stakeholder,' and 'resource efficiency' reflect key principles hovering around SR and CE. Terms like 'resource efficiency' reflects circular economy's alignment to basic principles, On the other hand, emphasis on the terms 'decision-making' and 'stakeholder' indicate the value of collaborative governance (Murray et al. 2017).

One rising theme of phrases like 'digital transformation' and 'innovation' in the word cloud highlights growing influence of technology in shaping sustainable practices. This is common through tools like Blockchain & IoT being used in business processes (Tseng et al. 2018). Thus, the word cloud as a whole represents multidisciplinary nature of SR and CE research, integrating environmental, technical and social dimensions.

Another visual representation is the *Thematic Map*, depicted in the Fig. 7. It represents four different *quadrants* and *themes* reflecting their *centrality and density.* The

Fig. 6 Word cloud. *Source* Biblioshiny

centrality represents the basicness of terms whereas density represents the quantum of literature or research already available on the topics.

The *Motor Themes* quadrant, located on the *upper right* quadrant, *indicating high density* and *high centrality* of the terms. The terms in this quadrant indicate well developed and highly connected research areas in the field. *Governance* and *firm performance* are both positioned as important components in the *Motor themes quadrant,* emphasising their crucial role in determining the future of sustainability.

The *Emerging Themes* quadrant positioned at the *lower-left* portion with *low centrality and low density,* indicating these themes are declining or at the very infant stage of emergence. Terms like *sustainability accounting* and *digital transformation* are reflected in this quadrant indicating the importance of sustainability-based accounting or reporting and use of new age technologies such as *blockchain, artificial*

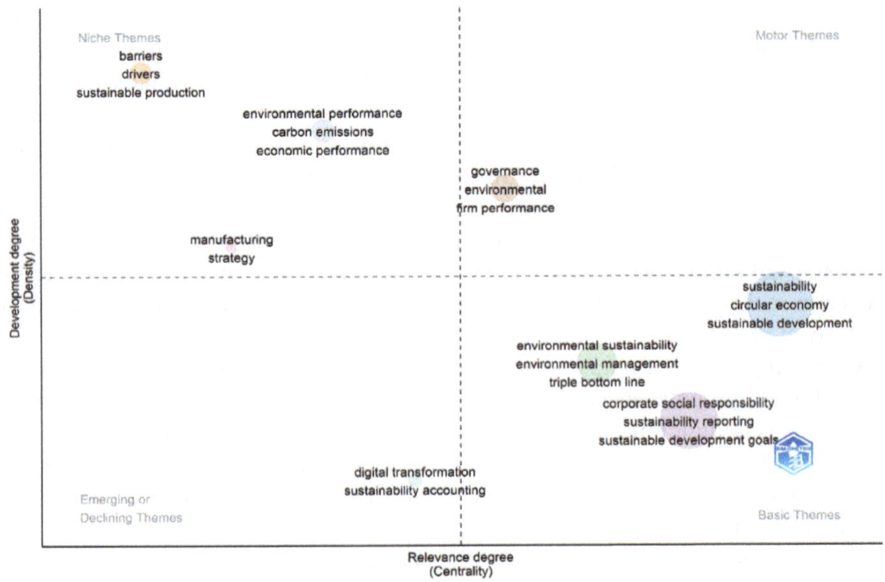

Fig. 7 Thematic map. *Source* Biblioshiny

intelligence, and the Internet of Things. But at the same time, low density indicates that research in these areas is still understudied.

The Niche themes quadrant is positioned at the *upper-left portion* of the graph, showing high density but low centrality. Phrases like *barriers, drivers, and sustainable production* exhibit *high density but low centrality.* This is an indication that these themes are well-developed internally, having rich conceptual frameworks, but are relatively isolated from other *thematic clusters*, due to which they *lack centrality.* These themes offer opportunities for future research based on integration of gaps observed among the research clusters. For example, research on *barriers and drivers* evaluate the factors that can hinder or promote sustainability transitions respectively. But due to lower centrality, these themes may not yet be influencing the broader aspects in sustainability like the ESG or Circular Economy.

Lastly, the *Basic themes* quadrant is positioned in the *lower-right* portion. This shows *low density but high centrality* of themes in picture. The *high centrality* suggests that these themes are common across different studies and thus, form the backbone of the research field. On the other hand, their *low density* indicates that comprehensive research on these themes may still be limited. Terms like the 'Circular *Economy (CE)', 'Sustainability Reporting (SR),' 'Triple bottom line (TBL) reporting', environmental management, Sustainable Development Goals (SDGs), and Corporate Social Responsibility (CSR)* indicate that they are the foundation of most studies revolving around CE and SR. Thus, it provides a decent beginning for scholars entering the field, offering opportunities for linking new ideas to well-established literature ***(RQ4)***.

In Summary, the *Thematic Map* reflects that the field of SR and Circular Economy is undergoing significant changes. It also reflects strategic insights into the structure and direction of the research field. It highlights the importance of areas of mature interest with potential ideas that can dive deep into organisational planning and governance.

5 Future Research Recommendations

Despite the expanding volume of literature on *Circular Economy (CE) and SR*, significant challenges continue to limit their smooth integration. One major challenge is the fragmented treatment of sustainability concepts where individual dimensions are often examined rather than an approach that cohesively address all the dimensions. The fragmentation is compounded by geographical and sector bias, with research centred on developed nations, particularly the European countries. There exists limited focus on developing and under-developing countries or even sectors such as agriculture or construction that necessitates comprehensive studies that strives into these regions. Additionally, CE's short-term benefits like resource optimization and waste reduction are well researched, but on the other hand, the long-term impacts on economic stability, growth and consumer behaviour are unexplored. This highlights the need of longitudinal studies in this area. Moreover, there is an absence of a standardise metrics that assesses both CE and SR practices. Most of the guideline's hover around qualitative and self-reporting data which serves as a limitation for comparison. Integration of technologies like the Blockchain, IoT, and AI can enhance transparency and comparison. *Lastly,* the role of policies and governance in fostering the CE and SR's under-researched. There are very few studies that examine how collaborative policy frameworks support circular practices and boost sustainability. Addressing these challenges though a comprehensive, interdisciplinary, and robust research is the key to align with the objectives of our bibliometric study to map the research gaps and guide future directions ***(RQ5)***.

6 Conclusion

The bibliometric analysis performed on the selected dataset of 1604 articles that were published during year 2000 to 2024 has established a clear intersection of the evolving landscape of CE and SR. Authors such as Lozano R., Ramos T. B., and Zhang Y., were each found prolific with nine publications each ***(RQ1)***. Together they have significantly influenced the research output as they have got the most citations for their works like '*A holistic perspective on Corporate Sustainability Drivers*' (1202 citations), '*Sustainable Supply Chain Management: A review of Trends and Future Directions*' (310 citations), and '*Circular Economy in Corporate Sustainability Reporting: A review of Organizational Approaches*' (290 citations).

Studies like these along with seminal contribution from authors like Murray et al. (2017); Hassini et al. (2012) and Manavalan et al. (2018) reflect the advancements that have taken place in the interdisciplinary frameworks and performance metrics. Leading journals, such as the '*Journal of Cleaner Production,*' and '*Sustainability*' have come out as key source platform ***(RQ2)*** with strong collaborative networks among European institutions, particularly in UK, and Netherlands ***(RQ3)***. The country collaboration map, the word cloud, and the thematic maps offer insights into the whole landscape of SR and CE. These visualisation structure also indicate the absence of integrated frameworks which serve as a limitation to the existing research. Thus, reflecting on the above findings, the study has mapped the evolution of CE and SR identifying robust foundations and gaps that seek policymakers, authorities and different stakeholders' attention. By laying these foundations, this chapter not only accumulates the existing literature but also advocates for a collaborative scholarly approach in future to transform the theoretical findings into practical and sustainable solutions.

References

Adams CA (2017) The sustainable development goals, integrated thinking and the integrated report. IIRC and ICAS, London

Aria M, Cuccurullo C (2017) Bibliometrix: an R-tool for comprehensive science mapping analysis. J Informet 11(4):959–975
Bocken NMP, Short SW, Rana P, Evans S (2016) A literature and practice review to develop sustainable business model archetypes. J Clean Prod 65:42–56
De Pascale A, Arbolino R, Szopik-Depczyńska K, Limosani M, Ioppolo G (2020) A systematic review for measuring circular economy: the 61 indicators. J Cleaner Prod 281:124942. https://doi.org/10.1016/j.jclepro.2020.124942
Drexhage J, Murphy D (2010) Sustainability reporting: from Brundtland to Rio 2012. Background Paper, United Nations
Ellen MacArthur Foundation (2013) Towards the circular economy vol. 1: an economic and business rationale for an accelerated transition. Ellen MacArthur Foundation, Cowes, UK
Falagas ME, Pitsouni EI, Malietzis GA, Pappas G (2008) Comparison of PubMed, Scopus, web of science, and google scholar: strengths and weaknesses. FASEB J 22(2):338–342
Geissdoerfer M, Savaget P, Bocken NMP, Hultink EJ (2017) The circular economy—a new sustainability paradigm? J Clean Prod 143:757–768
Ghisellini P, Cialani C, Ulgiati S (2016) A review on circular economy: the expected transition to a balanced interplay of environmental and economic systems. J Clean Prod 114:11–32
Global Reporting Initiative (GRI) (2016) Consolidated set of GRI sustainability reporting standards. Global Reporting Initiative, Amsterdam
Hahn R, Kühnen M (2013) Determinants of sustainability reporting: a review of results, trends, theory, and opportunities in an expanding field of research. J Clean Prod 59:5–21
IFRS Foundation (2021) IFRS foundation announces international sustainability standards board (ISSB) at COP26
Jaeger B, Upadhyay A (2020) Understanding barriers to circular economy: cases from the manufacturing industry. J Enterp Inf Manag 33(4):729–745
Kirchherr J, Piscicelli L, Bour R, Kostense-Smit E, Muller J, Huibrechtse-Truijens A, Hekkert M (2018) Barriers to the circular economy: evidence from the European Union. Ecol Econ 150:264–272
Kolk A (2004) A decade of sustainability reporting: developments and significance. Int J Environ Sust Dev 3(1):51–64
Kolk A (2010) Trajectories of sustainability reporting by MNCs. J World Bus 45(4):367–374
Korhonen J, Honkasalo A, Seppälä J (2018) Circular economy: the concept and its limitations. Ecol Econ 143:37–46
Marquis C, Toffel MW, Zhou Y (2016) Scrutiny, norms, and selective disclosure: a global study of greenwashing. Organ Sci 27(2):483–504
Murray A, Skene K, Haynes K (2017) The circular economy: an interdisciplinary exploration of the concept and application in a global context. J Bus Ethics 140(3):369–380
Rockström J, Steffen W, Noone K, Persson Å, Chapin FS III, Lambin E, Foley JA (2009) A safe operating space for humanity. Nature 461(7263):472–475
Sachs JD (2015) The age of sustainable development. Columbia University Press, New York
Singh H, Puri N, Mishra N (eds) (2024) Sustainability reporting and blockchain technology. Routledge, London. https://doi.org/10.4324/9781003378341
United Nations (2015) Transforming our world: the 2030 agenda for sustainable development. https://sustainabledevelopment.un.org/post2015/transformingourworld
Upadhyay A, Mukhuty S, Kumar V, Kazancoglu Y (2021) Blockchain technology and the circular economy: implications for sustainability and social responsibility. J Clean Prod 293:126130
World Commission on Environment and Development (WCED) (1987) Our common future. Oxford University Press, Oxford

Reimagining Carbon Accountability: Innovative Tools for Sustainable and Inclusive Growth

Krishnendu Ghosh, Veena Anand, and Trupti Dandekar Humnekar

Abstract The increasing concentration of carbon dioxide (CO_2), driven by forms of production and consumption across individuals, corporations, and institutions, has emerged as a serious concern for governments and global stakeholders due to its profound implications for climate stability. In response, the concept of the carbon footprint has evolved from a static emissions accounting metric into a dynamic strategic instrument for sustainability transformation. This chapter examines the methodological advancements, technological innovations, and sector-specific applications that have expanded the role of carbon footprint in advancing the Sustainable Development Goals (SDGs). Carbon footprint analysis is no longer limited to retrospective reporting; it has become an essential decision-support system that informs corporate strategy, public policy formulation, sustainable finance, and climate governance. By integrating data-driven insights with sustainability objectives, carbon footprint tools now serve as catalysts for considerable progress toward climate action and broader SDG integration.

Keywords Climate stability · Carbon footprint · Sustainability · Sustainable development goals · SDGs · Carbon footprint tools

---

K. Ghosh (✉)
Associate Professor, School of Management, Bennett University, Greater Noida, India
e-mail: kghosh.fin@gmail.com; krishnendu.ghosh@bennett.edu.in

V. Anand
Assistant Professor, Christ Academy Institute for Advanced Studies, Bangalore, India

Research Scholar, JAIN (Deemed-to-be) University, Bangalore, India

V. Anand
e-mail: veenav2@gmail.com

T. D. Humnekar
Professor, Faculty of Management Studies, Jain (Deemed-to-be) University, Bangalore, India
e-mail: truptidandekar@gmail.com

K. Kumar et al. (eds.), *Sustainable Value Creation: Corporate Strategies for SDGs Implementation*, World Sustainability Series,
https://doi.org/10.1007/978-3-032-24652-3_16

1 Introduction

The term "carbon footprint" was introduced in the early 1990s and has gained immense popularity over the last few years. This has become a buzzword worldwide in terms of sustainability and action against climate change. This is a tool to measure greenhouse gas emissions. The increasing level of CO_2, resulting from production and consumption by individuals, companies, organizations, and others, in the environment, has raised a major concern for the government and other stakeholders regarding the overall impact on the global climate and the Earth. Although it has been observed that the widespread use of this catchphrase is a main agenda item in various political and corporate forums worldwide, there is a dearth of clarification of this term in the scientific literature.

The concept of "carbon footprint" is not that new. It has evolved over almost two decades. To date, many researchers have tried to define the term based on their own understanding. Wiedmann and Minx (2008) have defined the term 'carbon footprint' as, "*The carbon footprint is a measure of the exclusive total amount of carbon dioxide emissions that is directly and indirectly caused by an activity or is accumulated over the life stages of a product.*" In another study, East (2008) has defined carbon footprint as, "*A direct measure of greenhouse gas emissions (expressed in tonnes of CO_2) caused by a defined activity. At a minimum, this measurement includes emissions resulting from activities within the control or ownership of the emitter and indirect emissions resulting from the use of purchased electricity.*"

By searching the term "carbon footprint" through the Scopus database, it has been observed that since 2006, the concept has started gaining momentum. In the line chart titled "Documents by Year," shown in Figure 1, the number of documents published annually from 1982 to 2026 is shown according to Scopus data. The horizontal axis represents years, marked in increments of four years, and the vertical axis represents the number of documents, ranging from 0 to 10,000 in increments of 2,000. The data points remain near zero from 1982 until around 2026, and then show a steady increase starting around 2007. The number of documents rises sharply after 2014, peaking near 8,000 around 2024, followed by a steep decline approaching zero by 2026. The chart indicates rapid growth in document publication in recent years, with a projected sharp drop at the end of the timeline.

The analysis of the Scopus database also provides us with the details about the research initiatives taken by different countries in this area. The graphical representation of the same has been depicted below. It is clearly visible that the United States is leading with its research initiatives in carbon footprint, followed by China, the United Kingdom, and India. Though countries such as Germany, Italy, Australia, Canada, Spain, and France have also contributed significantly to the area of carbon footprint (Fig. 2).

Though most of the definitions are related to the CO_2 emission, the concept of carbon footprint is not restricted to CO_2 emission only. It is based on the emission of greenhouse gases. Greenhouse gases (GHGs) are atmospheric gases that block heat, leading to the greenhouse effect that heats the Earth. While this is a natural

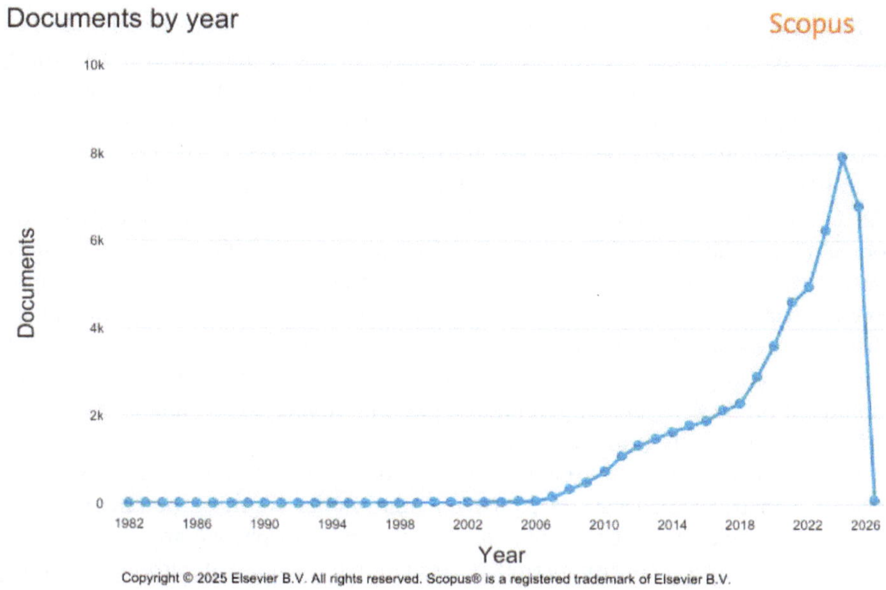

Fig. 1 Documents by year (*Source* www.scopus.com)

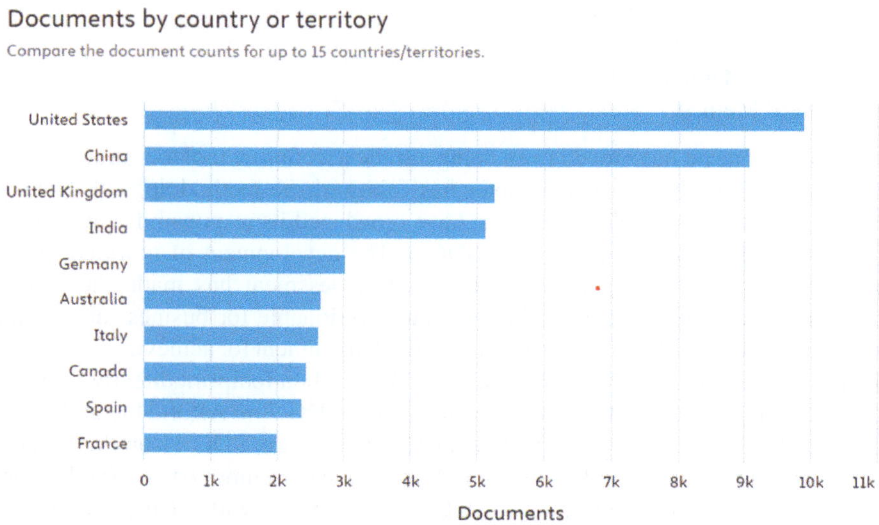

Fig. 2 Documents by country or territory (*Source* www.scopus.com)

and necessary process for maintaining a liveable temperature, human activities have dramatically increased the concentration of these gases, intensified the warming effect, and altered the global climate. Major GHGs include carbon dioxide (CO_2), methane (CH_4), nitrous oxide (N_2O), and fluorinated gases (like HFCs, PFCs, and SF6), which have various natural and human-related sources.

In 1988, the United Nations body, the Intergovernmental Panel on Climate Change (IPCC), was established by the United Nations Environment Programme (UN Environment) and the World Meteorological Organization (WMO) to provide policymakers with consistent scientific assessments on climate change, its implications, potential future risks, and available adaptation and mitigation options (Emmanuel 2019). At present, it has 195 member countries (Seventh Assessment Report—IPCC 2019). According to a new UN Environment Programme (UNEP) Report, the world should cut off 42% of annual greenhouse gas emissions by 2030 and 57% by 2035 (UN Environment Programme 2024).

Carbon accounting is the decisive tool in decarbonization strategies. With the help of a metric called carbon equivalent, the global warming potential (GWP) assesses how much heat a greenhouse gas traps in the atmosphere compared to carbon dioxide (CO_2) over a specific period. There are five methods of carbon accounting as mentioned below:

(a) The Physical-Unit Method,
(b) Supplier-Specific Method,
(c) Spend-Based Method,
(d) Average Method, and
(e) Hybrid Method.

Carbon footprint has moved from being a static accounting tool to a strategic tool for sustainability change. This chapter has followed the methodological, technological, and sectoral developments that have broadened its responsibility in driving the Sustainable Development Goals (SDGs). The development of more advanced analytics, artificial intelligence, and life cycle assessment has made the carbon footprint not only more accurate but also more actionable for business and policy decision-making. Carbon footprint is an essential instrument for achieving the United Nations Sustainable Development Goals (SDGs). Carbon footprint directly supports several SDGs, such as SDG 13 on Climate Action; SDG 7 on Affordable and Clean Energy; SDG 9 on Industry, Innovation, and Infrastructure; SDG 11 on Sustainable Cities and Communities, and SDG 12 on Responsible Consumption and Production.

Sustainable development aims to fulfil existing needs without negotiating the capability of future generations to meet their own needs. A carbon footprint is a method to measure total greenhouse gas emissions from an individual, activity, or product. Reducing carbon footprints is essential for sustainable development because excess emissions lead to climate change, threatening social development aspects like health, food, and energy security, and requiring a transition to low-carbon economies.

1.1 Connection Between Carbon Footprint and SDGs (Especially SDG 7, 9, 11, 12, 13)

The concept of carbon footprint is intricately linked to the Sustainable Development Goals (SDGs), particularly with SDG 13 on Climate Action; SDG 7 on Affordable and Clean Energy; SDG 9 on Industry, Innovation, and Infrastructure; SDG 11 on Sustainable Cities and Communities, and SDG 12 on Responsible Consumption and Production. To maximize the contribution of carbon footprint to the SDG's, concerted effort is required from all, starting from individuals, businesses, and governments. Understanding of climate change and its correlated carbon footprint is crucial for advancing environmentally responsible behaviours and advocating policy initiatives aimed at reducing greenhouse gas emissions.

Carbon Footprint and SDG 13

Climate change has become a real threat to the entire civilization. Sustainable Development Goal 13 (SDG 13) focuses on taking initiative for climate change awareness, imperative action to combat climate change planning and management, and its impacts, resilience, and adaptive capacity, as well as United Nations Framework Convention on Climate Change (UNFCCC) commitments. The effect of climate change is quite evident worldwide. In fact, the climate is changing much faster than anticipated due to the increasing level of greenhouse gases. To achieve net zero emissions by 2050, the carbon emissions need to fall by 45% by 2030 from the level of 2010 globally. As per the Sustainable Development Goals Report 2025, published by the United Nations, 2024 was marked as the hottest year in the last 175 years. The global average temperature has risen above the pre-industrial level by 1.55 °C. It has already started impacting the climate. Melting of glaciers, water security concerns in the long term, rising sea level, and acidification of the oceans are becoming irreversible impacts of climate change.

Carbon Footprint and SDG 7

SDG 7 is the United Nations' Sustainable Development Goal 7 on Affordable and Clean Energy, which aims to ensure access to modern energy for all, which is affordable, reliable, and sustainable. Reducing dependence on fossil fuels and raising the use of renewable energy sources is the prima facie objective. SDG 7 is focused on energy efficiency. The target is to strengthen the share of renewable energy in the global energy mix. With the help of more investments, advanced technology, and expanding infrastructure can be a way to counter climate change issues. Organizations, cities, and countries can use Carbon footprint to establish a baseline of their energy-related emissions. Guaranteeing consistent, accurate data across different regions, sectors, and supply chains remains challenging due to variations in measurement and reporting standards. A practice known as "greenwashing," some companies are using carbon footprint data to misrepresent their environmental efforts without making considerable changes.

Carbon Footprint and SDG 9

SDG 9 is the United Nations' Sustainable Development Goal 9 on Industry, Innovation, and Infrastructure, which aims to provide a sustainable future by structuring resilient, sustainable infrastructure, endorsing inclusive and sustainable industrialization, and encouraging innovation. Transport is one of the biggest sources of CO_2 emissions. Road, air, and sea transportation require serious and immediate attention to develop reliable infrastructure to reduce greenhouse gas emissions. Access to mobile broadband networks is another important factor, as almost 50% of the global population has yet to get the 5G broadband mobile network service. 4% of the global network still does not have mobile network access. The assimilation of carbon footprint into SDG 9 initiatives can help recognise and mitigate greenhouse gas emissions linked with industrial processes and infrastructure development, thereby fostering sustainable industrial growth. Liang et al. (2023) have shown the efficacy of the green infrastructure initiatives, green technology, green financing solutions, government's digital governance capacities enhancement, and in internet infrastructure and digital applications, filling in the gaps.

Carbon Footprint and SDG 11

Sustainable Development Goal 11, titled "sustainable cities and communities," aims to create sustainable cities and communities by reducing greenhouse gas emissions and enhancing urban resilience with the help of endorsing sustainable urban development and refining urban health and thermal comfort through green infrastructure and carbon appropriation strategies. Cities across the globe are major sources of greenhouse gas emissions and resource consumption. Carbon footprint is an essential tool for measuring carbon emissions and energy usage. This measurement can act as an indicator to formulate strategies in terms of adopting strategies and formulate policies to build resilient, safe, sustainable, and inclusive cities towards achieving the goals of SDG 11.

Carbon Footprint and SDG 12

UN's SDG 12, on Responsible Consumption and Production, which intends to "ensure sustainable consumption and production patterns". Carbon footprint acts as a critical tool in measuring and managing the environmental influences of consumption and production patterns, in this manner supporting the achievement of SDG

12. Sustainable management and effective use of natural resources, Sustainable consumption, and production patterns, reducing food wastage at the retail and consumer levels, sustainable practices, reduction, recycling, and reuse of waste are the key aspects through which the goals of SDG 12 can be achieved.

1.2 Evolution of Carbon Accounting from Traditional Reporting to Advanced Footprint

There is a paradigm shift in carbon accounting from traditional reporting to advanced footprint. Wright et al. (2011) have proposed a definition of carbon footprint as, "A measure of the total amount of CO_2 and CH_4 emissions of a defined population, system, or activity, considering all relevant sources, sinks, and storage within the spatial and temporal boundary of the population, system, or activity of interest. Calculated as CO_2e using the relevant 100-year global warming potential (GWP100)." The demand for carbon accounting, also known as greenhouse gas accounting, is growing rapidly worldwide. It is the process to measure the emission of greenhouse gases (GHGs) directly and indirectly by an individual or from business or organizational activities and understand their climate impact. Accordingly, the goals to reduce their emissions can be set. Carbon accounting is a complex process. In many countries, companies must report their emissions to their governments or to the Environmental Protection Agency. The World Resources Institute and the World Business Council for Sustainable Development have established the Greenhouse Gas Protocol, the most popular framework and international standard for carbon accounting. The GHG Protocol includes a suite of standards for different needs (Risco Morillo 2023), such as:

- The Corporate Accounting and Reporting Standard
- The Corporate Value Chain (Scope 3) Standard
- The Global Protocol for Community-Scale Greenhouse Gas Emission Inventories (GPC)
- The Mitigation Goal Standard
- The GHG Protocol Policy and Action Standard
- The Product Standard
- The GHG Protocol for Project Accounting.

2 Conceptual Framework

The world's most widely used greenhouse gas accounting standard, Greenhouse Gas Protocol, developed by the World Resources Institute (WRI) and the World Business Council for Sustainable Development (WBCSD), has classified the carbon emissions into three types, known as Scope 1, Scope 2, and Scope 3. The focus is to become carbon neutral by creating a GHG emissions inventory and adopting preventive measures for reducing carbon emissions and carbon footprint.

- Scope 1 emissions cover the emissions resulting from a set of operations from the direct sources owned and controlled by the organization itself. Burning fuel in a fleet of vehicles, fuel combustion in the planes, boiler, or furnace, etc. There are four categories of Scope 1 emissions, named, stationary combustion, mobile combustion, fugitive emissions, and process emissions.
- Scope 2 emissions, also known as indirect emissions, are caused by GHG emissions released in the atmosphere by the purchased energy, such as electricity, steam, heat, or cooling, from an organization's consumption.
- Scope 3 emissions, all indirect emissions not included in Scope 2, also known as the life-cycle emissions, are the most complex ones. It encompasses the emissions that occur in the value chain, both upstream and downstream, of the reporting company. Scope 3 emissions encompass the major proportion of any company's total emissions. Upstream activities consist of business travel emissions, waste generated in operations, transportation, and distribution emissions due to purchased goods and services, etc. Downstream activities include investment emissions, franchise emissions, leased assets emissions, use of old product emissions, etc.

For product and corporate value chain GHG accounting and reporting, the WRI and the WBCSD are now developing new standards by following the multi-stakeholder process across the globe to cover both the product, goods, and services, life cycle, and the corporate level value chain accounting and reporting guidelines. Life Cycle Assessment (LCA) is an important analysis to understand the GHG emissions and their effect on climate change. Carbon footprint deals with assessing the impact of a product on the environment from raw material extraction, manufacturing and processing, distribution and transportation, use, to end of life (disposal or recycling) stages. Carbon accounting is a crucial factor of the E, 'Environment', in ESG. It is the most important tool to achieve net-zero emissions by 2050.

3 Methodological Advances in Carbon Footprint

Across countries and organizations for measuring, reporting, and verifying GHG emissions, there exist certain international carbon accounting standards, such as Greenhouse Gas Protocol, ISO 14064, Partnership for Carbon Accounting Financials (PCAF), and PAS 2060, to meet the goals of the Paris Agreement (International Standards for Carbon Accounting | Minimum.com, n.d.).

For corporate accounting and reporting emissions, as an international standard, WRI and WBCSD created the GHG Protocol in 1998. The GHG Protocol has five principles: relevance, accuracy, consistency, completeness, and transparency. The GHG Protocol is the most extensively used for measuring and managing greenhouse gas emissions.

Based on the GHG Protocol, the International Organization for Standardization (ISO) created the ISO 14064 in 2006. These standards apply to the entities that measure, report, and manage greenhouse gas (GHG) emissions that already use ISO principles.

For the financial sector, to standardize the carbon accounting of financed greenhouse gas emissions, the Partnership for Carbon Accounting Financials (PCAF) framework was developed globally by using category 15 of the GHG protocol related to investment. The financed GHG are the carbon emissions produced by the entities or the processes financially supported by the Banks, Asset managers, Pension funds, and Other financial entities.

For accomplishing and displaying carbon neutrality, the British Standards Institution (BSI) provides a framework known as Publicly Available Specification (PAS) 2060. It provides a consistent methodology for businesses, governments, and other organizations to achieve Net Zero and align with global climate goals. Input–Output (I–O) Analysis and Hybrid Approaches have become popular nowadays in linking macroeconomic data with emission factors. The I–O Analysis tracks the direct emissions and the indirect emissions across different economic sectors. The Hybrid Approach is a mix of different carbon accounting methods. The Hybrid Approach combines the strengths of top-down I-O analysis with bottom-up, process-based LCA.

Traditional carbon accounting models are primarily static—they rely on historical data and periodic reporting (typically quarterly or annually). In contrast, dynamic and real-time footprint models utilize digital technologies, IoT sensors, and AI analytics to continuously monitor, calculate, and inform an organization's carbon emissions in real-time. Several studies have proposed various models. Gu et al. (2025), in their study to explore the patterns of city-wide energy-carbon nexus across some selected cities, have applied a novel infrastructure-oriented framework to understand the systemic zero-carbon transition actions. Holzapfel et al. (2025) talked about a three-stage research procedure via five different chains of custody (CoC) models in LCA to build awareness on the prospective challenges and work towards suitable solutions and requirements. Kelbel et al. (2025), in their study using an artificial neural network, have developed a scalable method for carbon footprint oriented on the Crisp-DM process to support a producing company's sustainability goals by integrating sustainability into portfolio management. Integration of water, energy, and material footprints with carbon footprint involves an integrated approach to recognize the interdependencies between resource use and GHG emissions. An LCA provides a comprehensive framework that offers us the details of the carbon emissions due to the consumption of water, energy, and materials.

4 Technological Tools and Innovations

4.1 Digitalization in Carbon Footprint

Digitalization has played a significant role in climate mitigation efforts, combining both intelligent technologies and the monitoring, reporting, and verification (MRV) systems in industries. Digitalisation is no longer restricted to the conventional mechanism of carbon accounting and has advanced by converting data management to an automated and open process. Traceability, effectiveness, and accountability are increased by digital ecosystems—the requirements for synchronizing regional activity with universal sustainability objectives (Li et al. 2025; Wang et al. 2025).

Wang's (2025) research mentions that digitalization greatly decreases urban CO_2 emissions by helping governments and industries get correct emissions information for effective policymaking (Wang et al. 2025). In the same lines, Li et al. (2025) explain that the digital economy helps in enhancing carbon emission efficiency across regions, depicting spatial synergy and reducing data latency in carbon management.

Blockchain for Supply Chain Traceability

Blockchain technology is playing a significant role in transforming supply chain carbon accountability. The distributed ledger of blockchain technology contributes to providing real-time confirmation of information related to sustainability at every stage of production. According to Farmonaut (2025) and Gupta (2025), each of these transactional events—manufacturing, transport, and storage—has a time stamp that reduces the risk of manipulation and greenwashing.

There have been case studies in recent times that highlight the incorporation of blockchain in eco-enterprise networks, which in turn increases emission responsibility by 45% and decreases compliance burden by 37%. This helps in acquiring greater transparency all over the global supply chains (Hind Alofaysan 2024). For example, Farmonaut's blockchain-powered agricultural traceability platform helps in connecting smallholder farmers to carbon credit markets by making the verified emission reductions tokenised, while simultaneously enabling consumers to get their hands on end-to-end sustainability data (Farmonaut 2025). The joining of blockchain, smart contracts, and IoT has made sustainability validation dynamic and almost real-time.

IoT Sensors for Real-Time Emission Monitoring

IoT sensors are going a long way in helping to monitor emissions, which in turn leads to accurate results and temporal resolution in carbon counting. IoT sensors use both fixed and mobile sensor modules in the form of those installed on vehicles, drones, etc., which help monitor real-time levels of CO, NO_2, etc. This real-time data is then processed through analytics that are AI-based, which help in calibration, visualization, and predictions (Pathak 2025).

Real-time policy reaction by municipal and industrial governments is allowed by LoRaWAN and 5G-enabled IoT-based hybrid networks. This is through evidence-based environmental management (Pathak 2025). This, in turn, helps in identifying promptly the emissions from industrial sites, which further helps urban city planners to take up dynamic traffic or any interventions in zooming. To sum up, IoT-enabled emission monitoring helps in aligning with the SDG governance requirements on the digital platform (Gupta 2025).

4.2 Artificial Intelligence and Machine Learning Applications

Predictive Modeling for Carbon Reduction Pathways

The future of reducing carbon footprint requires carbon forecasting and mitigation that can be helped by Artificial Intelligence (AI) and Machine Learning (ML). AI platforms can predict probable paths for emission reduction through multiple source datasets, such as satellite imaging, IoT sensors, etc. (Polagani 2025, n.d.). The research done by Polagani (2025) reveals that CNNs (convolutional neural networks), along with Sentinel-5P–5P and Landsat data, can help in providing information about CO_2 and methane concentration in industrial areas. This is a major improvement over conventional sampling with 70% accuracy.

Research by Farmonaut (2025) says that AI can also help in facilitating carbon budgeting for 1.5 °C trajectories with the help of proactive recalibration as and when new information arises. This is a key element in corporate SDG alignment and country NDC tracking.

4.3 Carbon Management Software and Platforms

The corporate climate responsibility requires advanced software ecosystems for carbon management, and there are three major platforms—Sphera, SimaPro, and GaBi- which are the major players in this area due to their intense methodology and integrated regulation.

Sphera

The 1st major platform, Sphera provides a varied sustainability management suite that encompasses environmental, health, safety, and sustainability (EHS&S) performance modules. As per (Sphera I Sustainability, Operational Risk Management and EHS Software, n.d.), its databases containing more than 20,000 verified process datasets help in real-time tracking of emissions and automated compliance reporting.

SimaPro

The 2nd major platform, SimaPro, brings together life cycle modelling capabilities and collaborative cloud-based computing through Synergy. Further, integration with Eco Invent 3.11 helps in scenario testing on the product chains. This is a major requirement for consultants and consumer packaged goods firms that have Science-Based Targets in mind. SDGs 7 and 9 have been mapped by Japan by applying SimaPro-based industrial LCAs to maximize the flow of energy (LCA Software Comparison in 2025: Best Tools Ranking—Devera 2025).

GaBi

The 3rd major platform, GaBi automation modules, brings together corporate ERP systems and emission databases that in turn, produce product carbon footprints (PCFs) in bulk. This helps industrial applications, more specifically automotive, aerospace, and heavy industry, due to its data set integration and support given to PACT v3. (LCA Software Comparison in 2025: Best Tools Ranking—Devera 2025). As per (Sphera I Sustainability, Operational Risk Management and EHS Software, n.d.), this guarantees both regulation compliance and helps corporates to correlate the environmental outcomes with the SDG goals.

4.4 Big Data and Remote Sensing for Sectoral Emission Tracking

Big data analysis and remote sensing provide the building blocks for current emission mapping by sectors. Satellite missions like Sentinel and Landsat provide high-resolution, multispectral information that AI models translate into carbon maps for large-scale monitoring (Polagani 2025).

In the oil and gas sectors, geospatial data management systems such as The Pyxis Project reconcile non-matching emission records through spatial indexing—significantly improving inventory accuracy. Big data pipeline integration enables cross-validation of emissions from various sources, and a comprehensive and transparent national register of emissions is generated. Remote sensing feeds are now merged with government NDC dashboards, with real-time authentication of the rate of renewable adoption and deforestation figures, directly linked with SDG 7 (Clean Energy) and SDG 13 (Climate Action) progress.

5 Sectoral Applications of Advanced Carbon Footprint

5.1 Energy and Power Sector: Decarbonization and SDG 7

The energy sector alone accounts for about 75% of the greenhouse gas emissions and is therefore the key to global SDG 7 achievement. Total Energies (2025) documents embedding IoT data into green generation networks to maximize production, with a 16.5% life-cycle emission reduction since 2015. AI-optimized grid management forecasts renewable uncertainty and avoids carbon-intensive balancing at power stations. Digital twins of the energy system model the potential for efficiency improvement from new technologies, promoting cross-sector decarbonization. Blockchain-secured carbon markets enable open tracking of renewable certificates, promoting fair low-carbon investments (Farmonaut 2025).

5.2 Transport and Mobility: Low-Carbon Transition and SDG 11

Transport generates close to one-quarter of worldwide CO_2 emissions, requiring sophisticated carbon footprint analysis. Telematics information integrated with IoT sensors makes it possible to perform real-time fleet emissions modeling and intelligent traffic forethought. According to, big data-based transport system goes a long way in enhancing route optimisation, minimising fuel use, and reducing carbon intensity. As per the GCC Association (2021), the foundation of SDG 11 to create inclusive and sustainable city systems is now intelligent mobility technologies. Carbon modelling integration with the infrastructure policies is enabled by sustainable mobility models.

5.3 Agriculture and Food Systems: Methane Tracking and SDG 2

The carbon footprint that is created by agriculture, particularly by methane from livestock and fertiliser, can be traced in real time by satellite imagery and soil monitoring. AI blockchains are used by platforms like Farmonaut to enable end-to-end visibility in agriculture value chains (Farmonaut 2025). SDG 2 and 12 have been mapped in India by using Farmonaut for tracing methane and fertiliser emissions that implement blockchain-based Agri-foot printing.

The AI platforms help to make carbon tokenised in areas where the reduction in emissions is financially available to the farmers. SDG 13 and 15 have been mapped in Sweden by utilizing satellite-linked AI models for forestry carbon apportionment monitoring. Moreover, the monitoring of NO_2 and Methane increases the accuracy of sustainable agriculture, which contributes directly to SDG 2 (Zero Hunger) in the form of maximising yields and minimising emissions. Small holders benefit from blockchain integration that opens access to the digital carbon markets.

5.4 Manufacturing and Industry: Sustainable Production and SDG 9 & 12

The sensors and LCA software in the form of SimaPro and GaBi that are based on IoT help in measuring carbon intensities and detecting hotspots in areas that highly involve energy-consuming production, for decarbonizing the process (LCA Software Comparison in 2025: Best Tools Ranking—Devera 2025). Circular manufacturing involving eliminating waste and optimising material use is assisted by Predictive analytics that is driven by AI. This aims at achieving SDG 9 (Industry, Innovation, Infrastructure) and SDG 12 (Responsible Consumption and Production). If carbon footprint tools can be embedded within digital manufacturing environments, sustainability can be guaranteed within all production decisions, which would help to connect every industrial process with climate results that can be quantified.

5.5 Urban Planning: Smart Cities and SDG 11

The resilience of smart cities can be improved if contemporary urban designs can use carbon footprint technology. Emission information can be gathered from mobility, energy infrastructure, and real estate with the help of combined IoT systems, which in turn update city-level carbon dashboards. (Wang et al. 2025). The aim of SDG 11 is inclusive and low-carbon urban development. There has been a decline in per capita emissions in forward-thinking municipalities with the help of AI-powered energy grids and green buildings' analytics. As per the GCC Association 2021, digitalisation makes city decarbonisation quantifiable and actionable.

6 Linking Carbon Footprint to SDG Monitoring

6.1 Mapping Footprint Metrics to SDG Indicators

SDG measures like unit of output emissions, renewable energy percentage, and air quality scores can be easily attained through carbon footprint measurements. SDG 7.2 is aimed at the adoption of renewable energy, and SDG 11.6 is aimed at urban carbon density. These metrics can be automatically mapped in national SDG dashboards by digital systems that allow real-time assessment. There are machine learning models that can match emission datasets with a single SDG goal.

6.2 *Corporate ESG Reporting and SDG Integration*

Efforts have been made, and this has resulted in corporate environmental, social, and governance reports converging with SDG monitoring systems. Spehera, as a platform, has incorporated LCA data within ESG reports, making them traceable and transparent as per the needs of investors and regulators (Sphera | Sustainability, Operational Risk Management and EHS Software, n.d.). Companies like Siemens and Unilever use predictive analytics to help link operational carbon targets to SDG goals 12 and 13 by making decisions based on data. The link between sustainability reporting and global policy initiatives is formed by the carbon management software.

6.3 *Nationally Determined Contributions (NDCs) and SDG Progress*

AI, blockchain, and remote sensing integration assure operational transparency as required by Nationally Determined Contributions (NDCs). SDG 13 (Climate Action) accordance with national climate targets, can be revised by carbon MRV systems based on these technologies.

Nations that use common data platforms to make emission reductions and clean energy expansion authentic are inspiring cross-border partnerships. Uniform reporting structures and progress records that can be verified can be provided by online partnerships. Digital ecosystems can help convert carbon footprint to anticipatory governance from reactive accounting.

7 Challenges and Limitations

In spite of significant methodological and digital advancements, several challenges and limitations remain within the domain of carbon footprint. These challenges limit the accuracy, comparability, and usability of carbon accounting as an impetus for sustainable change.

7.1 Data Quality and Availability Issues

Sound data is still the foundation of precise carbon footprint calculation. Nevertheless, data weaknesses persist, especially across developing economies, small businesses, and informal economies, hindering thorough examination. Non-standardized reporting, transparency deficiencies in value chains, and variability in emission factors among databases also impede cross-sector comparison. Lack of immediate, validated emission information by suppliers frequently compels organizations to make do with secondary or proxy data, inflicting estimation errors.

7.2 Complexity of Scope 3 and Worldwide Supply Chains

Scope 3 emissions, including upstream and downstream operations, generally represent more than 70% of an organization's overall carbon footprint (n.d.). Estimating these emissions over intricate, multi-tier supply chains continues to be one of the most critical methodological issues. Data quality variation, alternative accounting boundaries, and supplier proprietary limitations challenge the validity of Scope 3-foot printing. Multinational companies are also challenged to standardize methodologies across varying geographical and regulatory environments.

7.3 Gaps in Standardization Between Regions and Industries

While standards such as the Greenhouse Gas (GHG) Protocol and ISO 14064 exist internationally, the application of these standards is not uniformly consistent across countries and industries. Countries and industries interpret and use these standards differently, leading to different reporting structures and baselines of emissions. This lack of harmonization restricts cross-comparability and dilutes the global benchmarking of carbon performance.

7.4 Risk of Greenwashing

The growing corporate focus on carbon disclosure has inadvertently created "greenwashing", the misleading representation of environmental performance to achieve reputational or market benefit (Delmas and Burbano 2011). If organizations make selective disclosure of limited emissions information or use unverifiable carbon offsets, it undermines public confidence and weakens the transformative purpose of carbon footprint. Strong verification mechanisms, third-party audits, and open disclosure thus become essential to uphold the integrity of climate reporting systems.

8 Future Directions

The development of carbon footprint is moving into a revolutionary stage characterized by the convergence of digital technologies, policy innovation, and behavioural change. The future of research and practice needs to move beyond measurement to adopt strategic integration through economic systems, social education, and technological ecosystems.

8.1 Towards Net-Zero and Beyond: Carbon-Negative Pathways

As the global discourse emphasizes bringing net-zero emissions by 2050, new strategies support carbon-negative pathways, where the removal of atmospheric carbon surpasses emissions. Carbon capture, utilization, and storage (CCUS) technologies, bio chip manufacturing, and direct air capture (DAC) systems are prime facilitators of this shift (Fawzy et al. 2020). Integrating carbon-negative solutions into country-level decarbonisation plans can fast-forward towards climate neutrality and stretch further to net-positive ecological restoration.

8.2 Integration with Circular Economy Models

Carbon footprint in the future will have to integrate closely with circular economy principles, such that waste reduction, resource recovery, and closed-loop production systems become key (Geissdoerfer et al. 2017). Synthesis of carbon accounting with material flow and resource efficiency analysis allows companies to target hotspots and align emission reduction with economic resilience. Merging circularity metrics and carbon data analytics has the potential to redefine sustainability performance measurement across corporate and policy contexts.

8.3 Carbon Literacy and Education for Achievement of SDGs

Carbon literacy, knowledge about carbon sources, impacts, and responses, is critical for enabling people and organizations to make a difference towards Sustainable Development Goals (Oziewicz 2023). Integrating carbon literacy throughout education curricula, vocational training programs, and public campaigns improves behavioural change and responsibility. Business schools and universities, in general, have a critical role in developing future-proof leaders who can move carbon data into strategic action.

8.4 Policy, Business, and Academic Opportunities

Collaborative environments between policymakers, companies, and universities will decide the next level of carbon footprint maturity. Policymakers can link carbon indicators into budget and trade policies, and companies can use digital innovations for open reporting and emission minimization. Academia, on the other hand, can develop methodological strength and interdisciplinary research that connects climate science, economics, and behavioural research (Bai et al. 2025).

9 Conclusion

Carbon footprint has become a key facilitator of SDG attainment, connecting climate action (SDG 13) with sustainable industry (SDG 9), sustainable consumption (SDG 12), renewable energy (SDG 7), and sustainable cities (SDG 11). However, unlocking its true potential depends on overcoming long-standing challenges in data quality, standardization, and transparency of institutions.

The way ahead will need to be based on international cooperation and innovation. Governments will need to standardize and invest in open-access data platforms; companies need to embed carbon metrics into strategic decision-making frameworks; and universities need to push interdisciplinary research at the intersection of technology, social science, and environmental policy. All these players working together can lock in carbon footprint from measurement to transformative action, propelling the world towards a resilient, carbon–neutral, and ultimately regenerative future. Carbon footprint has been taken from a mere accounting exercise to an active tool to achieve SDG goals through technology enablers such as digitalisation, blockchain, IoT, big data, and LCA software. These technologies help stakeholders like local farmers and global corporates by enabling accurate tracking of emissions, predictive modelling of decisions, and open book reporting. All these technologies prove that the carbon footprint is pivotal for sustainable development. It is no longer only for compliance, but it is an active tool based on data that has resilience to climate and global equity.

Technology enablers, such as digitalization, blockchain, IoT, AI, big data, and LCA software, have taken carbon footprint beyond an accounting exercise to an active tool for advancing SDG progress. Through enabling accurate tracking of emissions, predictive modeling of decisions, and open book reporting, these technologies equip stakeholders at all levels—local farmers to global corporates—with the ability to make sustainability objectives operational.

The convergence of these technologies guarantees that carbon footprint is at the heart of sustainable development: no longer an institution of compliance, but a living, data-based tool for resilience to climate and global equity.

References

Alofaysan H (2024) The effect of environmental smart technology and renewable energy on carbon footprint: a sustainability perspective from the MENA region. Energies 17(11):2624–2624. https://doi.org/10.3390/en17112624

Bai C, Du X, Chua SY (2025) The green potential of the digital economy in balancing economic growth and carbon emissions. Sci Rep 15(1). https://doi.org/10.1038/s41598-025-04616-x

Delmas MA, Burbano VC (2011) The drivers of greenwashing. California Manage Rev 54(1):64–87. https://doi.org/10.1525/cmr.2011.54.1.64

East AJ (2008) What is a carbon footprint? An overview of definitions and methodologies. In: Vegetable industry carbon footprint scoping study—Discussion papers and workshop, 26 September 2008. Horticulture Australia Limited, Sydney

Emmanuel O (2019, August 9) UN warns against rising world food insecurity from unprecedented climate change. National Daily Newspaper. https://nationaldailyng.com/un-warns-against-rising-world-food-insecurity-from-unprecedented-climate-change/

Farmonaut® (2025, July 12) Blockchain for supply chain traceability in 2025. Farmonaut®. https://facebook.com/farmonaut; https://farmonaut.com/blogs/blockchain-for-supply-chain-traceability-in-2025

Fawzy S, Osman AI, Doran J, Rooney DW (2020) Strategies for mitigation of climate change: a review. Environ Chem Lett 18(18):2069–2094. https://doi.org/10.1007/s10311-020-01059-w.

Geissdoerfer M, Savaget P, Bocken NMP, Hultink EJ (2017) The circular economy—a new sustainability paradigm? J Clean Prod 143(1):757–768. https://doi.org/10.1016/j.jclepro.2016.12.048

Goal 7 | Department of Economic and Social Affairs (n.d.) Sdgs.un.org. http://sdgs.un.org/goals/goal7

Gu Y, He Y, Tong K, Sun K, Chen W-Q (2025) Spatial and temporal patterns of urban energy-carbon nexus associated with multiple infrastructure provisioning sectors. Resour Conserv Recycl 223:108532. https://doi.org/10.1016/j.resconrec.2025.108532

Gupta IAK (2025) Blockchain-based supply chain optimization for eco-entrepreneurs: enhancing transparency and carbon footprint accountability. Int J Environ Sci 11(17s):2025

Holzapfel P, Blizniukova D, Bach V, Finkbeiner M (2025) Chain of custody modelling in carbon footprinting and life cycle assessment. Int J Life Cycle Assess 30(5):980–993. https://doi.org/10.1007/s11367-025-02452-9

https://www.omdena.com/blog/ai-in-carbon-management-strategies-businesses-202

https://www.novapublishers.com/catalog/product_info.php?products_id=5999

International standards for carbon accounting | Minimum.com (n.d.) https://www.minimum.com/resources/international-carbon-accounting-standards

International Energy Agency (2024) Net zero by 2050: tracking clean energy progress 2024. IEA Publications

Kelbel N, Yang P, Riesener M, Keuper A, Schuh G (2025) Fast carbon footprinting in complex product portfolios using machine learning approaches. Lect Notes Prod Eng 430–437. https://doi.org/10.1007/978-3-031-86893-1_47

LCA Software Comparison in 2025: Best Tools Ranking—Devera (2025). Devera.ai. https://www.devera.ai/resources/lca-software-comparison-in-2025-best-tools-ranking

Li Y, Chen B, Guo L, Kang J (2025) Analysis of the impact of the digital economy system on carbon emissions and carbon footprint from the perspective of high-quality development. Front Environ Sci 13. https://doi.org/10.3389/fenvs.2025.1443405

Liang X, Liu M, Huang Y (2023) Studying financial development with low-carbon architecture development and green technological innovation: sustaining SDG-9. Environ Sci Pollut Res 31(2):2009–2025. https://doi.org/10.1007/s11356-023-31155-7

Oziewicz MC (2023) What is climate literacy? *Climate Literacy in Education, 1*(1), 34–38. https://doi.org/10.24926/cle.v1i1.5240

Pathak, S. P. (2025). Real-time air quality monitoring using low-cost sensors—IoT system. *International Journal Of Scientific Research In Engineering and Management, 9*(5), 1–9. https://doi.org/10.55041/ijsrem47226

Polagani SS (2025) AI for carbon emissions monitoring: Computer vision and remote sensing for automated carbon emissions tracking. World J Adv Res Rev 26(2):2324–2334. https://doi.org/10.30574/wjarr.2025.26.2.1847

Risco Morillo, Á. (2023). *Recycling of plastics of waste from electrical and electronic equipment by microwave-assisted pyrolysis* [Thesis]. https://ebuah.uah.es/dspace/bitstream/handle/10017/60476/Thesis%20Alvaro%20Risco%20Morillo.pdf?isAllowed=y&sequence=1

Seventh Assessment Report—IPCC (2019) Ipcc.ch; IPCC. https://www.ipcc.ch/assessment-report/ar7/

Sphera | Sustainability, Operational Risk Management and EHS Software (n.d.) Sphera. https://sphera.com

Sustainable transport | Department of Economic and Social Affairs (2016) Un.org. http://sdgs.un.org/topics/sustainable-transport

Sustainability and Climate 2025 Progress Report (n.d.) https://totalenergies.com/system/files/documents/totalenergies_sustainability-climate-2025-progress-report_2025_en.pdf

Transparency in transition: the state of corporate environmental disclosure (2024)

UNEP (2024) Nations must close huge emissions gap in new climate pledges and deliver immediate action, or 1.5°C lost. UN Environment. https://www.unep.org/news-and-stories/press-release/nations-must-close-huge-emissions-gap-new-climate-pledges-and

United Nations Environment Programme (2024) Emissions Gap Report 2024: Broken record—temperatures hit new highs. UNEP, Nairobi

Wang T, Li R, Zhang Q, Sun S (2025) Digitalization and urban carbon emissions: unraveling the mechanisms of agglomeration economics. J Environ Manage 387:125855–125855. https://doi.org/10.1016/j.jenvman.2025.125855

Wiedmann T, Minx J (2008) A definition of 'carbon footprint. In: Pertsova CC (ed) Ecological economics research trends. Nova Science Publishers, Hauppauge, Chapter 1, pp 1–11

Wright LA, Kemp S, Williams I (2011) "Carbon footprinting": towards a universally accepted definition. Carb Manage 2(1):61–72. https://doi.org/10.4155/cmt.10.39

World Business Council for Sustainable Development (2023) Beyond value chain mitigation: driving climate action. WBCSD, Geneva

A Conceptual Framing of Ethical Reflexivity and Sustainable Value Creation for Building Sustainable Business Models: A Theoretical Exploration

Sanjana Santra, Kishore Kumar, and Nitendra Kumar

Abstract The question of applying ethical reflexivity through defined parameters and mechanism in Sustainable Value Creation (SVC) for businesses is imperative. Studies are segregated on either of the two in the context of building Sustainable Business Models (SBM). A wide corpus of literature has attempted to design business models that are contextually sustainable with the implications of integrity and ethics. However, very few studies have focused on the institutional feedback loop through theoretical exchange of the overlaying elements that might define the pathways of ethical integration in the SBMs. This chapter bridged this gap by drawing theories from literature to analyze its components parallelly, analyzing ethical considerations and SVC separately on SBM to understand the distinctions and interrelations between the two concepts. The chapter acts as a comparative analysis between six major theories in this domain. While discussing the intersecting elements through this process, the study develops a conceptual framing to analyze the interrelations in depth. The chapter also presents business cases to analyze and support the developed concepts. The study presents a consolidated understanding for academics and business stakeholders to design and implement SBM strategies through opportunities and inclusive strategies. The study actively engages with the business strategies, analyzing the importance of ethical reflexivity in corporate social responsibility, which will help the stakeholders capture the important aspects of the sustainability parameters. These insights would guide the policymakers, managers and the other decision-makers effectively embed sustainability into the organization's business functions and meet the stakeholders' expectations.

Keywords Sustainable business models · Sustainable value creation · Ethics · Stakeholder theory · Triple bottom line

S. Santra (✉) · K. Kumar
School of Management, Bennett University, Greater Noida, Uttar Pradesh, India
e-mail: santrasanjana1995@gmail.com

N. Kumar
Department of Decision Sciences, Amity Business School, Amity University, Greater Noida, India

K. Kumar et al. (eds.), *Sustainable Value Creation: Corporate Strategies for SDGs Implementation*, World Sustainability Series,
https://doi.org/10.1007/978-3-032-24652-3_17

1 Introduction

When the two words, 'Ethics,' and 'Value' are used interchangeably, it is often mistaken in understanding what are the parameters that make them contextually different (Fischer 2004; Schwartz 2005). While ethics have a heavy history with cultural, philosophical and also legal definitions, value is more relaxed with changing its agenda from person to person (Yuan et al. 2023). In the organizational context, creating a sustainable value creation however has a set of understanding that can be generalizable and implementable to sustain uniform business practices (Alshukri et al. 2024; Kaith and Sachdeva 2024). In this implementation process, the tool that comes into use is the 'Ethics,' through which the idea of 'value' is generated. The ethical principles help the organizations shape the beliefs and vision, that define the strategies and policies (Mert et al 2025). In designing these policies, ethics determine the extent of rightness or wrongness in a particular conduct (Peterson 2022; Singh et al. 2024; Joselin et al. 2025). Hence, ethics can be stated as values or standard of behavior, defined by a set of moral principles, by an organization (McGrath and Keith 2022; Agrawal 2025). In the present time, the biggest ethical agenda of this century is achieving the Sustainable Development Goals (SDGs), by creating the original triple bottom line contribution through social, economic and environmental benefits (Nogueira et al. 2022; Srivastava et al. 2018; Tripathi et al. 2022). However, when it comes to the meaning of corporate sustainability or enterprise-level sustainability, the motivation might seem confusing to some managers. For some, it is a way of doing the business at ease, with lower cost and risk aversion methods, while some see it only as the part of Corporate Social Responsibility (CSR) (Wang and Yan 2023; Srivastava et al. 2024). In some firms, the agenda to meet sustainability goals come at the price of shareholders' value in the favor of public good (Hart and Milstein 2003). Therefore, there are varying definitions of sustainable value creation among business organizations (Méndez-León et al. 2022). Bocken et al. (2014) defined as creating value through sustainable business models (SBM) that can reduce the negative impacts from society and environment. Richardson (2005) proposes three pillars of the concept of value in business models: i. the value proposition, consisting of target customer, offering and differentiation, ii. value creation and delivery system with required resources, assets, processes and position in a value network and iii. value capture system with financial and competitive strategic models. Evans et al. (2017) presented a qualitative framework with different forms of value for each stakeholders explaining the concept of value from opportunity cost parameters and analyzing it in multiple forms. The Sustainable Value Analysis Tool (SVAT) has been designed with the help of multiple cards that flow from product life cycle to Value Captured (VC) to Uncaptured (VU), Opportunities (VO) and Opportunity Assessment (VOA) with a step-by-step guide. Manninen et al. (2024) defined a framework for sustainable value creation by synthesizing the fragmented SBM literature, which presents SVC elements in detail and builds basic conceptualization of SVC. However, the study identifies it as a subject of business ethics and sustainability but does not articulate the idea of ethics in any of its argument. Therefore, it is safe to say that even though the existing literature has identified the concept of SVC has ethical implications but hardly any study attempted to focus solely on articulating ethical parameters in SVC with theoretical identification and clear propositions.

While the question of ethics is inevitable in creating sustainable model of businesses and especially to design corporate social responsibility strategies, the parameters of ethical considerations are not conceptualized under same umbrella. The elements of ethics are often scattered in different literature body of stakeholder responsibility, justice and fairness, accountability and corporate responsibility, values and integrity and moral legitimacy (Prencipe 2025; Cosa 2024; Srivastava et al. 2021). However, when one tries to understand 'Ethics' and 'Sustainable Value Creation' (SVC) together, the literature becomes scanty and scattered. This might mislead oneself into believing that there might be less significance of understanding ethical implications in sustainable value creation, however it is a more complex phenomenon that has been observed by the present authors. The contemporary journals on strategic management discipline focuses deeply on value creation alone and the strategies that are in place to achieve the feat. Parallelly journals on sustainable business practices talk about the impact of the factors responsible in creating SVC in different segments such as SMEs, entrepreneurial leadership and CSR. On the other hand, ethics journal extensively talk about the justice, inclusivity, morality et cetera with the performed business practices. Very few studies have been identified which have bridged the gap between all these three points. Also, understanding ethics might be a little difficult to map through empirical tests, explicit theories and quantifying mechanisms (Singh et al. 2025a; Prakash and Tiwari 2025; Kumar et al. 2025). Hence, what sustainable value equates to ethics is more of a logic-based understanding that needs conceptual argumentation. Also, the concept of sustainable value creation has been widely discussed in financial management studies, corporate strategies and performance management (Jordão et al. 2022). Ethics in this sense requires a separate base of articulation, grounded in moral philosophy in business practices. Thereby, the concept of ethics has been associated in a fragmented manner with either surface-level meanings or indirect assumptions of businesses following prescribed methods of doing CSR, following business laws and policies with justified performances through management levels. However, the emphasis is on finding ethical pillars in SVC, which remains conceptually under-explored.

This chapter aims to bridge this gap by presenting a conceptual underpinning with methodical definition of ethics in SVC, by drawing instances from literature from different disciplines. The authors aim to present a conceptual understanding of ethical reflexivity, identifying the parameters of ethical implication in SVC practices. The search does not confine to only ESG reports, governance papers and managerial guidelines but also attempts to gauge the theories that have been used to conceptualize ethical dimensions in academic research. To accomplish this, the study first synthesizes the academic discussions on business ethics in SBM and then compare with the elements of SVC within SVM literature. Hence, the research questions in this study are as follows:

RQ1. What are the ethical dimensions in SVC practices?

RQ2. How does ethical implications shape SVC for building SBM?

As outcomes of assessing these research questions, we analyze the elements of ethics and SVC parallelly in SBM literature to present the interactional elements under a single framing. It will help the elements to come under a qualitative visual network that helps the future researchers identify the constructs and items for developing tools for evaluating ethical aspects in sustainable business models. The study significantly contributes to both the Ethics and SBM literature with a consolidated conceptualization, backed by theoretical grounding. Practically, it may help the CSR practitioner, managers and other mentioned stakeholders of SBM in understanding the role of ethics in the strategic decision-making process. The elements of justice pillars, mentioned in this chapter categorizes the distinctive features of sustainable value that are backed by ethical support. It will present a clear picture to the stakeholders understanding the opportunity cost and risk associated with every value element in creating a business model sustainable in nature.

The chapter is structured as follows. In the Sect. 1, the authors present a general introduction, explaining the problem statement and context of the study. This section is followed by identifying theoretical frameworks from business ethics, SVM and SVC literature, to identify the common grounds and elements that negotiate between these areas in Sect. 2. In Sect. 3, the authors develop the concepts with theoretical underpinnings and presents with the strategic processes. Section 4 presents a comparative case analysis with three selected cases to analyze the applicability of the ethical parameters, with stating concluding remarks in Sect. 5.

2 Theories on Ethics and Sustainable Value Chain in Creating Sustainable Business Models

This section briefly discusses the relevant theories, revolving around ethics in sustainable business model on one hand and SVC on another. Then drawing theoretical elements from both sides, the analysis attempts to identify the intersecting elements to understand what are the constructs that may explain how value is created in sustainable business models through ethical practices. The study focuses on three primary theories from both the domains to keep the analysis focused on the research questions. In ethics side, the most prominent theories used by the academics are, i. Stakeholder Theory (Freeman and Phillips 2002), iii. Integrative Social Contracts Theory (Donaldson and Dunfee 1994), and iii. Care Ethics (Gilligan 1982; Tronto 1999; Singh et al., 2025b). From SVC side, we have identified the foundational Triple Bottom Line by Elkington (1997) that is widely used by every SVC model in any organization. Secondly, the Shared Value Theory (Kramer and Porter 2011), and Stakeholder Value Creation framework by Harrison et al. (2010) emerge as the relevant theories that are taken into consideration in multiple empirical papers on SBM literature. The agenda of this section is not to discuss the theories in detail but to analyze the intersecting elements through a comparative lens.

Any business is just not about the profits, but the people who are involved at any end of its operation. The importance of benefiting stakeholders is the primary concern of any business that makes a particular business model sustainable in nature. In this argument, the ethical implications hold three basic questioning imperatives, 'why,' 'for whom' and 'from whom.' Why be ethical is the principal pillar that holds a business moral in place. For whom the business needs to be ethical involves all the stakeholders and from whom inevitably reverts to the decision-makers, managers and shareholders. Stakeholder Theory in this ground opens the avenues of bringing fundamental designing of activities, bringing 'the universal attribution of moral personality' (Taylor 1989). Stakeholder theory opens the ethical stakeholder management who help in creating sustainable business models. Stakeholders' interests help the businesses understand new sustainable innovative ideas, green supply chain and logistics management, designing green products and creating new market needs (Moratis et al. 2018). This point becomes more applicable in the SBM practices in global south, in the developing economics, where ethics not just only remain confined to the business stakeholders but also opens to the society showcasing how a business can create significant social impact with its value creation mechanisms. Integrative Social Contracts Theory (ISCT) adheres to that global norm with a local focus. In the theories of business ethics and organizational behaviour the concept of integrity plays multiple roles. Gosling and Huang (2009) have stated the importance of integrity in ethical theory showcasing its importance in creating wholeness, consistency and authenticity. These attributes are also defined in SVC framework by many authors. Parry and Proctor-Thomson (2002) argue that a lack of integrity puts an organization at risk. In this sense 'ethical charismatic leadership' also has been defined as one of the most important characteristics to answer the 'from whom' question, as the managers are required to have the ethical leadership characteristics to build an authentic form of leadership for the organization's good (Endrissat et al. 2007). In these leadership categories, the ethics of care is also a vertical that managers follow to create a sustainable form of business models. Nicholson and Kurucz (2019) suggest the ethics of care can ultimately redirect the leadership towards more sustainability-driven business outcomes. The study articulated a Relational leadership for Sustainability, in which Nicholson and Kurucz (2019) present a framework on the 'ethics of care,' that has multiple dimensions, ecological, social and economic. The 'caring-for' and 'caring-about' verticals in that study shows the concern through which care has been centered as the fundamental point of ethical value in sustainable businesses. These points directly cater to the present study's argument of categorizing the parameters of three ethical questions.

The above discussion has its clear reflection on our analysis of the theories in SVC. The Triple Bottom Line (TBL) theory has been very widely used in this aspect. TBL also works very well with the stakeholder theory. In the discussion of SVC through business models, the discourse shifts to sustainability-oriented streams. Sustainability-driven concepts are based on creating values and change through business operations and organizational behavior. Lüdeke-Freund et al. (2020) have discussed these interactions between TBL and Stakeholder Theory in detail while proposing a framework through valid assumptions. The study asks very fundamental questions on what value is and what are the resources, for whom it is created, how it is created and ultimately who captures this value. Our proposition at this point is to put more emphasis on the 'who' defines the value and what are the sources of it. We argue that these questions inevitably connect us with the ethical concerns. These are the areas where the question of ethics arises prominently. Value is defined by the stakeholders in a form of 'value pluralism'. They are the satisfiers with the capability to define the sources of value and at the end they are the ones who allocate the perceived value within the stakeholders benefit itself. In this segment of 'who' particularly the analysis of ethical patterns can be situated and identified as the opportunity cost, willingness and beneficiary agendas are determined. This point also brings us closer to the Shared Value Theory (SVT). Lüdeke-Freund et al. (2016) in a study explained the interrelations between SBM and shared value creation in an extensive manner. In this report, they have identified the concept of shared value coincide closely with CSR, corporate sustainability, business ethics and value-based leadership. As per their observation, the point of sustainability and CSR becomes a 'window dressing' by the organizations in creating societal impact. Kramerand Porter's (2011) study bridges the gap between these two, understanding the trade-offs between the organizations' self-interest and the imperatives of societal progress. Schaltegger et al. (2016) analyzed what a business model for sustainability must have. A company's value creation possibilities depend on the utilization of resources to benefit all the stakeholders and the shared value stand upon the ethical ground that has to be a collective understanding too. Another study by Freudenreich and Lüdeke-Freund (2020) bridges this understanding ultimately with creating an association between the stakeholder theory and shared value creation. This study situated 'value creation' in an intersecting point between the business model perspective and stakeholder theory perspective, where they have segmented the 'what' and 'how' to sustainable business models and 'with' and 'for whom' to stakeholder theory. One of their primary areas of exploration is to understand the different types of value created for different stakeholder's interest. The study also argues the value exchange process through a definite framing of concepts aligning the corporate sustainability ambitions and stakeholder expectations. The present study argues that in this intersecting point to facilitate the stakeholder's value creation ethics is the fundamental tool that may hold all the elements together. In this aspect, the final theory from SVC side, emerged to be the Stakeholder Value Creation by Harrison et al. (2010). This is very evident in this aspect as their study identifies the principles of distributional, procedural and interactional justice being the key determinants. These are the conditions upon which the ethical judgment of any stakeholder is performed. It has the potential to avoid the causal ambiguity and path dependence associated in the value creation process by the stakeholders. In this argument a point of competitive advantage has been argued to be sustainable that emerges from 'managing-for-stakeholders' approach. Here, the authors of the present study concludes this section with an interactional understanding between all the theories and how the elements involved in all the verticals. Associated theories can be merged to understand how the base of the SBMs are standing on the superstructure of Stakeholders' priorities, context-sensitivity for individual businesses, societal impact and collective vision of ensuring justice. All these elements comprise to create a truly inclusive and sustainable business model.

3 Developing the Concept of Ethical Reflexivity on the Context of Sustainable Business Models

Based on the above discussion we have followed the shared understanding between the theories and assumptions to develop the concept of ethical reflexivity in a form of feedback loop between SVC and ethical reflexivity (ER). This section briefly discusses the concept without reiterating the already discussed propositions. Keeping the institutional feedback loop as the anchoring point between the two concepts SVC and ER, the discussion proceeds with involving important theories in place. As we started this study with one standpoint that ethics and value are often used interchangeably across contexts, the institutional feedback keeps this interchangeability in perspective and operationalizes the exclusiveness of both the terms. This emerges from SBM, as it can institutionalize both the value propositions and ethical perspectives. Similarly, it may also dilute the principles. In this aspect the concept of value creation in SBM is not inherently 'sustainable' until it is ethically guided. When we argue that ethical guidance channelizes the sustainability in a business model, we propose three central questions, that we discussed before. The 'why ethical?' 'For Whom' and 'From Whom.' These parameters may work as the epistemic layers of ethical reflexivity in an organization. Why be ethical asks the fundamental questions about the societal, environmental and economic development of all the stakeholders. Creating a lasting societal impact along with keeping the business' interest in check bring us to the TBL theory where why ethical questions are needed to be answered in all these aspects. This is where the major step of value-creation has been defined, as we attempt to answer the 'for whom' and 'from whom' questions. Stakeholder inclusion is the major point of 'for whom' pillar, because a business model can only be sustainable when all its stakeholders' inclusive and collective benefit is considered. Also, the stakeholders in charge of decision-making process are the ones who decide this inclusiveness and its process. The stakeholders collectively decide how a contextual value proposition would be determined in a particular business model. Here, the theory of Stakeholder Value Creation is applicable, as it keeps the value exchange process in place through keeping a balance between the stakeholders' expectations and creating a pluralistic agreement of defining the values.

From here, we position the channels through which the co-created value is channelized through the three lines that we mentioned. On the other hand, here is the node, where a shared value is created by evaluating the impacts of a certain business model. Here comes the Shared Value Theory that puts these channels to more sustainable outcomes. Businesses may leverage this position to design the CSR strategies and sustainability imperatives. Keeping this point active here, we look at the SVC side. Stakeholder Theory is very widely applied in SVC literature, exploring the negotiations between the stakeholders' intent, benefit and interests that drive the major decision-making processes in an organization. Here, we pointed few major stakeholders from the organizations side, as the objective of this study is to explore the creation mechanism of value creation in the business models. As we discussed the ethical leadership through care principles, the theory of Care Ethics is situated here. Ethical leadership in managers is sustained by relationship-building, caring for the organization as well as the society, which contributes to relational leadership in managers. This cumulatively generates the ethical perspectives and integrity in decision-making. As this standpoint contributes to an integrative position of a business impacting the people, profit and planet simultaneously, it directly relates with the ISCT positioning. Keeping the profitability in check to run the business sustainably, but at the same time through a legitimate inclusion strategy for all stakeholders at any point, and performing the ecological responsibility, a sustainable business model can be created that serve to all the parameters of TBL. Here, at the concluding end, one more perspective that is needed to keep in observation is to tie any loose end in a the trade-offs between the self-interests of the organization and the society's betterment to strike the balance. At this point the feedback loop is complete, and it is safe to say, that here all the boxes can be ticked for designing an Ethically Shared Value Creation strategy for making business models profitable but also sustainable and inclusive in nature.

4 Analyzing Cases on Ethical Reflexivity and Sustainable Value Creation in Business Models

In this section, we present a brief comparative discussion on three cases to support and analyze the developed concepts. First case is the organization called, Natura & Co., the renowned consumer goods company from Brazil, second is Naspers, the multinational technology company from South Africa and Mercado Libre (ML) from Argentina. These cases are illustrated to understand the sustainability parameters with a focus on value creation and ethical reflexivity. These cases are selected for its multinational visibility and strong presence in developing economies. Also, these companies value CSR practices and have established themselves as practicing corporate sustainability norms. Our analysis tries to trace the elements of ethical reflexivity from these cases to test our conceptual framing in this context. The analysis has been performed collectively by the authors, and the findings were shown to a case study expert for review. This way the study achieves internal agreement and external validity through discussion and review insights. Also, all the information collected from the three units of analysis are triangulated for improved validity (Creswell and Miller 2000). However, due to the limited structural scope of the present study, the major elements of our findings are discussed in brief here.

How Companies follow Ethical Reflexivity: A Critical Comparative Case Analysis

In creating sustainable business models, Natura & Co has been a 'best-in-class' consumer goods company operating in multiple countries. Certified as one of the first B-Companies, it is also a part of the Dow Jones Sustainability Index since 2014 (S & P Global 2019). Not only the company has been able to integrate sustainability in their business models but also practices ethical considerations consciously. The stakeholders state that their success in global expansion can be credited to their commitment towards sustainability ethics. Their increasing sophistication in considering the need for ethical parameters are affluent that not only involve the internal stakeholders but also keeps the interest of all external stakeholders too. Their core purpose is to preserve the biodiversity through the operating channels and ecosystems. The company prioritized the indigenous communities with an integration of biodiversity protection to source the goods through a benefit sharing mechanism and agreements. This has been also practiced by Naspers who have established a group ethics in all their compliance framework to create a sustainable, shared trust governance. Both the platforms have internal traceability to track bio-ingredients to meet circular economy goals. Natura & Co. has set a long-term sustainability goals keeping with an investment in green packaging and also through a refill system that is biodegradable. In this process multiple jurisdictions are engaged to ensure a regular review of these adaptation practices in governance and reporting. In this aspect these companies have strengthened their supply chain channels to benefit all the TBL pillars and kept an integrative approach through accountability mechanisms. On the contrary case, Mercado Libre has been criticized for its inequality and malpractices in hiring systems. Even though Mercado Libre literally translates to 'free market,' their stakeholder engagement especially in the lower level of management, has not remained fair consistently. The company is managed by algorithmic insights which could have ensured more clarity; however, it resulted into labor broking practices (Atzeni 2023). This violates the 'From Whom' parameter where the internal system of the organization failed to co-create the shared value association. Even though the company promises a dispute-resolution mechanism through transparent process in consumer and seller protection benefiting the external stakeholders, but their union representation and formal regulatory frameworks have seldomly protected the workers. The workers are put into stressful, tiring and physically harmful situations that go far beyond the SVC agendas. The supervisors of the company are stated as 'toxic' in nature, and this violates the very first mile of the company's ethics in production system. The workers in the company have reported their job instability, lack of induvial and collective rights with a systemic and structural oppression.

In this context, Natura & Co. has ensured that suppliers are to be followed business practices with integrity, transparency and commitment towards sustainability imperatives. Their adherence to transparent accounting and reporting systems in all assets, transactions and liabilities with statutory accounting standards have met the sustainability compliance. They also have a zero-tolerance policy towards corruption and labor rights violation. Any bribery or extortion observed by the company results into strict disciplinary steps and regular audits are performed in this regard. Their ethical reflexivity practices remain shared through all stakeholder to keep a lawful transparency ensuring compliances with anti-money laundering laws, economic and trade sanctions law. Also, they meet the 'for whom' parameter by ensuring responsible sourcing following the sustainable code of conducts that are collectively shared and available le through the whole supply chain with a responsibility dedicated to all suppliers, sub-suppliers and even temporary agencies. While ML has taken a leverage of the loopholes in legal rights of the labor unions, Natura has kept a strict policy towards the people and planet by ensuring an inclusion of the local, national and international environment laws and labor rights. With an improvement in energy efficiency, regenerative farming, treatment of wastewater and reducing carbon footprint through responsible agricultural practices, measures to reduce carbon emissions from raw materials and finished goods. Natura meets all the ethical reflexivity parameters by encouraging the suppliers in a fair and equitable benefits with respecting genetic resources of a community, place and its people. The transparency in shared knowledge strengthens the due diligence capabilities and committing towards the ability to trace conflicting points in the entire supply chain and logistics. Natura has been called the business of 'aesthetics and ethics,' for its strategies to build the kind of SBM that truly prioritizes sustainable agendas in all the pillars of development. Like their tagline *'bem-estar bem,'* which roughly translates to 'wellbeing done well,' their business strategies are grounded in real-data insights and meeting more ambitious sustainability goals than its peers.

While Natura has ensured transparent reporting practices, Naspers have attempted but struggled with the monitoring and evaluation of environmental sustainability parameters. Naspers reported an approximate 44% generated waste being recycled, resulting in reducing carbon footprint of an estimated 1949 tons, however it was not measured in any industrial benchmark of reporting regulations (Vos and Reddy 2014). Even though this problem exists in the company, but Naspers still meet few of the criteria from our framework with its commitment to sustainability issues. As Naspers is a constituent of JSE Socially Responsible Investment Index, the company could have higher level of disclosure practices that are shared through all stakeholders' levels. One reason could be the massive scaling of the company across South Africa, with an emphasis on moving from ideology to mainly financial sustainability (Teer-Tomaselli and Tomaselli 2019). Also, very similar to ML, Naspers also have been criticized for its poor practices. The company has repeatedly opposed the efforts to improve precarious working conditions, inadequate wages and also poor health and safety. The Shareholder Association for Research and Education (SHARE) has reported concerns about the work life of the delivery workers. Even though their CEO has reassured the shareholders that workers' rights are important for the company, however SHARE anticipates that the controversial labor practices will only spread and contribute under his leadership. However, the company has recently brought a huge focus on creating sustainable value by engaging stakeholders as reported in their recent report. As stakeholders they have mentioned society along with our planet, industry bodies, investors and lenders, media and governance. Their 'purpose of engagement' section served the parameter of 'why ethical' question with a clear addressal of 'for whom' and 'from whom' distinctions. They have prioritized the stakeholder theory in all TBL pillars understanding the need of sustainability parameters in all its business operations. The reason for selecting the case of Naspers is to illustrate that even though the historical trajectory of a company may remain problematic, but measures can be taken to address it and adopt the pathways to balance the global scale with a local grounding. The report explicitly mentions the creation of a value chain through sustainability measures by adopting Global Reporting Initiatives (GRI) with European Sustainability Reporting Standards (ESRS) topics to map, WEF risk reports and Sustainability Accounting Standards Board standards. Through these assessment methodologies the company has attempted to strike the balance between both the Financial and Impact Perspectives. In this aspect they have prioritized zero-emission deliveries, data-driven climate-risk assessment strategy with two key emission categories, corporate emissions and portfolio emissions. Both the emissions categories are subjected to stay within Science Based Target Initiative's (SBTi) goal of 1.5 degree centigrade. In their new report a claim towards fair use of AI, cyber-resilience and data privacy has been committed to benefit the end-users and the consumers. This also brings our attention towards their newly designed social vertical, in which the company prioritized social imperatives to achieve social sustainability goals through which renewed measures have been mentioned to improve all stakeholders' benefit and ease. Lastly, the company's ethics and integrity vertical promises to provide competition compliance, anti-money laundering and counter-financing terrorism policies to be strictly adhered with anti-bribery and anti-corruption agenda. With clear sustainability statements reported, the organization has committed to business conduct with integrity in social, ethical and sustainability committee charter, human rights policy, Diversity, Equity and Inclusion (DEI). Hence, this report testified all the elements of our conceptual framing, analyzing every requirement of the theories discussed.

5 Concluding Remarks

The chapter has rigorously engaged in theorization of ethical reflexivity in business practices, in the context of sustainable value creation measures in building sustainable business models. The findings iterate the major constructs in this regard, exemplifying the categories and elements borrowed from major theories from the associated domains. The chapter aims to understand how SBMs are created through the institutional feedback loop between the stakeholders to engage in the sustainability agenda translating theory to practice. In this attempt, the chapter not only develops a conceptual framing but also applies the interactive elements in real-life cases. The cases are analyzed comparatively to understand how the intersecting elements result into practical outcomes. Two motivations drive our analysis, first, to explore how corporate sustainability visions change from compliance ethics to practical implementation, prioritizing the innovation and stakeholders' engagement throughout the process. Second, how the internal and external stakeholders' function through a transparent integrative manner to perform decision making through ethical reflexivity. Three parameters in this aspect have been proposed by the authors that may be taken into consideration while designing the shared ethical agreements for value creation in an organization. The cases are analyzed keeping these objectives under consideration. While Natura & Co.'s positioning in sustainability has emerged as a pro-SVC organization, keeping all its sustainability parameters in check through inclusion of all stakeholders, Mercado Libre has been criticized for its warehouse management policies and labor rights. Mediating between the two cases, Naspers has been identified as the organization, adopting adequate measures to build a sustainable form of business that has not been there before. The immense criticism led to its refined polices, stakeholders' engagement, care ethics and moral integrity, the organization can be an emerging case of equitable, sustainability-sensitive business models with real impact created in all the pillars of sustainable development (Engster 2011). The chapter critically analyzes these points by theoretical underpinning and critical comparative case analysis.

The chapter opens new avenues of future scopes of research in this domain. The concepts can be adapted and tested across different cases from different contexts and industries. Depending on the nature of the business the study can identify more elements that may not have emerged in our analysis. Also, cases from developed nations can also be undertaken to understand how labor rights, indigenous knowledge sharing mechanisms and local diversity is considered and included in the ethical perspectives. The external side of the stakeholders are the major variable when it comes to the sustainability agenda and the goal of creating social and environmental impact. In that sense, the insights from this chapter may provide propositions that can elevate the theories to new nuances. The internal stakeholders of the companies can also get beneficial insights from the case analysis, understanding how sustainability governance structures are to be designed and how to engage the stakeholders in all its process of business operations. The study has been limited to a brief analysis of the cases' ethical considerations only and could detail more about the sustainability agenda integrated with each element. This gap can be fulfilled in the future studies and bringing one theory at a time with one sustainability metrics in a particular ethical standpoint. In this way the study provides important theoretical and practical implications for both academic and industrial research in understanding how sustainable business models is created keeping the ethics and value creation in its core focus.

References

Agrawal D (2025) A comprehensive review of investing and financing for sustainable tourism projects. Environ Sustain Governance Insights 1(1):53–70

Alshukri T, Ojekemi OS, Öz T, Alzubi A (2024) The interplay of corporate social responsibility, innovation capability, organizational learning, and sustainable value creation: does stakeholder engagement matter? Sustainability 16(13):5511

Atzeni M (2023) The labour process and workers' rights at Mercado Libre: hiding exploitation through regulation in the digital economy. Work in the Global Economy 3(2):181–200

Bocken NMP, Short SW, Rana P, Evans S (2014) A literature and practice review to develop sustainable business model archetypes. J Cleaner Prod 65:42–56

Cosa M (2024) Equity integration in corporate social responsibility: analyzing stakeholder, legitimacy, and social activism dynamics. In: CSR, governance and value. Springer Nature Singapore, pp 133–154

Creswell JW, Miller DL (2000) Determining validity in qualitative inquiry. Theory into Pract 39(3):124–130

Donaldson T, Dunfee TW (1994) Toward a unified conception of business ethics: integrative social contracts theory. Acad Managem Rev 19(2):252–284

Elkington J (1997) The triple bottom line. Environ Managem: Read Cases 2:49–66

Endrissat N, Müller WR, Kaudela-Baum S (2007) En route to an empirically-based understanding of authentic leadership. Europ Managem J 25(3):207–220

Engster D (2011) Care ethics and stakeholder theory. In: Applying care ethics to business. Springer Netherlands, pp 93–110

Evans S, Fernando L, Yang M (2017) Sustainable value creation—from concept towards implementation. In: Sustainable manufacturing: challenges, solutions and implementation perspectives. Springer International Publishing, pp 203–220

Fischer J (2004) Social responsibility and ethics: clarifying the concepts. J Business Ethics 52(4):381–390

Freeman RE, Phillips RA (2002) Stakeholder theory: a libertarian defense. Business Ethics Quarter 12(3):331–349

Freudenreich B, Lüdeke-Freund F, Schaltegger S (2020) A stakeholder theory perspective on business models: value creation for sustainability. J Business Ethics 166(1):3–18

Gilligan C (1982) In a different voice. Harvard University Press

Gosling M, Huang HJ (2009) The fit between integrity and integrative social contracts theory. J Business Ethics 90(Suppl. 3):407–417

Harrison JS, Bosse DA, Phillips RA (2010) Managing for stakeholders, stakeholder utility functions, and competitive advantage. Strategic Managem J 31(1):58–74

Hart SL, Milstein MB (2003) Creating sustainable value. Acad Managem Perspect 17(2):56–67

Jordão RVD, de Almeida VR, Novas J (2022) Intellectual capital, sustainable economic and financial performance and value creation in emerging markets: the case of Brazil. The Bottom Line 35(1):1–22

Joselin D, Singh A, Tiwary KS (2025) The nexus between CSR practices and financial performance: empirical evidence from the manufacturing sector in India. Environ Sustain Governance Insights 1(1):1–17

Kaith A, Sachdeva G (2024) Bibliometric insights into green human resource management: a connection with business ethics and values. PURUSHARTHA-A J Managem Ethics Spirituality 17(1):51–63. https://doi.org/10.21844/16202117104

Kramer MR, Porter M (2011) Creating shared value: FSG. Boston, MA, USA, Vol. 17

Kumar K, Singh R, Choudhary M, Sharma A (2025) Climate change mitigation strategies in environmentally sensitive industries. In: Climate change and social responsibility, vol 25. Emerald Publishing Limited, pp 185–204

Lüdeke-Freund F, Rauter R, Pedersen ERG, Nielsen C (2020) Sustainable value creation through business models: the what, the who and the how. J Business Models 8(3):62–90

Lüdeke-Freund F, Massa L, Bocken N, Brent A, Musango J (2016) Business models for shared value. Network for Business Sustainability South Africa

Manninen K, Laukkanen M, Huiskonen J (2024) Framework for sustainable value creation: a synthesis of fragmented sustainable business model literature. Managem Res Rev 47(1):99–122

McGrath SK (2022) Ethics, morality, values, principles and beliefs. In: Speaking management: how to spot language traps and resolve contested management terms. Springer Nature, Singapore, pp 197–244

Méndez-León E, Reyes-Carrillo T, Díaz-Pichardo R (2022) Towards a holistic framework for sustainable value analysis in business models. Business Strategy Environ 31(1):15–31

Mert G, Akkaya B, Palazzo M, Micozzi A, Ferri A, Notari F (2025) Investigating the relationships among organizational policy, strategic planning performance, and business ethics. Int Entrepreneurship Managem J 21(1):1–22
Moratis L, Melissen F, Idowu SO (2018) Sustainable business models: principles, promise, and practice. Springer
Nicholson J, Kurucz E (2019) Relational leadership for sustainability: building an ethical framework from the moral theory of ethics of care. J Business Ethics 156(1):25–43
Nogueira E, Gomes S, Lopes JM (2022) The key to sustainable economic development: a triple bottom line approach. Resources 11(5):46
Parry KW, Proctor-Thomson SB (2002) Perceived integrity of transformational leaders in organisational settings. J Business Ethics 35(2):75–96
Peterson M (2022) Moral rightness comes in degrees. J Amer Philosop Assoc 8(4):645–664
Prakash A, Tiwari AK (2025) The role of institutional factors in achieving SDG 16: a thematic review of drivers of corporate anti-corruption action. Environ Sustain Governance Insights 1(1):34–52
Prencipe A (2025) Accountability between compliance and legitimacy: rethinking governance for corporate sustainability. Sustainability 17(20):9305
Richardson JE (2005) The business model: an integrative framework for strategy execution. SSRN Working Paper
Schaltegger S, Hansen EG, Lüdeke-Freund F (2016) Business models for sustainability: origins, present research, and future avenues. Organizat and Environ 29(1):3–10
Schwartz MS (2005) Universal moral values for corporate codes of ethics. J Business Ethics 59(1):27–44
Singh R, Kumar K, Khan S (2024) A comprehensive view of artificial intelligence (AI)–based technologies for sustainable development goals (SDGs). In: Artificial intelligence enabled management: an emerging economy perspective, vol 183. 9783111172408-012
Singh R, Sharma A, Kumar K (2025a) Navigating carbon neutrality: strategies for sustainable future and environmental challenges. In: Climate neutrality through smart eco-innovation and environmental sustainability. Cham, Springer Nature Switzerland, pp 69–80
Singh R, Tiwary KS, Kumar K, Kumar A (2025b) Empowering a sustainable future: blockchain's role in economy, organizations, and finance towards smart circular economy. In: Blockchain technologies for smart circular economy and organisational sustainability. Cham, Springer Nature Switzerland, pp 271–283
Srivastava SK, Anshul A, Pathak P, Bansal J (2018) Responsible consumption for curbing food wastage: an exploratory enquiry. PURUSHARTHA-A J Managem Ethics Spirituality 11(2):104–116
Srivastava SK, Singh S, Srivastava M (2021) Food waste at household and social gatherings: drivers and possible remedies. PURUSHARTHA-A J Managem Ethics Spirituality 14(2):117–133
Srivastava SK, Sinha AK, Kumar S (2024) Workplace spirituality empowered by spiritual leadership to promote organizational effectiveness: a bibliometric analysis. PURUSHARTHA-A J Managem Ethics Spirituality 17(1):1–22. https://doi.org/10.21844/16202117101
Taylor C (1989) The diversity of goods. In: Clarke SG, Simpson E (eds) Anti-theory in ethics and moral conservatism. State University of New York Press, pp 223–240
Teer-Tomaselli R, Tomaselli K, Dludla M (2019) Peripheral capital goes global: naspers, globalisation and global media contraflow. Media Culture and Soc 41(8):1142–1159
Tripathi V, Singh K, Parihar S, Srivastava SK (2022) A bibliometric analysis on supply chain collaboration. PURUSHARTHA-A J Managem Ethics Spirituality 15(2):66–103. https://doi.org/10.21844/16202115205
Tronto JC (1999) Care ethics: moving forward. Hypatia 14(1):112–119
Vos BS, Reddy TL (2014) The use of environmental sustainability reporting by South African companies to promote environmental sustainability. In: Actuarial society of South Africa's 2014 convention

Wang Q, Yan W (2023) Your behaviour reflects your risk attitude: the influence of CEOs' insurance behaviours on corporate social responsibility. British J Managem 34(3):1584–1605
Yuan L, Chia R, Gosling J (2023) Confucian virtue ethics and ethical leadership in modern China. J Business Ethics 182(1):119–133

Repair-Oriented Branding: Reframing Consumer Participation in Sustainable Fashion Through Brand-Enabled Product Longevity

Kandarp Singh, Devika Agrawal, and Teena Bagga

Abstract The research on sustainable fashion has largely focused on ethical production processes and end-of-life waste management, providing little information on the role of brands in sustainable fashion use phases. At the same time, a new generation of fashion brands is promoting consumer engagement in repairing garments through brand-enabled repair initiatives, indicating a paradigm shift from disposal-oriented sustainability to product longevity-oriented sustainability through use-phase interventions. Although increasingly relevant in practice, this paradigm shift has yet to be adequately theorized in sustainable marketing and branding scholarship. This conceptual paper proposes the concept of Repair-Oriented Branding (ROB) as a new branding paradigm that explains how brands can strategically engage consumers in repair as a means to achieve sustainability. Based on Service-Dominant Logic, circular economy thinking, sustainable consumption theory, product attachment, and brand relationship theory, this paper argues that repair practices redefine the conventional transactional relationship between brands and consumers by turning passive consumers into active product life managers. A conceptual framework is proposed to demonstrate how brand-enabled repair initiatives can encourage consumer engagement in repair activities, build product attachment, reduce replacement rates, and maximize product longevity, thus achieving circular economy outcomes while simultaneously deepening long-term brand-consumer relationships. The paper also proposes a set of propositions for future empirical research on sustainability strategies through repair practices. By shifting the focus of repair practices from a post-purchase service to a branding strategy, this paper provides a use-phase-oriented perspective to the sustainable fashion and sustainable branding literature and provides practical

K. Singh (✉)
School of Management, Bennett University, Greater Noida, India
e-mail: Kandarp.singh@bennett.edu.in

D. Agrawal
Institute of Business Management, GLA University, Mathura, India
e-mail: devikaagrawal.phd@gmail.com

T. Bagga
Department of Management Studies, Jamia Millia Islamia, Delhi, India
e-mail: tbagga@jmi.ac.in

K. Kumar et al. (eds.), *Sustainable Value Creation: Corporate Strategies for SDGs Implementation*, World Sustainability Series,
https://doi.org/10.1007/978-3-032-24652-3_18

guidance for brands interested in translating sustainability into practice beyond the production and disposal phases.

Keywords Repair-oriented branding · Sustainable fashion · Product longevity · Circular economy · Consumer co-creation · Sustainable consumption · Product attachment

1 Introduction

Sustainability has emerged as a key defining word in the global fashion world. In fact, within the past two decades, a lot of talk has gone into concepts such as ethical sourcing, green materials, reduction of waste, recycling, and the slow fashion ethos of sustainability in fashion production and consumption (Fletcher 2014; Gwilt 2014; Joy et al. 2012). In fact, concepts such as circular economy and sustainable consumption have largely influenced our thinking about how fashion brands can engage in lower forms of environmental degradation and sustainability (Bocken et al. 2016; Jackson 2005). As such, the most common theme within sustainable fashion has largely revolved around how fashion products are produced and what happens to them in post-consumption (Kemi and Zilahy 2025; Rotimi et al. 2021).

However, another equally significant stage of a product's life cycle; its use period remains relatively undervalued both theoretically as well as empirically (Rehman et al. 2024). In similar contexts related to circular economies, the strategy of grinding down the so-called resource loops, hence attempts to increase the lifespan of products that have been seen as a crucial step for reaching sustainability goals (Stahel 2010; Bocken et al. 2016). Within the existing literature on sustainable consumption patterns, it has been found that the extension of product life and the reduction of replacement rates are essential in managing sustainability risks from a societal perspective (Rehman et al. 2024; Cooper 2005; Jackson 2005). However, in the existing literature on the fashion industry, the production phase is rarely analyzed, with special attention not being paid to post-consumer strategies for clothing products (Kemi and Zilahy 2025; Schiaroli et al. 2024; Niinimäki 2013).

The emerging trends of a small number of fashion brands suggest there is a real shift in how sustainability is being considered. Rather than focusing on recycling or persuading consumers to buy less, these brands are actively promoting repair of clothing items (Michel et al. 2019). For instance, brands such as Nudie Jeans, Decathlon, and Eileen Fisher are collaborating with two-way repair into their brand ideology (Michel et al. 2019; Laitala and Klepp 2018). This is beyond offering an additional service to the customer; it is a comprehensive change in how brands think of their brand-consumer relationship. Thus, while repair in this context is obviously a practical activity, it is also assumed to be a strategic means through which consumers can be encouraged to practice sustainability. As research into product attachment and emotionally durable design has identified, a person's emotional attachment to a repaired or maintained object is increased when they repair and maintain products, thus making them less likely to replace them (Chapman 2005; Mugge et al. 2005). Thus, repair is a relational practice as well.

Additionally, theory in marketing and branding provides some insight into the change. The Service-Dominant Logic perspective highlights the value co-creation concept between firms and customers (Vargo and Lusch 2004), while the brand relationship perspective underlines the long-term relationship between people and brands (Fournier 1998). However, there has been little investigation into how the brand relationship perspective relates to maintenance practices in sustainable fashion in recent branding literature. The goal is to explore and establish the concept of Repair-Oriented Branding (ROB), which is a form of branding that encourages, informs, motivates, and normalizes repair practices for consumers as a means of achieving sustainability through the concept of circular economies (Hernandez et al. 2020), sustainable consumption (Moalem and Mosgaard 2021), Service-Dominant Logic (Hernandez et al. 2020), and brand relationships (Amend et al. 2022). With regards to this, it is possible that repair can help change the original relationship between a brand and a consumer, which is based on a one-time transaction (Korsunova et al. 2023). In other words, it's not just that sustainability means sustainable production and then sustainable end-of-life recycling; sustainability means sustainable lifespan usage. Repair acts as a bridge between branding and user engagement and sustainability, providing a new way to think about sustainable marketing (Hernandez et al. 2020).

2 Sustainable Fashion and Its Current Theoretical Boundaries

One thread that has managed to take root in sustainability, design, and marketing is sustainable fashion. Initially, sustainable fashion was centered on green processes, ethical sourcing, sustainable resources, and lessening the footprint of processes (Fletcher 2014; Gwilt 2014). Subsequent developments on sustainable fashion have attempted to address issues such as over-consumption, fast fashion, and its damaging effects on the environment (Joy et al. 2012). These studies have played a vital role in sensitizing scholars and practitioners on issues regarding sustainable development. Concurrently with all this, the circular economy has presented itself as a new and better system of economic organization, better than the linear 'take-make-waste' model (Sariatli 2017; Merli et al. 2017) Circular thinking focuses on closing and slowing resource loops, and as such, recycling, reuse, and remanufacturing are key concepts, as are new forms of business (Bocken et al. 2016; Stahel 2010). In the fashion industry, it has materialized particularly in the form of take-back schemes, secondary markets, upcycling, and textile recycling, driven by crucial industry reports (Ellen MacArthur Foundation 2017).

As a consumer, research into sustainable consumption behavior has highlighted the importance of behavioral change, such as responsible buying, reducing consumer behavior, and ethical behavior (Jackson 2005; Sheth et al. 2011). The basic idea is that sustainability cannot happen without changing consumer behavior, such as changing the way consumers buy and consume as well as reducing the frequency of replaces. For instance, this has been the basis for the rise of slow fashion and anti-consumerist behavior (Niinimäki 2013). Still, upon perusing the literature more deeply, one can observe that sustainability interventions predominantly fall into one of two phases: production or post-consumption (Dao and Martinez 2024). Ethical materials, clean tech, and labor practices impact how a product is made, whereas recycling, resell cultures, and product take-back programs impact what happens after the product is no longer needed. There is little research on the space between product purchase and product disposal, especially the duration of time a product is used (Shi et al. 2020).

Research in product-service systems and product longevity has shown that it is by stretching how long a product lasts that one attains sustainability. Objects that acquire emotional attachments and people who become attached to an object tend to keep it longer instead of replacing it quickly. However, these ideas have not been folded neatly into sustainable fashion or branding studies as intentional or designable tools for firms. Therefore, theories of sustainable fashion often simplify sustainability to reduced purchase frequency, enhanced purchase quality, or responsible disposal of products, with relatively less focus on how brands can engage with the usage phase to extend the life of garments. The current text establishes a large theoretical gap. Although the literature on the circular economy recognizes the need to slow down material flow, and the literature on sustainable consumption focuses on reducing replacement, there is relatively less progress on how brands can implement these goals through consumer behaviors such as repair.

The presence of this gap indicates the need for an alternative theory of sustainable fashion and marketing that has not been explored adequately with respect to how brands can build systems that promote the care and maintenance of current garments. This theoretical niche is where the theory of ROB can be developed.

3 Emerging Practice: Brands Normalizing Repair Culture

The most apparent change presently underway in the fashion industry, after the emphasis on ethical production and recycling at the end of a product's life, is the extension of product life through repair. Brands today are no longer content with encouraging recycling or upcycling but are now integrating repair as a fundamental behavior of consumers. This approach is consistent with the circular economy model, which gives greater priority to the conservation of resources through the extension of product life rather than focusing solely on recycling (Stahel 2010; Bocken et al. 2016).

Today, the most prominent brands in the industry, such as Patagonia, Nudie Jeans, Decathlon, and Eileen Fisher, are integrating repair into their business models through repair points in-store, free lifetime repairs, repair workshops, repair tutorials on the web, and repair kits. As a result, repair has moved from being a secondary afterthought to sales to becoming a fundamental part of brand identity and consumer engagement. These actions demonstrate brands assuming responsibility beyond the purchase and into the use phase of a product's life cycle. Research data indicates that an extension in product use is one of the most powerful but unused strategies for enhancing sustainability (Cooper 2005). Research on emotionally durable design and attachment to products have confirmed that when consumers maintain a product through repair and maintenance activities, they bond with it; consequently, user attachment reduces the probability that they would discard a product (Chapman 2005; Mugge et al. 2005). From a perspective of sustainable consumption patterns, the environmental benefits rely less on buying new things in an ethical way and more on reducing the frequency with which we replace things in the first place (Jackson 2005; Sheth et al. 2011). Accordingly, repair services bring this into practice and eliminate any barriers that encourage people to discard clothing that is damaged.

This is along the lines of Service-D Dominant Logic, which accepts the common viewpoint that value is created together with, and through interaction between, firms and users (Vargo and Lusch 2004). The repair, thus, requires consumers to take a proactive role in preserving value and makes consumers move beyond being merely consumers to being survival partners. The brand relationship theory extends this further and suggests that such interactions also enhance consumer relationships (Fournier 1998). Although evidence of the practice of repair is visible, there is limited theory written about it, and what there is, settles unevenly across the academic fields of Design Studies, Sustainable Consumption, and Circular Economy. In fact, discussions about repair appear to center on durability, frugality, and the resurgence of craftsmanship (Laitala and Klepp 2018), rather than as something companies build into their brand approaches as an opportunity for sustainable co-engagement with consumers. That is, when leading fashion organizations make repairing a product an empirical case that conjures up new concepts, it illustrates how concepts related to the circular economy (Stahel 2010), product longevity (Cooper 2005), emotional longevity (Chapman 2005), co-creating (Vargo and Lusch 2004), and brand relationship concepts (Fournier 1998) integrate in practice as part of repairing. Understanding this integration provides the foundation of theory that Repair-Oriented Branding (ROB) uses in the next section.

4 Theoretical Foundations for Repair-Oriented Branding (ROB)

The increasingly popular brand-enabled repair in fashion theory synthesizes concepts from marketing theory, sustainability studies, and design research. These concepts have mainly developed independently of one another; however, synthesis of these concepts helps us to understand Repair-Oriented Branding (ROB) as a mode of sustainable marketing in its own right.

Service-Dominant Logic initiates this trend with the idea of redefining the traditionally understood concept of value created exclusively in the product, towards its co-creation in the relationship between brands and consumers (Vargo and Lusch 2004). In the process of repair initiatives, the consumer is not only the receiver of the value provided by the brand but becomes an actor in co-creating the utility of the garment themselves. By providing consumers with repair outlets provided by the brand, they become co-creators of the value in the utility of the garment beyond its lifespan. The idea of circular economy reinforces this approach by highlighting the importance of closing and slowing down resource loops, as proposed by (Bocken et al. 2016; Stahel 2010). A large amount of literature on fashion sustainability focuses on end-of-life approaches that emphasize either recycling or reuse, and methods that aim to extend the lifespan of products to slow down the consumption cycle are often overlooked. Stahel's performance economy approach argues that the best way to achieve product sustainability is to keep products in use for as long as possible, and this can be achieved through repair activities.

Alternative viewpoints can be derived from studies on sustainable consumption, which argue that sustainable consumption is necessary and that the rate of product replacement is a key determinant (Jackson 2005; Sheth et al. 2011). The current literature on this topic is primarily focused on attitudes and intentions, and it is unclear how this can be implemented in practice. Repair activities show how conscious consumption can be reflected in brand actions by reducing obstacles to use. Knowledge from brand relationship theory and product attachment studies provides an understanding of the importance of brands in relation to strategic repair practices. Brand relationship theory defines brand attachment as an outcome of repeated interactions with the consumer (Fournier 1998). Studies on emotionally durable design and product attachment show that consumer investment in product maintenance is an important factor in attachment and reducing the chances of product replacement (Chapman 2005; Mugge et al. 2005). Repair work provides these repeated interactions, thus strengthening both garment attachment and brand attachment.

Overall, all of these theories line up to show that there is more to repair than just fixing something. While co-creating emphasizes the role of the consumer as an active collaborator, the circular economy emphasizes the sustainability of longevity as a measure of sustainability in particular. Sustainable consumption points to the behavioral change of not replacing things as frequently, and brand relationship theory clearly maps the effects of a conversation leading to a relationship. ROB is the crossroads of all of these ideas as it is a branding approach towards sustainability that includes the consumer as a participant during use.

5 Conceptualizing Repair-Oriented Branding (ROB)

The previous sections illustrate that much of sustainable fashion scholarship has zeroed in on responsible production and how to handle waste at the end of a garment's life. At the same time, new industry practices are signaling a shift: a growing focus on keeping clothes usable longer by repairing them. Drawing on Service-Dominant Logic, the circular economy, sustainable consumption, and brand relationship theory, this paper proposes a Repair-Oriented Branding (ROB) concept as unique. ROB explains in detail how brands can enact sustainability during the use phase of products.

Repair-Oriented Branding is a branding practice wherein companies intentionally design systems, messaging, and cultural signals that spur and assist consumer-driven repair toward an extension of a product's lifetime. In line with Service-Dominant Logic, the repair service changes the role of the consumer from a passive value receiver to an active co-creator of value, thereby contributing to the retention of value in the garment over time (Vargo and Lusch 2004; Prahalad and Ramaswamy 2004; Srivastava et al. 2021). At the same time, the repair service also adheres to the principles of the circular economy, which aims to achieve resource circularity through extended use, in addition to post-consumer recycling (Stahel 2010; Bocken et al. 2016; Srivastava et al. 2018). From the point of view of sustainable consumption, repair services lower the speed of replacement of commodities, (Jackson 2005; Sheth et al. 2011; White et al. 2019). In addition to the ethical stance of consumers, the notions of repairability and ownership behavior (ROB) provide insights into how brands can create systems that make long-term use possible and desirable. Repair services promote emotional durability, which in turn increases the attachment to the product and decreases early discarding (Chapman 2005; Mugge et al. 2005; Belk 1988). The recurring interactions made possible by repair services are expected to improve the consumer-brand relationship (Fournier 1998; Keller 2001), thus connecting sustainability and relationship marketing (Singh et al. 2025b; Prakash and Tiwari 2025; Kumar et al. 2025).

In conclusion, ROB transforms the branding paradigm from value creation at the point of purchase to value maintenance throughout the product life cycle, thus positioning the consumer as the protector of longevity and brands as facilitators of this protection.

6 Conceptual Framework of Repair-Oriented Branding

Here, ROB is considered a system where the brand incorporates sustainability into the wear stage of the clothing. Repair is not understood as a singular marketing tactic, but rather as a continuous and reinforcing process that helps create the sense of "normality" concerning sustainable use. At its heart is the concept that the brand can create repair-enabled systems to define the life of clothing that the consumer adopts. Product longevity research indicates that extending product life may be the most powerful thing a brand can do for sustainability (Singh et al. 2024; Agrawal 2025; Joselin et al. 2025). ROB is discussed as extending product life, as opposed to "durable" product development and "ethically" shopping, and therefore repairing, clothing and other consumer items.

As for brands, infrastructure for repair such as services, tools, training, and info, does not merely shift units, it extends their use. These brand strategies engage consumers in repair, making maintenance a value-creation activity consistent with Service-D Dominant Logic. In this way, consumers engaging in repair and maintaining products are beneficial to the product's functionality and meaning-making, transforming them from consumers to steward-like consumers who are involved in the product's lifespan. This creates products with higher durability and fewer replacement needs, consistent with sustainable consumption objectives that focus on extending product use rather than repeated acquisition (Singh et al. 2025a). This, in turn, contributes to slowing down resource flows: a major concept behind the circular economy theoretical framework.

ROB: Notably, circularity arrives with ROB at the point of use, rather than the end of product life.

This framework has also proposed a relational feedback mechanism, and a greater number of repair-centered touch points for brand and customer will enhance their relationship over time (Fournier 1998; Keller 2001). This will further enhance the willingness of consumers to mend goods over time, building on a sustainable cycle of longevity and sustainability already initiated by ROB. ROB has provided a closed-loop branding system where co-creation, attachment, sustainable consumption, and circular economy concepts and outcomes create a feedback mechanism for each other, giving rise to a number of propositions as suggested by this framework in the future.

P1. Brand-enabled repair systems have a positive effect on consumer engagement with garment repair activities.

P2. Consumer engagement with repair activities has a positive effect on product attachment and emotional durability.

P3. Higher product attachment reduces consumers' intention to replace garments prematurely.

P4. Reduced replacement behavior leads to increased product lifetimes.

P5. Increased product lifetimes result in slowed resource cycles and positive circular economy outcomes.

P6. Consumer engagement with repair activities increases the strength of long-term consumer-brand relationships.

P7. Stronger consumer-brand relationships further increase consumers' engagement with repair behavior.

By outlining these relationships, the ROB conceptual framework places Repair-Oriented Branding on the map as a tool that enables brands to operationalize abstract sustainability goals into real-world consumer behavior, hence making product longevity the key mediator of environmental and relationship value creation.

7 Theoretical Contributions of Repair-Oriented Branding

This paper makes four key theoretical contributions. First, it expands the sustainable fashion literature by considering not only the creation and disposal of products but also their use. Following on from existing literature that has explored the ethical considerations of recycling and sustainable fashion (Fletcher 2014; Niinimäki 2013; Tripathi et al. 2022), this paper shows how the longevity of products can be achieved through repair. Second, it makes a theoretical contribution to the circular economy by showing how the concept of slowed resource loops (Stahel 2010; Bocken et al. 2016), can be realized through branding, rather than other ways of thinking about the concept. Third, it contributes to sustainable consumption research by showing that sustainable behavior can indeed be facilitated by means of systems created by brands rather than by the ethical intentions of consumers (Jackson 2005; White et al. 2019; Srivastava et al. 2024). Fourth, it will enhance our understanding of branding and relationship marketing theory by linking related ideas such as product attachment, emotional durability, and co-creation to sustainability, demonstrating how repair can foster consumer-brand relationships (Fournier 1998; Chapman 2005).

8 Managerial Implications

ROB is a guide for fashion brands to go beyond superficial sustainability claims and to actively influence the use of their garments. Instead of investing only in sustainable sourcing, material development, and reporting, ROB claims that the actual environmental and relational added value is created when managers stop striving for a maximum number of sales within a limited timeframe and focus on extending the life of a garment. First, repair should be woven into the brand's retail and service framework, not booked as a separate after-sales add-on. Drawing on product-service system ideas, brands can move from pure supply of products to support ongoing use in-store through repair services, tools, training, and transparent informational support. This turns stores into sites of maintenance and continued involvement, rather than pure transaction points. Second, ROB emphasizes the role of consumer education and guidance. Sustainable consumption research has established that to change behaviour, awareness is not enough: it requires practical assistance (Jackson 2005; White et al. 2019). In practice, this can be carried out by brands, integrating repair tutorials, workshops, and care guidance into the brand experience—making repair both desirable and approachable.

Third, ROB calls for a pivot in brand storytelling. Traditional fashion branding tends to chase novelty and newness, but repair-oriented brands would spotlight durability, longevity, and the garment's life journey. This resonates with research on emotionally durable design and product attachment that found well-maintained products assume personal significance and are kept longer (Chapman 2005; Mugge et al.

2005). So, the value messages need to shift from celebrating replacement to celebrating endurance. Fourth, ROB underscores the relational worth of repair works. In the language of relationship marketing, the act of repair creates repeated instances of contact between consumers and brands. These, in turn, foster loyalty and participation in long-term relationships (Fournier and Keller 1998). Far from undermining revenue streams, this deeper involvement also invites supplementary services, memberships, and community-based brand affiliations.

Essentially, what ROB does is to call on managers to think differently in terms of how we measure sustainability. So, it's not just about measuring how much recycled material a product uses or how much footprint reduction we effect. Rather, it's also important that fashion brands think in terms of how long products last, how often products are repaired, or how much effort customers are putting into product servicing. This is because such measurements are in greater line with the concept/theory of the circular economy, as broadly researched here (Stahel 2010; Bocken et al. 2016). With the adoption of Repair-Oriented Branding, fashion leaders can thus operationalize the otherwise abstract sustainability narrative in day-to-day consumers.

9 Future Research Directions

Repair-Oriented Branding (ROB) has identified a number of promising avenues for future research in sustainable marketing, consumer behavior, branding, and the literature on the circular economy. ROB is based on current industry practice and a variety of theoretical approaches and can be used to support theory development and empirical/cross-cultural research. One promising avenue for research lies in the area of consumers and their behavior in response to brand-led repair initiatives. Current research has shown that product attachment and emotionally durable design can extend the life of products (Chapman 2005; Mugge et al. 2005). Future research might investigate consumer attitudes towards repair services offered by brands, the factors that influence participation in repair activities, and the impact of participation in repair activities on the decision to replace. Using a combination of quantitative and qualitative research approaches, researchers could investigate the impact of repair initiatives on consumer behavior and the extension of the life of clothing and other products. Also of interest is the impact of repair participation on brand outcomes. Using the framework of brand relationship theory, repeated interactions can build brand attachment and loyalty (Fournier 1998; Keller 2001; Kaith and Sachdeva 2024). Research might investigate whether customers who participate in repair activities show greater brand trust and loyalty than customers who do not participate in repair activities.

ROB unlocks the doors to behavioral mechanisms that cause sustainable consumption behaviors. Although most sustainable consumption research focuses on motives for ethics and sustainable consumption (Jackson 2005; Sheth et al. 2011; White et al. 2019), it appears that brand system design can affect behavior through mechanisms other than ethics and sustainable consumption; for instance, experiments can investigate behavioral mechanisms for infrastructural support and sponsorship for repairing instead of replacing a product. Cross-cultural studies may uncover how ROB norms vary across societies. In certain societies, repair may be deeply ingrained, and it may be viewed very unfavorably in others. Such studies may investigate the impact of such attitudes on ROB's effectiveness.

Future studies have the potential to investigate the economic viability of ROB initiatives. Utilizing literature about Product-Service Systems (PSS) such as that developed by researchers Mont (2002) and Armstrong et al. (2016), studies have the potential to investigate the financial viability of ROB initiatives for fashion brands. Finally, ROB could be extended to other product categories beyond fashion—electronic devices, furniture, etc., where repair could potentially lengthen product lifespan and eco-sustainability. This would thus make ROB a universal sustainable brand strategy rather than a fashion-centric one. By addressing these research directions, future scholarship can empirically validate and extend the theoretical contributions of Repair-Oriented Branding, strengthening its role within sustainable marketing literature.

10 Conclusion

This paper aims to address an important niche in the literature of sustainable fashion and marketing, which is concerned with the lack of consideration of the longevity or length of usage of products. While most studies were more focused on ethical production, green products, and even product recycling, there has been a lack of conceptual consideration of strategies for fashion marketers to promote product longevity by focusing on consumers. At exactly this point, however, actual industry practices were observed, indicating that numerous fashion brands were encouraging consumers to repair products, prolonging product longevity. Grounded in Service-Dominant Logic, circular economy, sustainable consumption, product attachment, and brand relationship theory, the present research proposes a novel concept in the literature: Repair-Oriented Branding (ROB), which refers to the ability of brands to enact sustainability in the use stage through consumer participation. This study was conducted as another exploration of the position of repair in brand strategy, where it is argued to be able to play a part in generating value co-creation, reducing resource turnover, and increasing the relationship between a brand and consumers. The theory also discusses how brand-based repair systems can be used to facilitate people repairing their clothing, developing emotional ties with their clothing, delaying discarding them, and providing new opportunities for circularity and sustainability. In this new perspective of sustainable fashion, there is a move away from production and end-of-life patterns, and more focus is given to product usage periods and product lifespan as the primary driver for creating environmental and relational values. We see through this discussion of sustainable fashion that, while repair is related to new approaches to sustainability, it is also seen as a tool for reconsidering and looking at branding within the circular economy. From this perspective of seeing consumers as product custodians and brands as enablers, Repair-Oriented Branding takes its well-earned place as being of prime importance within sustainable fashion marketing. The message is clear: sustainable fashion needs to be thought about beyond just production and end-of-life patterns, but also in relation to how long products are used for and how long they can be used for.

References

Agrawal D (2025) A comprehensive review of investing and financing for sustainable tourism projects. Environ Sustain Governance Insights 1(1):53–70

Amend C, Revellio F, Tenner I, Schaltegger S (2022) The potential of modular product design on repair behavior and user experience—Evidence from the smartphone industry. J Cleaner Prod.https://doi.org/10.1016/j.jclepro.2022.132770

Armstrong CM, Niinimäki K, Kujala S, Karell E, Lang C (2016) Sustainable product-service systems for clothing: exploring consumer perceptions of consumption alternatives in fashion. J Cleaner Prod 97:30–39. https://doi.org/10.1016/j.jclepro.2014.01.046

Belk RW (1988) Possessions and the extended self. J Consum Res 15(2):139–168. https://doi.org/10.1086/209154

Bocken NMP, de Pauw I, Bakker C, van der Grinten B (2016) Product design and business model strategies for a circular economy. J Indus Prod Eng 33(5):308–320. https://doi.org/10.1080/21681015.2016.1172124

Chapman J (2005) In: Emotionally durable design: objects, experiences and empathy. Earthscan

Cooper T (2005) Slower consumption: reflections on product life spans and the "throwaway society." J Indus Ecol 9(1–2):51–67. https://doi.org/10.1162/1088198054084671

Dao J, Martinez C (2024) Defining sustainable clothing use: a taxonomy for future research. Int J Consum Stud.https://doi.org/10.1111/ijcs.13033

Ellen MacArthur Foundation (2017) A new textiles economy: redesigning fashion's future. Ellen MacArthur Foundation.

Fletcher K (2014) In: Sustainable fashion and textiles: Design journeys. 2nd ed. Routledge

Fournier S (1998) Consumers and their brands: developing relationship theory in consumer research. J Consum Res 24(4):343–373. https://doi.org/10.1086/209515

Gwilt A (2014) In: Fashion design for living: An introduction to sustainable fashion. Routledge

Hernandez RJ, Miranda C, Goñi J (2020) Empowering sustainable consumption by giving back to consumers the 'Right to Repair'. Sustainability 12(3). https://doi.org/10.3390/su12030850

Jackson T (2005) Motivating sustainable consumption: a review of evidence on consumer behaviour and behavioural change. In: Sustainable development research network

Joselin D, Singh A, Tiwary KS (2025) The nexus between CSR practices and financial performance: empirical evidence from the manufacturing sector in India. Environ Sustain Govern Insights 1(1):1–17

Joy A, Sherry JF, Venkatesh A, Wang J, Chan R (2012) Fast fashion, sustainability, and the ethical appeal of luxury brands. Fash Theory 16(3):273–296. https://doi.org/10.2752/175174112X13340749707123

Kaith A, Sachdeva G (2024) Bibliometric insights into green human resource management: a connection with business ethics and values. PURUSHARTHA-A J Managem Ethics Spirituality 17(1):51–63. https://doi.org/10.21844/16202117104

Keller KL (2001) Building customer-based brand equity. Market Managem 10(2):14–19

Kemi A, Zilahy G (2025) Sustainable post-purchase behaviour of consumers of fashion textile products. Discover Sustain 6.https://doi.org/10.1007/s43621-025-00888-5

Korsunova A, Heiskanen E, Vainio A (2023) Consumer decision-making on repair in a circular economy: a process model based on experiences among young adults and stakeholders in Finland. J Cleaner Prod. https://doi.org/10.1016/j.jclepro.2023.137052

Kumar K, Singh R, Choudhary M, Sharma A (2025) Climate change mitigation strategies in environmentally sensitive industries. In: Climate change and social responsibility, vol 25. Emerald Publishing Limited, pp 185–204

Laitala K, Klepp IG (2018) Clothing repair practices: findings from consumer research. Sustainability 10(7):1–16. https://doi.org/10.3390/su10072203

Merli R, Preziosi M, Acampora A (2017) How do scholars approach the circular economy? A systematic literature review. J Clean Prod 178:703–722. https://doi.org/10.1016/j.jclepro.2017.12.112

Michel G, Feori M, Damhorst M, Lee Y, Niehm L (2019) Stories we wear: promoting sustainability practices with the case of Patagonia. Family and Consum Sci Res J.https://doi.org/10.1111/fcsr.12340

Moalem R, Mosgaard M (2021) A critical review of the role of repair cafés in a sustainable circular transition. Sustainability.https://doi.org/10.3390/su132212351

Mont O (2002) Clarifying the concept of product–service system. J Cleaner Prod 10(3):237–245. https://doi.org/10.1016/S0959-6526(01)00039-7

Mugge R, Schoormans JPL, Schifferstein HNJ (2005) Design strategies to postpone consumers' product replacement: the value of a strong person–product relationship. J Prod Innov Managem 22(6):525–537. https://doi.org/10.1111/j.1540-5885.2005.00147.x

Niinimäki K (2013) In: Sustainable fashion: new approaches. Aalto University

Prahalad CK, Ramaswamy V (2004) Co-creation experiences: the next practice in value creation. J Interact Market 18(3):5–14. https://doi.org/10.1002/dir.20015

Prakash A, Tiwari AK (2025) The role of institutional factors in achieving SDG 16: a thematic review of drivers of corporate anti-corruption action. Environ Sustain Governance Insights 1(1):34–52

Rehman M, Petrillo A, Barrios M, Forcina A, Baffo I, De Felice F (2024) Sustainable fashion: mapping waste streams and life cycle management. J Cleaner Prod.https://doi.org/10.1016/j.jclepro.2024.141279

Rotimi E, Topple C, Hopkins J (2021) Towards a conceptual framework of sustainable practices of post-consumer textile waste at garment end of lifecycle: a systematic literature review approach. Sustainability.https://doi.org/10.3390/su13052965

Sariatli F (2017) Linear economy versus circular economy: a comparative and analyzer study for optimization of economy for sustainability. Visegrad J Bioecon Sustain Developm 6:31–34. https://doi.org/10.1515/vjbsd-2017-0005

Schiaroli V, Fraccascia L, Dangelico R (2024) How can consumers behave sustainably in the fashion industry? A systematic literature review of determinants, drivers, and barriers across the consumption phases. J Cleaner Prod.https://doi.org/10.1016/j.jclepro.2024.144232

Sheth JN, Sethia NK, Srinivas S (2011) Mindful consumption: a customer-centric approach to sustainability. J Acad Market Sci 39(1):21–39. https://doi.org/10.1007/s11747-010-0216-3

Shi X, Li F, Chumnumpan P (2020) The use of product scarcity in marketing. Eur J Marketing 54(2):380–418

Singh R, Kumar K, Khan S (2024) A comprehensive view of artificial intelligence (AI)–based technologies for sustainable development goals (SDGs). In: Artificial intelligence enabled management: an emerging economy perspective, vol 183. 9783111172408-012

Singh R, Sharma A, Kumar K (2025a) Navigating carbon neutrality: strategies for sustainable future and environmental challenges. In: Climate neutrality through smart eco-innovation and environmental sustainability. Cham, Springer Nature Switzerland, pp 69–80

Singh R, Tiwary KS, Kumar K, Kumar A (2025b) Empowering a sustainable future: blockchain's role in economy, organizations, and finance towards smart circular economy. In: Blockchain technologies for smart circular economy and organisational sustainability. Cham, Springer Nature Switzerland, pp 271–283

Srivastava SK, Anshul A, Pathak P, Bansal J (2018) Responsible consumption for curbing food wastage: an exploratory enquiry. PURUSHARTHA-A J Managem Ethics Spirituality 11(2):104–116

Srivastava SK, Singh S, Srivastava M (2021) Food waste at household and social gatherings: drivers and possible remedies. PURUSHARTHA-A J Managem Ethics Spirituality 14(2):117–133

Srivastava SK, Sinha AK, Kumar S (2024) Workplace spirituality empowered by spiritual leadership to promote organizational effectiveness: a bibliometric analysis. PURUSHARTHA-A J Managem Ethics Spirituality 17(1):1–22. https://doi.org/10.21844/16202117101

Stahel WR (2010) In: The performance economy. 2nd ed. Palgrave Macmillan

Tripathi V, Singh K, Parihar S, Srivastava SK (2022) A bibliometric analysis on supply chain collaboration. PURUSHARTHA-A J Managem Ethics Spirituality 15(2):66–103. https://doi.org/10.21844/16202115205

Vargo SL, Lusch RF (2004) Evolving to a new dominant logic for marketing. J Market 68(1):1–17. https://doi.org/10.1509/jmkg.68.1.1.24036

Watson M, Shove E (2008) Product, competence, project and practice: DIY and the dynamics of craft consumption. J Consumer Cult 8(1):69–89

White K, Habib R, Hardisty DJ (2019) How to shift consumer behaviors to be more sustainable: a literature review and guiding framework. J Market 83(3):22–49. https://doi.org/10.1177/0022242919825649

MIX
Papier aus verantwortungsvollen Quellen
Paper from responsible sources
FSC® C105338

If you have any concerns about our products,
you can contact us on
ProductSafety@springernature.com

In case Publisher is established outside the EU,
the EU authorized representative is:
Springer Nature Customer Service Center GmbH
Europaplatz 3, 69115 Heidelberg, Germany

Printed by Libri Plureos GmbH
in Hamburg, Germany